Periodic Solutions of Hamiltonian Systems and Related Topics

NATO ASI Series

Advanced Science Institutes Series

A series presenting the results of activities sponsored by the NATO Science Committee, which aims at the dissemination of advanced scientific and technological knowledge, with a view to strengthening links between scientific communities.

The series is published by an international board of publishers in conjunction with the NATO Scientific Affairs Division

A Life Sciences B Physics	Plenum Publishing Corporation London and New York
C Mathematical and Physical Sciences	D. Reidel Publishing Company Dordrecht, Boston, Lancaster and Tokyo
D Behavioural and Social Sciences E Engineering and Materials Sciences	Martinus Nijhoff Publishers Dordrecht, Boston and Lancaster
F Computer and Systems Sciences G Ecological Sciences H Cell Biology	Springer-Verlag Berlin, Heidelberg, New York, London, Paris, and Tokyo

Series C: Mathematical and Physical Sciences Vol. 209

Periodic Solutions of Hamiltonian Systems and Related Topics

edited by

P. H. Rabinowitz

Department of Mathematics,
University of Wisconsin-Madison, U.S.A.

A. Ambrosetti

Scuola Normale Superiore, Pisa, Italy

I. Ekeland

Université de Paris 9 Dauphine, Paris, France

and

E. J. Zehnder

Department of Mathematics,
Rühr University, Bochum, F.R.G.

D. Reidel Publishing Company

Dordrecht / Boston / Lancaster / Tokyo

Published in cooperation with NATO Scientific Affairs Division

Proceedings of the NATO Advanced Research Workshop on
Periodic Solutions of Hamiltonian Systems and Related Topics
Il Ciocco, Italy
13-17 October 1986

Library of Congress Cataloging in Publication Data

NATO Advanced Research Workshop on Periodic Solutions of Hamiltonian Systems (1986: Pisa, Italy)
Periodic solutions of Hamiltonian systems and related topics.

(NATO ASI series. Series C, Mathematical and physical sciences; vol. 209)
"Proceedings of the NATO Advanced Research Workshop on Periodic Solutions of Hamiltonian Systems, Il Ciocco, Italy, 13–17 October, 1986"– T.p. verso.
"Published in cooperation with NATO Scientific Affairs Division."
Includes index.
1. Hamiltonian systems–Congresses. 2. Periodic functions–Congresses. I. Rabinowitz, Paul H. II. North Atlantic Treaty Organization. Scientific Affairs Division. III. Title. IV. Series: NATO ASI series. Series C, Mathematical and physical sciences; vol. 209.
QA614.83.N38 1986 515.3′52 87–16276
ISBN 90–277–2553–5

Published by D. Reidel Publishing Company
P.O. Box 17, 3300 AA Dordrecht, Holland

Sold and distributed in the U.S.A. and Canada
by Kluwer Academic Publishers,
101 Philip Drive, Assinippi Park, Norwell, MA 02061, U.S.A.

In all other countries, sold and distributed
by Kluwer Academic Publishers Group,
P.O. Box 322, 3300 AH Dordrecht, Holland

D. Reidel Publishing Company is a member of the Kluwer Academic Publishers Group

---

Printed in The Netherlands

TABLE OF CONTENTS

PARTICIPANTS

Antonio Ambrosetti, Scuola Normale Superiore, Piazza dei Cavalieri, 56100 Pisa, Italy

Abbas Bahri, 22 Rue Petrelle, 75009 Paris, France

Vieri Benci, Istituto di Matematiche Applicate, Università di Pisa, Pisa, Italy

D. Bennequin, IRMA, 7 rue R. Descartes, Strasbourg, France

A. Capozzi, Dipartimento di Matematica, Campus Universitario, Bari 70125, Italy

Marc Chaperon, Centre de mathématiques, Ecole Polytechnique, 91128 Palaiseau Cedex, France

A. Chenciner, Department of Mathematics, University of Paris VII, Place Jussieu, 75230 Paris 05, France

Frank H. Clarke, Centre de recherches mathématiques, Université de Montréal, C.P. 6128, Station A, Montréal (Quebec) Canada H3C 3J7

Vittorio Coti Zelati, SISSA, Strada Costiera 11, 34014 Trieste, Italy

G. F. Dell'Antonio, Department of Mathematics, University of Rome, La Sapienza, Italy

R. Devaney, Department of Mathematics, Boston University, Boston, Massachusetts 02215

Ivar Ekeland, Université Paris Dauphine, Ceremade, 75775 Paris, Cedex 16, France

J. Font, Department d'Equacions Funcionals, Facultat de Matematiques, University of Barcelona, Avda. Jose Antonio 585, Barcelona 7, Spain

Donato Fortunato, Dipartimento di Matematica, Università, 70125 Bari, Italy

Giovanni Gallavotti, Dipartimento di Matematica, II[a] Università di Roma, Via Raimondo, 00173 Roma, Italia

Mario Girardi, Dipartimento Matematico, dell'Universita di Roma I, I - 00185 Roma, Italy

C. Greco, Dipartimento di Matematica, Università di Bari, 70125 Bari, Italy

M. Gromov, IHES, Bures-sur-Yvette 91440, France

M. Herman, Ecole Polytechnique, Department of Mathematics, Plateau de Palaiseau, 91128 Palaiseau Cedex, France

H. Hofer, Department of Mathematics, Rutgers University, New Brunswick, New Jersey 08903

J. Mallet-Paret, Division of Applied Mathematics, Brown University, Providence, Rhode Island 02912

G. Mancini, Department of Mathematics, University of Trieste, Trieste, Italy

John N. Mather, Department of Mathematics, Princeton University, Fine Hall - Washington Road, Princeton, New Jersey 08544

Michele Matzeu, Dipartimento Matematico, dell'Universita di Roma I, I - 00185 Roma, Italy

Richard Moeckel, School of Mathematics, University of Minnesota, Minneapolis, Minnesota 55455

Jürgen Moser, Mathematik, ETH-Zentrum, 8092 Zürich, Switzerland

D. Offin, University of Missouri, Mathematical Science Building, Columbia, Missouri 65211

Paul H. Rabinowitz, Mathematics Department, University of Wisconsin-Madison, Madison, Wisconsin 53706

B. Ruf, Department of Mathematics, University of Köln, Köln, Federal Republic of Germany

L. Sanchez, C.M.A.F., Av. Prof. Gama Pinto, 2 - 1699 Lisbon Codex, Portugal

E. van Groesen, Department of Applied Mathematics, University of Twente, 7500 AE Enschede, The Netherlands

Michel Willem, Institut Mathematique, 2 Ch. du Cyclotron, B-1348 Louvain-la-Neuve, Belgium

E. Zehnder, Mathematical Institute, Ruhr University, Bochum, Federal Republic of Germany

CONTRIBUTORS

Antonio Ambrosetti, Scuola Normale Superiore, Piazza dei Cavalieri, 56100 Pisa, Italy

Abbas Bahri, 22 Rue Petrelle, 75009 Paris, France

Vieri Benci, Istituto di Matematiche Applicate, Universitá di Pisa, Pisa, Italy

A. Capozzi, Dipartimento di Matematica, Campus Universitario, Bari 70125, Italy

Marc Chaperon, Centre de mathématiques, Ecole Polytechnique, 91128 Palaiseau Cedex, France

Frank H. Clarke, Centre de recherches mathématiques, Université de Montréal, C.P. 6128, Station A, Montréal (Quebec) Canada H3C 3J7

Vittorio Coti Zelati, SISSA, Strada Costiera 11, 34014 Trieste, Italy

Marco Degiovanni, Scuola Normale Superiore, Piazza dei Cavalieri, 7, I 56100 Pisa, Italy

G. F. Dell'Antonio, Department of Mathematics, University of Rome, La Sapienza, Italy

Ivar Ekeland, Université Paris Dauphine, Ceremade, 75775 Paris, Cedex 16, France

Andreas Floer, Courant Institute of Mathematical Sciences, New York University, 251 Mercer Street, New York, New York 10012

Donato Fortunato, Dipartimento di Matematica, Università, 70125 Bari, Italy

Giovanni Gallavotti, Dipartimento di Matematica, II[a] Universitá di Roma, Via Raimondo, 00173 Roma, Italia

Fabio Giannoni, Dipartimento di Matematica, Università di Roma Tor Vergata, Via Orazio Raimondo, I 00173 Roma, Italy

Mario Girardi, Dipartimento Matematico, dell'Universita di Roma I, I - 00185 Roma, Italy

C. Greco, Dipartimento di Matematica, Università di Bari, 70125 Bari, Italy

M. Gromov, IHES, Bures-sur-Yvette 91440, France

H. Hofer, Department of Mathematics, Rutgers University, New Brunswick, New Jersey 08903

Antonio Marino, Dipartimento di Matematica, Università, Via Buonarroti, 2, I 56100 Pisa, Italy

John N. Mather, Department of Mathematics, Princeton University, Fine Hall - Washington Road, Princeton, New Jersey 08544

Michele Matzeu, Dipartimento Matematico, dell'Universita di Roma I, I - 00185 Roma, Italy

Richard Moeckel, School of Mathematics, University of Minnesota, Minneapolis, Minnesota 55455

Jürgen Moser, Mathematik, ETH-Zentrum, 8092 Zürich, Switzerland

Arturo Olvera, Departament d'Equacions Funcionals, Facultat de Matematiques, Universitat de Barcelona, Gran Via 585, Barcelona 08007, Spain

Paul H. Rabinowitz, Mathematics Department, University of Wisconsin-Madison, Madison, Wisconsin 53706

Carles Simó, Departament d'Equacions Funcionals, Facultat de Matematiques, Universitat de Barcelona, Gran Via 585, Barcelona 08007, Spain

E. van Groesen, Department of Applied Mathematics, University of Twente, 7500 AE Enschede, The Netherlands

Michel Willem, Institut Mathematique, 2 Ch. du Cyclotron, B-1348 Louvain-la-Neuve, Belgium

E. Zehnder, Mathematical Institute, Ruhr University, Bochum, Federal Republic of Germany

PREFACE

This volume contains the proceedings of a NATO Advanced Research Workshop on Periodic Solutions of Hamiltonian Systems held in Il Ciocco, Italy on October 13-17, 1986. It also contains some papers that were an outgrowth of the meeting. On behalf of the members of the Organizing Committee, who are also the editors of these proceedings, I thank all those whose contributions made this volume possible and the NATO Science Committee for their generous financial support. Special thanks are due to Mrs. Sally Ross who typed all of the papers in her usual outstanding fashion.

Paul H. Rabinowitz
Madison, Wisconsin
April 2, 1987

PERIODIC SOLUTIONS OF SINGULAR DYNAMICAL SYSTEMS[1]

Antonio Ambrosetti
Scuola Normale Superiore
Piazza dei Cavalieri
56100 Pisa, Italy

Vittorio Coti Zelati
SISSA
Strada Costiera 11
34014 Trieste, Italy

ABSTRACT. The paper contains a discussion on some recent advances in the existence of periodic solutions of some second order dynamical systems with singular potentials.

The aim of this paper is to discuss some recent advances in the existence of periodic solutions of some second order dynamical systems with singular potentials.

More precisely, we look for T-periodic solutions of N-dimensional systems like

$$\ddot{x} + V_x(t,x) = 0 \tag{*}$$

where Ω is an open subset of $\mathbb{R}^N$, $x \in \Omega$, $V \in C^1(\mathbb{R} \times \Omega, \mathbb{R})$ (for many results in §1 we will assume $V \in C^2$), V is T-periodic in t and V_x denotes the gradient $\left(\frac{\partial V}{\partial x_i}\right)_{i=1,\ldots,N}$. A specific feature is that V has singularities, in the sense that $V(t,x) \to -\infty$ (or $+\infty$) as $x \to \partial\Omega$.

The paper is divided in 2 sections. In §1 we deal with the case in which, roughly,

$$\Omega = \mathbb{R}^N - \{0\} \tag{Ω1}$$

and, either

$$V(t,x) \to -\infty \quad \text{as} \quad x \to 0, \quad \text{uniformly in} \quad t \tag{1^-}$$

or

---

[1]Supported by Min. P.I. (40%), Gruppo Naz. "Calcolo delle Variazioni".

P. H. Rabinowitz et al. (eds.), Periodic Solutions of Hamiltonian Systems and Related Topics, 1–10.

$$V(t,x) \to +\infty \quad \text{as} \quad x \to 0, \quad \text{uniformly in} \quad t\,. \tag{1^+}$$

In both the cases it is assumed that $V(t,x) \to 0$ as $|x| \to \infty$. Potentials of the above type remind the gravitational ones, the main, substantial, difference being that our approach works provided V behaves like $\mp|x|^{-\alpha}$, with $\alpha \geq 2$, near $x = 0$.

In §2 we suppose that:

$$\Omega \text{ is a bounded, convex, open subset of } \mathbf{R}^N \tag{$\Omega 2$}$$

and, again, that $V(t,x)$ diverges either to $-\infty$ or to $+\infty$ as $x \to \bar{x} \in \partial\Omega$ (uniformly).

The results of §1 review those of [2], [8] as well as [10]; that ones of §2 review those of [1], [9]. We refer to [1,2,8,9] for more details and other references.

Notations. The following notations will be used in the rest of the paper:

- for $x,y \in \mathbf{R}^N$, (x,y) denotes the Euclidean scalar product and $|x|$ the corresponding norm;

- $L^p(0,T;\mathbf{R}^N)$, resp. $H^{k,p}(S^1,\mathbf{R}^N)$ denote the usual Lebesgue (resp. Sobolev) spaces. By S^1 we mean $\mathbf{R}/[0,T]$

- for $u \in L^2(0,T;\mathbf{R}^N)$ we set $|u|_2^2 = \int_0^T |u(t)|^2 dt$; for $u \in H^{1,2}(S^1,\mathbf{R}^N)$ we set $\|u\|^2 = |u|_2^2 + |\dot{u}|_2^2$;

- if $\Omega \subset \mathbf{R}^N$ is an open set, Λ stands for $\Lambda^1(\Omega) = \{u \in H^{1,2}(S^1,\mathbf{R}^N) : u(t) \in \Omega\}$.

- If E is a B-space, A an open subset of X and $f \in C^1(A,\mathbf{R})$ we indicate by $f'(u)$ the F-derivative of f at $u \in A$. If E is a Hilbert space, the same symbol $f'(u)$ will indicate the gradient of f at u.

- If $f : A \to \mathbf{R}$, $f^a := \{u \in A : f(u) \leq a\}$.

- If X is a topological space, $H_q(X)$ denotes the q-th homology group of X.

§1.

Here we study (*) assuming (Ω1).

Before stating various kind of results, it is convenient to indicate the functional setting, which is the same throughout this section.

Let $\Lambda = \Lambda^1(\Omega)$ and, for $u \in \Lambda$, let

$$f(u) = \int_0^T \{\frac{1}{2} |\dot{u}|^2 - V(t,u)\}dt . \tag{1}$$

Clearly $f \in C^1(\Lambda,\mathbf{R})$ (f is C^2 if V is C^2). Moreover, it is well known that the critical points of f on Λ correspond to (classical) T-periodic solutions of (*).

The main specific features of (1) will be described by discussing the following result, which is a particular case of a more general statement of W. B. Gordon [10].

Theorem 1. Let $\Omega = \mathbf{R}^2 - \{0\}$ and suppose $V \in C^1(\Omega,\mathbf{R})$ is independent on t and satisfies:

$$V(x) \leq \text{cost.} \quad \forall x \in \Omega . \tag{2}$$

In addition, suppose

$$\left.\begin{array}{l} \text{there is } U \in C^1(\Omega,\mathbf{R}), \text{ with } U(x) \to -\infty \text{ as} \\ x \to 0, \text{ such that } V(x) \leq -|\nabla U(x)|^2 \text{ for } x \in \Omega . \end{array}\right\} \tag{SF}$$

Then, for all $T > 0$ (*) has a non-constant T-periodic solution. □

Idea of the proof. By (2) it follows that f is bounded from below on Λ. In order to find a minimum for f, one needs: (i) to control the behavior of f at the boundary $\partial\Lambda$ of Λ; (ii) to show that f is coercive. Condition (SF) is employed to overcome (i). In fact one shows:

Lemma 2. Suppose (SF) holds and let $\bar{u} \in \partial\Lambda$. Then $f(u) \to +\infty$ as $u \to \bar{u}$. □

As for (ii), one first notices that the trivial estimate

$$f(u) \geq \frac{1}{2} |\dot{u}|^2 - \text{const.} \tag{3}$$

does not suffice for the coercivity, in general. However, in the present case, since $\Omega = \mathbf{R}^2 - \{0\}$, one can argue as follows. Let Λ_0 be the subset of Λ consisting of all paths p which wind around the origin. For $p \in \Lambda$ one easily shows that $\|p\| \leq \text{const.} |\dot{p}|_2$. This, jointly with (3), implies f is coercive on Λ_0. Then the theorem follows, taking the minimum of f on Λ_0. □

Remarks 3. a) Condition (SF) (= "Strong Force") is violated in the case of Keplerian potentials. In fact, if $V(x) = -|x|^{-\alpha}$, then (SF) holds whenever $\alpha \geq 2$.

b) In order to extend Theorem 1 to higher dimensions (i.e. $N \geq 3$) a stronger condition on Ω is needed: Ω must possess a loop p_0 such that: for all $c > 0$ there exists a compact $K_c \subset \mathbf{R}^N$ containing every loop p homotopic to p_0 in Ω, with length $\leq c$.

c) The results of [10] have been extended in [5] to cover dynamical systems with a forcing term. □

Our next result is concerned with the case in which ($\Omega 1$) and (1^-) hold. We will follow [2].

As pointed out in the preceding Remark 3-b), the arguments of Theorem 1 cannot be carried over when $\Omega = \mathbb{R}^N - \{0\}$, $N \geq 3$. Actually, in general, f is neither coercive nor attains the minimum on Λ. For example, if $V(x) = -|x|^{-2}$, then $f(u) > 0$ on Λ; on the other hand, if $x_n \in \Omega$, $|x_n| \to \infty$, then $f(x_n) \to 0$.

Another way to express the above fact is to say that the (PS) (= Palais and Smale) compactness condition fails, in general.

To overcome this lack of compactness, the idea is to use Morse theory, taking advantage of the fact that $H_*(\Lambda)$ is infinite.

Roughly, the procedure is the following: under some mild assumption on the behavior of V as $|x| \to \infty$ (see (4) below), one shows that (PS) holds on the sets $\{f \geq \varepsilon\}$ for all $\varepsilon > 0$.

Second, one evaluates in a direct way the homology $H_*(f^\varepsilon)$ for $\varepsilon > 0$ small enough and shows that it is finite.

Usual arguments of Morse theory permit to conclude, proving the existence of critical points of f, possibly not minima.

Let us point out that, in any case, we shall assume (SF).

To be specific, we will discuss in the following one situation only. Other cases are studied in [2]; see also Remarks 7 below.

Let us suppose $V \in C^2(\mathbb{R} \times \Omega; \mathbb{R})$ is T-periodic in t, satisfies (SF) and:

$$\left.\begin{array}{l} V(t,x) \to 0,\ V_x(t,x) \to 0 \quad \text{as} \quad |x| \to \infty \quad \text{uniformly in } t; \\ \\ \text{and} \\ \\ \exists r > 0 : V(t,x) < 0 \quad \forall |x| \geq r,\ \forall t \,. \end{array}\right\} \quad (4)$$

Let us point out that, if (SF) and (4) hold, then ($V \leq$ const. and hence) f is bounded from below on Λ. The following lemma has a fundamental role.

Lemma 4. Under the above assumptions, one has:

1°) $\exists \varepsilon^* > 0$ and $r^* > 0$ such that $f^\varepsilon = \Gamma_1^\varepsilon \cup \Gamma_2^\varepsilon \ \forall 0 < \varepsilon < \varepsilon^*$,

where: $\Gamma_1^\varepsilon = \{u \in f^\varepsilon : |\int_0^T u| < r^*\}$ and

$$\Gamma_2^\varepsilon = \{u \in f^\varepsilon : |\int_0^T u| > r^*\}.$$

2°) (PS) holds on $\{f \geq \varepsilon\}\ \forall \varepsilon > 0$ and on $\Gamma_1^\varepsilon\ \forall 0 < \varepsilon < \varepsilon^*$. □

As a consequence of Lemma 4-1°), f^ε consists of two disjoint components Γ_1^ε, which could be possibly empty, and Γ_2^ε. This latter

---

(*)We say that (PS) holds on the set $X \subset \Lambda$, if every sequence $u_n \in X$ such that $f(u_n)$ is bounded and $f'(u_n) \to 0$ has a converging subsequence.

contains the large constants and causes the failure of (PS). In spite of this, it is possible to evaluate the homology groups $H_*(f^\varepsilon)$, for $\varepsilon > 0$ small, by an "ad hoc" argument. Essentially one argues as follows: letting $Pu = \frac{1}{T}\int_0^T u$ and $w = u - Pu$, consider the projection on $\mathbb{R}^N$:

$$\Pi(t,u) = tw + Pu, \qquad t \in [0,1] .$$

If $0 < \varepsilon < \varepsilon^*$, then one shows that $\exists \varepsilon' < \varepsilon$ such that $\Pi(t,u) \in \Gamma_2^\varepsilon$ for all $t \in [0,1]$ and all $u \in \Gamma_2^{\varepsilon'}$. Hence, combining a retraction of Γ_2^ε onto $\Gamma_2^{\varepsilon'}$ (along the steepest descent flow) with the projection Π, one obtains that Γ_2^ε can be retracted onto a sphere $S_R^{N-1} = \{x \in \mathbb{R}^N : |x| = R\}$, R large.

More precisely the following lemma holds:

Lemma 5. Suppose that, for some $\varepsilon \in]0,\varepsilon^*]$, f has no critical points in Γ_2^ε. Then:

$$H_*(\Gamma_2^\varepsilon) = H_*(S_R^{N-1}) . \qquad \square$$

We are now in position to state:

Theorem 6. Suppose $\Omega = \mathbb{R}^N - \{0\}$, $N \geq 2$, and let $V \in C^2(\mathbb{R} \times \Omega, \mathbb{R})$ be T-periodic in t, and satisfy (SF) and (4). Then (*) has infinitely many T-periodic solutions.

Idea of the proof. By Lemma 4-1°) we deduce:

$$H_*(f^\varepsilon) = H_*(\Gamma_1^\varepsilon) \oplus H_*(\Gamma_2^\varepsilon), \qquad \forall 0 < \varepsilon < \varepsilon^* .$$

If f has only a finite number of critical points on Λ, an application of Lemma 5 says that $\exists \delta \in]0,\varepsilon^*]$ such that $H_q(\Gamma_2^\delta) = \{0\}$ $\forall q \neq 0, N-1$. Moreover, since (PS) holds on Γ_1^ε (Lemma 4-2°), $H_*(\Gamma_1^\varepsilon)$ is finite, too. These two facts imply:

$$H_q(f^\delta) = \{0\} \qquad \forall q \ \text{large} .$$

Now, still assuming that f has finitely many critical points, we use Lemma 4-2°) and (SF) to infer, via the usual deformation along the steepest descent flow, that

$$H_q(f^\delta) = H_q(\Lambda) \qquad \forall q \ \text{large} .$$

Since (Ω1) implies $H_*(\Lambda)$ is infinite, we find a contradiction, and the theorem follows. $\square$

Remark 7. a) The same result holds if (Ω1) is replaced by the requirement that $\mathbb{R}^N - \Omega$ is compact;

b) by suitable modifications one can handle the case in which, instead of (4), V satisfies:

$$\left.\begin{array}{l} V(t,x) \to 0 \ \text{ as } \ |x| \to \infty, \ \text{ uniformly in } \ t, \ \text{ and} \\ \exists r_1 > 0 \ \text{ such that } \ (V_x(t,x),x) < 0 \ \ \forall |x| \geq r_1, \ \forall t \ . \end{array}\right\} \qquad (4')$$

Also in such a case one proves (*) has infinitely many T-periodic solutions;

c) if V is independent on t, the same arguments can be used. Assuming V satisfies (SF), either (4) or (4') and that the set $\{x \in \Omega : \nabla V(x) = 0\}$ is compact, it follows that the autonomous system

$$\ddot{x} + \nabla V(x) = 0$$

has infinitely many, distinct, non-constant, T-periodic solutions, for all given $T > 0$;

d) some results concerning the case $(\Omega 1)$-(1^-) have been obtained by C. Greco [11] by a different approach. □

The last part of this section is devoted to study the case (1^+). Our discussion will follow [8].

First, few words are in order concerning the (SF) condition, which shall be still assumed (on $-V$, of course). In the present case, such a condition is less unnatural: for example, the "effective potential" of the Kepler problem is just $V(r) = -\frac{k}{r} + \frac{M^2}{2r^2}$, $r = |x|$, k and M positive constants, (cf. [4, page 38]) which satisfies (SF).

The result we want to expose is the following:

Theorem 8. Suppose $\Omega = \mathbb{R}^N - \{0\}$ and let $V \in C^2(\mathbb{R} \times \Omega, \mathbb{R})$ be T-periodic in t and satisfy:

$$\left.\begin{array}{l} V(t,x) \uparrow 0 \ \text{ as } \ |x| \to \infty, \ \text{ uniformly in } \ t \ ; \\ V_x(t,x) \to 0 \ \text{ as } \ |x| \to \infty, \ \text{ uniformly in } \ t \ ; \\ (V_x(t,x),x) \leq \text{const.} \quad \forall x \in \Omega, \ \forall t \ . \end{array}\right\} \qquad (5)$$

Suppose, further, that $-V$ verifies (SF). Then (*) has at least one T-periodic solution.

Idea of the proof. The arguments are similar to those used in proving Theorem 6, the main difference being that now f is unbounded below on Λ.

First of all, one shows, as in Lemma 4-2°), that (PS) fails to hold only at the level $c = 0$ (namely that each sequence $u_n \in \Lambda$ such that $f(u_n) \to c \neq 0$ and $f'(u_n) \to 0$ has a convergent subsequence).

This allows us to prove that, if - by contradiction - f has no critical points on Λ, one has:

$$H_q(\Lambda) = H_q(f^{\varepsilon}) \quad \forall q,\ \varepsilon > 0 \tag{6}$$

and

$$H_q(\Lambda) = H_q(f^{-\varepsilon}) \quad \forall q,\ \varepsilon > 0\ . \tag{6'}$$

On the other hand, it is possible to show, similarly as in Lemma 5, that (for $\varepsilon > 0$ small)

$$H_q(f^{\varepsilon}) = H_q(f^{-\varepsilon}) \oplus H_q(S^{N-1})\ . \tag{6"}$$

Clearly (6), (6') and (6"), for $q = N - 1$, give rise to a contradiction. □

Remark 9. Also here one can study the autonomous system:

$$\ddot{x} + \nabla V(x) = 0\ . \tag{7}$$

Assuming V has only a finite number of nondegenerate stationary points, it can be shown that $\exists T_0 > 0$ such that $\forall T \geq T_0$ (7) has one T-periodic, non-constant, solution.

§2.

In this section we will deal with bounded, convex potential wells Ω. For simplicity, we will assume $0 \in \Omega$.

Our first result is taken from [1]. We suppose $V \in C^2(\Omega,\mathbb{R})$ satisfies:

$$\left.\begin{array}{l} V \text{ is strictly convex, } V(0) = 0 = \min_{\Omega} V,\ V(x) \to +\infty \\ \text{as } x \to \partial\Omega, \text{ uniformly, and } \exists\theta \in]0, \frac{1}{2}[\text{ such that} \\ V(x) \leq \theta(x,\nabla V(x)) \text{ near } \partial\Omega\ . \end{array}\right\} \tag{8}$$

The idea developed in [1] is to substitute the search of critical points of f on $\Lambda^1(\Omega)$, as in the preceding section, with a dual approach, according to the Dual Variational Principle of Clarke-Ekeland [6,7]. More precisely, letting

$$G(y) = \max_{x\in\Omega}\{(x,y) - V(x)\}\ ,$$

(8) implies $G \in C^2(\mathbb{R}^N,\mathbb{R})$ is convex and $|G(y)| \leq c|y|$ for $|y|$ large.

Let

$$E = \{u \in L^1(0,T;\mathbb{R}^N) : \int_0^T u = 0\}$$

and

$$Lu = v \quad \text{iff} \quad -\ddot{v} = u, \qquad v \in E .$$

Define, for $u \in E$

$$\Phi(u) = \int_0^T [G(u) - \frac{1}{2}(u,Lu)]dt .$$

Since $G \in C^1(\mathbb{R}^N,\mathbb{R})$ and $G(y) \cong |y|$ as $|y| \to \infty$, it follows that $\Phi \in C^1(E,\mathbb{R})$. Let $u \in E$ be such that $\Phi'(u) = 0$. Then $\exists \xi \in \mathbb{R}^N$ such that $-Lu + \nabla G(u) = \xi$. Setting $x = \nabla G(u)$ one has $u = \nabla V(x)$ as well as:

$$x(t) = Lu + \xi ; \tag{9}$$

hence $-\ddot{x} = u$ and thus;

$$-\ddot{x} = \nabla V(x), \quad \text{weakly} .$$

Remark that $x(t)$ is A.C. (indeed $x \in W^{2,1}$) and $x(t) \in \bar{\Omega}$ for all t. Let $I \subset [0,T]$ be an interval such that $x(t) \in \Omega$ $\forall t \in I$. By the conservation of energy one finds

$$E(t) := \frac{1}{2}|\dot{x}(t)|^2 + V(x(t)) = c \qquad \forall t \in I \tag{10}$$

with c constant independent on I. It follows $V(x(t)) \lesssim c$ and this readily implies that $x(t)$ cannot reach the boundary $\partial\Omega$, i.e. $x(t) \in \Omega$ $\forall t \in [0,T]$ and is a (classical) T-periodic solution of (*).

The advantage of this approach is that one can employ the usual critical point theory. In fact, using also the last condition in (8), one shows that Φ satisfies (PS) on E and an application of the Mountain-Pass theorem [3] leads to find a (non-trivial) critical point of Φ and, as consequence, a non-constant T-periodic solution of (*).

Another advantage is that it is now possible to show that T is the real (minimal) period of x. Precisely the following result holds:

Theorem 10. Suppose (Ω2) holds and V satisfies (8). Moreover, assume that $\exists k > 0$ such that

$$(V''(x)y,y) \geq k|y|^2 \quad \forall x \in \Omega, \ \forall y \in \mathbb{R}^N .$$

Let ω_N be the greatest eigenvalue of $V''(0)$ and $T_0 := (2/\omega_N)^{1/2}$. Then $\forall T \in]0,T_0[$ (*) has a T-periodic solution having T as minimal period. □

The preceding approach can be applied to obtain other existence results concerning conservative systems with non-convex potentials as well as to study some classes of Hamiltonian Systems; see [1] for details.

Let us remark that no (SF) condition is assumed in Theorem 10. Now, if this is not surprising in the case where $V(x) \to +\infty$ as $x \to \partial\Omega$ (see the discussion preceding Theorem 10, in particular concerning (10)), different is the situation if we want to study potentials $V(x) \to -\infty$ as $x \to \partial\Omega$. In fact, in such a case it is possible to find solutions of (*) which reach the boundary $\partial\Omega$.

In spite of this, it is a remarkable fact that the Dual approach permits to obtain solutions of (*) lying in Ω without assuming any (SF) condition.

Roughly, let $-V$ satisfy (8) and let us employ the Dual Variational approach. Again a critical point u of Φ corresponds, through (9), to a (weak) solution $x(t)$ of (*), and $x(t) \in \bar{\Omega}$. The idea is now to notice that the critical points of Φ give rise to "smooth" solutions. In fact, by (9), it follows that $x \in W^{2,1}$ and hence $\dot{x}(t)$ is an A.C. function. At this point, one uses again (10): since $|\dot{x}(t)|$ is bounded, the conservation of energy yields $V(x(t))$ is bounded, too. Thus $x(t) \in \Omega$ $\forall t \in [0,T]$.

The above arguments can be carried over in greater generality and leads to the following result contained in [9].

Theorem 11. Suppose (Ω2) holds with $0 \in \Omega$. Let $V \in C^1(\Omega;\mathbb{R})$ be such that:

(i) $V(x) \to -\infty$ as $x \to \partial\Omega$ uniformly;

(ii) $\exists m > 0$ such that $\frac{1}{2} m|x|^2 - V(x)$ is strictly convex;

(iii) $\exists \theta \in]0, \frac{1}{2}[$ such that $V(x) \geq \theta(x, \nabla V(x))$ near $\partial\Omega$.

Let $h(t)$ be smooth and T-periodic. Then the system

$$-\ddot{x} = \nabla V(x) + h(t)$$

has a T-periodic solution $x(t)$, with $x(t) \in \Omega$ $\forall t$.

REFERENCES

1. A. Ambrosetti and V. Coti Zelati, 'Solutions with minimal period for Hamiltonian systems in a potential well', to appear in Annales I.H.P. "Analyse non lineare".
2. A. Ambrosetti and V. Coti Zelati, 'Critical points with lack of compactness and singular dynamical systems', to appear in Annali Mat. Pura Applicata.
3. A. Ambrosetti an P. H. Rabinowitz, 'Dual variational methods in critical point theory and applications', J. Funct. Anal. **14** (1973), 349-381.
4. V. I. Arnold, Mathematical methods of classical mechanics, Springer Verlag, 1980.
5. A. Capozzi, C. Greco and A. Salvatore, 'Lagrangian systems in presence of singularities', preprint University of Bari, Italy, 1985.
6. F. Clarke, 'Periodic solutions of Hamiltonian inclusions', J. Diff. Equat. **40** (1981), 1-6.

7. F. Clarke and I. Ekeland, 'Hamiltonian trajectories having prescribed minimal period', Comm. Pure Appl. Math. **33** (1980), 103-116.
8. V. Coti Zelati, 'Conservative systems with effective-like potentials', to appear in Nonlinear Anal. TMA.
9. V. Coti Zelati, 'Remarks on dynamical systems with weak forces', preprint SISSA, Trieste, Italy, 1986.
10. W. B. Gordon, 'Conservative dynamical systems involving strong forces', Trans. Amer. Math. Soc. **204** (1975), 113-135.
11. C. Greco, 'Periodic solutions of a class of singular Hamiltonian systems', preprint University of Bari, 1986.

PSEUDO-ORBITS OF CONTACT FORMS

Abbas Bahri
22 Rue Petrelle
75009 Paris
France

This is a brief summary of a paper to appear, where I developed some tools in order to study the Weinstein conjecture [1]. This conjecture states that any contact vector-field on a compact contact manifold (M^{2n-1},α) has a periodic orbit, provided $H^1(M^{2n-1},\mathbf{Z}) = 0$. This conjecture has been seen to hold in case M^{2n-1} may be embedded in $\mathbf{R}^{2n}$, with an embedding i such that $i^*\omega = d\alpha$, where ω is the standard symplectic form in $\mathbf{R}^{2n}$. The proof in the $\mathbf{R}^{2n}$-case, after the results of Birkhoff and Seifert, starts with the work of Paul H. Rabinowitz and A. Weinstein, who studied the case where M is a starshaped or convex hypersurface in $\mathbf{R}^{2n}$. Recently, C. Viterbo proved the conjecture under the general assumption $i^*\omega = d\alpha$. There is, at this precise moment, a gap between the full conjecture and the known results; hopefully, it will be filled soon.

What I present here is essentially the description of a phenomenon I encountered while studying this conjecture in its abstract framework. I have limited the presentation of my results to the case $n = 2$, i.e. M is three dimensional.

All the proofs are in [2].

The conclusion of this paper is centered around a clear presentation of the notion of critical points at infinity, which I have been led to introduce starting from this work.

1. THE DATAS - THE FUNCTIONALS

Given (M,α), ξ is the Reeb vector-field of α, i.e.

$$\alpha(\xi) \equiv 1; \quad d\alpha(\xi,\cdot) \equiv 0 . \qquad (1.1)$$

Let $H^1(S^1;M)$ be the space of H^1-loops on M. On $H^1(S^1;M)$, there are some "natural" functionals whose critical points are orbits of ξ; namely:

P. H. Rabinowitz et al. (eds.), Periodic Solutions of Hamiltonian Systems and Related Topics, 11–56.

$$\int_0^1 \alpha_x(\dot{x})dt; \qquad \int_0^1 \alpha_x(\dot{x})^2 dt \ . \tag{1.2}$$

The gradient of these functionals along a variation z in $T_xH^1(S^1;M)$ is:

$$\int_0^1 d\alpha_x(\dot{x},z)dt; \quad 2\int_0^1 \alpha_x(\dot{x})\Big(\frac{d}{dt}\alpha_x(z) - d\alpha_x(\dot{x},z)\Big)dt \tag{1.3}$$

and we have the following Proposition.

Proposition 1: Critical points of $\int_0^1 \alpha_x(\dot{x})^2 dt$ of strictly positive energy are periodic orbits to ξ. Critical points of $\int_0^1 \alpha_x(\dot{x})dt$ of non-zero energy are, after reparameterization, periodic orbits to ξ.

Such a proposition leaves some hope to study the related variational problems on $H^1(S^1;M)$ and try to find an existence mechanism from Morse theory.

Nevertheless, the variational problems are very much ill-posed and in some sense, are useful in showing that a functional is not enough to derive the existence of critical points. Indeed, both functionals we are considering bear very bad features, which may be summed up as follows:

1. The gradients are not Fredholm.
2. The second variation at a critical point has infinite Morse index.
3. The level sets have the same homotopy type.
4. The Palais-Smale condition is not satisfied.

In order to see 1., one writes for instance the gradients in local coordinates. 2. also is almost immediate. 4. will be discussed in the sequel, at least for the restriction of these variational problems to a submanifold of the loop space. The arguments are more complicated to see 3. Nevertheless, there is a result by S. Smale [3], which shows 3. partly:

Theorem (Smale [3]): Let α be a contact form on M. Let $\mathcal{L}_\alpha = \{x \in H^1(S^1;M) \text{ s.t. } \alpha_x(\dot{x}) \equiv 0\}$. The injection of $\mathcal{L}_\alpha$ in $H^1(S^1;M)$ is then a weak homotopy equivalence.

Denoting now:

$$J(x) = \int_0^1 \alpha_x(\dot{x})^2 dt \ , \tag{1.4}$$

$$J^a = \{x \in H^1(S^1;M) \text{ s.t. } J(x) < a\} \ . \tag{1.5}$$

Smale's theorem asserts that J^{∞} and J^0 are homotopy equivalent. Although this does not prove 3., it is very reasonable after having studied the problem, to conjecture that $J^a \simeq J^b$ for any a and b. Such a result, which indeed holds, has nevertheless not very much interest, but in giving another bad feature of these functionals.

With 1., 2., 3. and 4., no variational theory is in fact possible and some thought shows that either of these forbids the use of variational arguments:

Indeed, if the gradients (or pseudo-gradients) are not Fredholm (at least in some weak sense), no Morse lemma is provided; and therefore one does not know anymore if a given critical point induces or not a difference of topology in the level sets.

If the Morse index of a critical point is infinite, and if the gradient is Fredholm, it is known that this critical point does not induce a difference of topology at least if the space of variations (here $H^1(S^1;M)$) is modelled on a Hilbert or a reflexive Banach space. As the gradient is not Fredholm here, we cannot derive such a result.

The homotopy type of the level sets is another key-feature. Usually, in order to find critical points, one tries to find variations in the homotopy type of these level sets. Here, there are none; and would there be some, we would not be able, for instance because of 1., or 2., or 4., to derive any existence result.

Lastly, the Palais-Smale condition is also a key tool as can be seen on very simple functions from $\mathbf{R}$ to $\mathbf{R}$, for instance.

Therefore, Proposition 1 is useless.

The remaining possibilities are either to find other functionals in order to study the problem, other methods (see, for instance, Paul Rabinowitz approach in case M is starshaped in $\mathbf{R}^{2n}$, or more recent developments by C. Viterbo); or to restrict the variations, i.e. to study the variational problems on suitable submanifolds of the loop space(*).

Following Smale's theorem, there are "natural" submanifolds of the loop space where it is tempting to consider the variational problem. Namely, considering:

$$v \text{ a vector-field in } \ker \alpha; \text{ which we assume to be non-singular.} \tag{1.6}$$

We introduce:

$$\beta = d\alpha(v,\cdot)\ , \tag{1.7}$$

$$\mathcal{L}_\beta = \{x \in H^1(S^1;M) \ \text{ s.t. } \ \beta_x(\dot{x}) \equiv 0\}\ , \tag{1.8}$$

(*)The variational framework presented here is a joint work of D. Bennequin and the author.

$$C_\beta = \{x \in H^1(S^1;M) \quad \text{s.t.} \quad \beta_x(\dot{x}) \equiv 0; \\ \alpha_x(\dot{x}) = C^t > 0\} . \tag{1.9}$$

From now on, we denote:

$$J(x) = \int_0^1 \alpha_x(\dot{x})^2 dt, \quad \text{if the space of variations is } \mathcal{L}_\beta , \tag{1.10}$$

$$J(x) = \alpha_x(\dot{x}) = \int_0^1 \alpha_x(\dot{x})dt, \quad \text{if the space of variations is } C_\beta . \tag{1.11}$$

There are arguments in favor of such a choice of restricted variations, which are summed up in the following proposition:

<u>Proposition 2</u>:

1) Generically on v, the singularities of $\mathcal{L}_\beta/M$ lie on finitely many <u>geometric</u> curves. If β is a contact form, $\mathcal{L}_\beta/M$ is a submanifold of the loop space.
2) C_β is, generically on v, a submanifold of the loop space.
3) The critical points of J on $\mathcal{L}_\beta/M$, of <u>finite Morse index</u>, are periodic orbits of ξ. The same holds true on C_β.

Proposition 2 leaves some hope to be able to find those periodic orbits by restricting the variational problem to $\mathcal{L}_\beta$ or C_β. In fact, in case β may be chosen to be a contact form (which imposes restrictions on v), the topology of $\mathcal{L}_\beta$ is known, through Smale's theorem, to be that of the loop space on M. To relate this to a more classical framework, we simply say here that if M is a convex hypersurface in $\mathbb{R}^{2n}$, then such a choice of v is possible (see [2] for more details).

We will rediscuss points 1 to 4 for these new variational problems later on. For the moment, we start a description of the dynamics of α along v.

2. THE PENDULUM EQUATION

From now on, we assume, for sake of simplicity, that β is itself a contact form. By transversality of β and α, we then have:

$$(\beta \wedge d\beta)(\alpha \wedge d\alpha) > 0 . \tag{2.1}$$

We normalize v so that:

$$\beta \wedge d\beta = \alpha \wedge d\alpha . \tag{2.2}$$

Let w be the Reeb vector-field of β. We set:

$$\alpha(w) = \bar{\mu}; \quad d\alpha([\xi,[\xi,v]],[\xi,v]) = \tau \tag{2.3}$$

and we have:

$$\left.\begin{aligned} \bar{\mu} &= d\alpha(v,[v,[\xi,v]]) \\ d\bar{\mu}(\xi) &= \bar{\mu}_\xi = d\alpha([\xi,v],[v,[\xi,v]]) \end{aligned}\right\} \tag{2.4}$$

where $d\bar{\mu}(\xi)$ is the value of the differential of $\bar{\mu}$ on ξ.

There is a geometric property of the dynamics of α along v which we single out now:

Let x_0 be a point of M and let φ_s be the one-parameter group generated by v. We choose at x_0 two vectors in $T_{x_0}(M)$, denoted $e_1(0)$ and $e_2(0)$ such that

$$\alpha \wedge d\alpha(v,e_1(0),e_2(0)) < 0 . \tag{2.5}$$

In particular, $(v,e_1(0),e_2(0))$ is a basis of $T_{x_0}(M)$. Let:

$$e_1(s) = D\varphi_s(e_1(0)); \quad e_2(s) = D\varphi_s(e_2(0)) . \tag{2.6}$$

The plane $P_s = \{z \in T_{x_s}(M) \text{ s.t. } \alpha_{x_s}(z) = 0\}$, where $x_s = \varphi_s(x_0)$, intersects transversally the plane $\operatorname{span}(e_1(s),e_2(s))$ along a direction:

$$u(s) = \alpha_{x_s}(e_1(s))e_2(s) - \alpha_{x_s}(e_2(s))e_1(s) . \tag{2.7}$$

We then have the following:

<u>Proposition 3</u>: $u(s)$ rotates monotonically (strictly) from $e_1(s)$ to $e_2(s)$ when s increases.

Proposition 3 expresses exactly the fact that α is a contact form. Due to this rotation property, we introduce the two following definitions:

<u>Definition 1</u>: $x_s = \varphi_s(x_0)$ is a <u>coincidence point</u> of x_0 if $u(s)$ has completed, in the $(e_1(s),e_2(s))$ frame, k rotations, $k \in \mathbf{Z}$, between time 0 and s. x_s is an <u>oriented coincidence point</u> if $k \equiv 0 \ (2)$.

Therefore, if x_s is a coincidence point of x_0,

$$(\varphi_s^*\alpha)_{x_0} = \lambda(s,x_0)\alpha_{x_0}; \quad \lambda \neq 0 . \tag{2.8}$$

$\lambda(s,x_0)$ is positive if x_s is an oriented coincidence point of x_0.

<u>Definition 2</u>: Let x_s be a coincidence point of x_0. x_s is said to be <u>conjugate</u> to x_0 if $\lambda(s,x_0) = 1$.

As can be checked, these notions of coincidence and conjugate points are intrinsic; i.e. they depend only on α and v.

Starting from here, we have a two-fold discussion. On one hand, there is a discussion on the existence of coincidence and conjugate points which we will complete later on. On the other hand, there are the analytical aspects of these definitions. To summarize these analytical aspects, we first point out the following second order differential equation:

Denoting L_v the Lie-derivative of forms along v, we have three one-forms $\alpha, L_v\alpha, L_v^2\alpha$ which vanish on v. They must therefore be dependent and we have the differential equation:

$$a(x)(L_v^2\alpha)_x + b(x)(L_v\alpha)_x + c(x)\alpha_x \equiv 0 \,. \tag{D0}$$

Recalling that:

$$\beta \wedge d\beta(\xi,v,w) = d\beta(\xi,v) = (L_v^2\alpha)(\xi) = \alpha \wedge d\alpha(\xi,v,w)$$
$$= d\alpha(v,w) = \beta(w) = 1 \,, \tag{2.9}$$
$$d\alpha(v,\xi) \equiv 0 \,.$$

We derive:

$$c(x) \equiv -a(x) \,. \tag{2.10}$$

Recalling that:

$$\left.\begin{aligned} &\alpha(w) = \bar{\mu} \,, \\ &(L_v\alpha)(w) = \beta(w) = 1 \,, \\ &(L_v^2\alpha)(w) = d\beta(v,w) = 0 \,. \end{aligned}\right\} \tag{2.11}$$

We have:

$$b(x) + c(x)\bar{\mu} \equiv 0 \,. \tag{2.12}$$

Therefore, (D0) yields:

$$L_v^2\alpha + \alpha - \bar{\mu}L_v\alpha \equiv 0 \,. \tag{D1}$$

This is the first pendulum equation that we are encountering. In case $M = S^3$ and $\alpha = \alpha_0$ is the standard contact form on S^3 with v defining a Hopf fibration of S^3 over S^2, $\bar{\mu} \equiv 0$ and we have the exact pendulum equation.

Next, we may write (D1) as a transport equation of forms along v, in a selected basis. Namely, introducing:

$$\gamma = d\beta(v,\cdot) \ . \tag{2.13}$$

We may write along a v-orbit x_s the transport equation of forms in the frame $(\alpha_{x_0}, -\gamma_{x_0}, \beta_{x_0})$. Let

$$\left.\begin{aligned} u^*(0) &= \begin{bmatrix} A^*(0) \\ B^*(0) \\ C^*(0) \end{bmatrix} \text{ be the coordinates of a one-form at } x_0 \\ u^*(s) &= \begin{bmatrix} A^*(s) \\ B^*(s) \\ C^*(s) \end{bmatrix} \text{ be the coordinates of } \varphi^*_{-s}u^*(0) \text{ at } x_0 \ . \end{aligned}\right\} \tag{2.14}$$

We then have:

<u>Proposition 4</u>:

$$\left.\begin{aligned} \dot{u}^*(s) &= -{}^t\Gamma_{(x_s)}u^*(s) \\ u^*(0) &= \begin{bmatrix} A^*(0) \\ B^*(0) \\ C^*(0) \end{bmatrix} \end{aligned}\right\}$$

where

$$\Gamma(x) = \begin{bmatrix} 0 & 0 & -1 \\ 0 & 0 & \bar{\mu}_\xi \\ 1 & 0 & \bar{\mu} \end{bmatrix}$$

or else

$$\left.\begin{aligned} \dot{A}^*(s) &= C^*(s) + \bar{\mu}_\xi B^*(s) \\ \dot{C}^*(s) &= -A^*(s) - \bar{\mu}C^*(s) \\ \dot{B}^*(s) &= 0 \ . \end{aligned}\right\}$$

In particular, if $B^*(0) = 0$, i.e. if the initial data belong to $\text{span}\{\alpha_{x_0}, \beta_{x_0}\}$, then $B^*(s) = 0$ and the differential equation reduces to:

$$\left.\begin{aligned} \dot{A}^*(s) &= C^*(s) \\ \dot{C}^*(s) &= -A^*(s) - \bar{\mu}(x_s)C^*(s) \end{aligned}\right\} \tag{D2}$$

once again the pendulum equation. Coincidence and conjugate points to x_0 are then described as follows:

Proposition 5: Considering the initial data $\begin{bmatrix}1\\0\\0\end{bmatrix}$, x_s is a coincidence point of x_0 if $u^*(s) = \begin{bmatrix}A^*(s)\\0\\0\end{bmatrix}$; x_s is conjugate to x_0 if $u^*(s) = \begin{bmatrix}1\\0\\0\end{bmatrix}$. Therefore, conjugate points are related to periodic solutions of (D2).

Through these analytical descriptions of coincidence and conjugate points, we will encounter them in the previously defined functional framework, while studying the Palais-Smale condition.

We discuss now the existence of coincidence and conjugate points: There are two opposite situations which are relevant to this discussion, which we summarize in two opposite hypotheses:

Hypothesis A1: α turns well along v; i.e. every x_0 on M has a coincidence point distinct from itself. In this case, we say that (α,v) is elliptic.

Hypothesis A2: v belongs to a codimension one foliation transverse to α.

It is proven in [2] that a large class of contact forms (presumably all of them) admit a transverse foliation $\tilde{\gamma}$. If v belongs then to $\ker \tilde{\gamma} \cap \ker \alpha$, no point x_0 of M can possibly admit another coincidence point distinct from itself. (A complete rotation forces $\tilde{\gamma}$ and α to coincide somewhere in between.)

In this situation, the variational problem, as well as the topology of C_β, have drastically different features from the hypothesis A1 case. This will be described in a later publication.

In case (A1) holds, there are of course infinitely many coincidence points to each given point x_0 of M and the question which is left open is the existence of conjugate points. In this framework, we introduce:

Definition 3: Under (A1), let $\gamma^i : M \to \mathbf{R}$ be the function mapping $x_0 \in M$ to the ith-time $s = \gamma^i(x_0)$ such that x_s is a coincidence point of x_0 $(i \in \mathbf{Z})$. Let $f^i : M \to M$ be the diffeomorphism of M mapping x_0 on $x_{\gamma^i(x_0)}$. Let $\mu_i(x_0)$ be the coefficient of $(\varphi^*_{\gamma^i(x_0)}\alpha)_{x_0}$ on α_{x_0}, i.e.:

$$(\varphi^*_{\gamma^i(x_0)}\alpha)_{x_0} = \mu_i(x_0)\alpha_{x_0} \ .$$

Conjugate points are defined by the equation:

$$\mu_i(x_0) = 1 \ . \tag{2.15}$$

Therefore, points admitting a conjugate point distinct from themselves lie generically on unions of hypersurfaces $\Sigma_i = \{x \in M_3 \text{ s.t. } \mu_i(x) = 1\}$. An interesting situation is provided here by $(S^3, \alpha = \lambda\alpha_0)$ where $\lambda \in C^\infty(S^3, \mathbf{R}^{+*})$; then all points admitting a conjugate point distinct from themselves lie on $\Sigma = \{x \in S^3 \text{ s.t. } \lambda(x) = 1\}$. If x belongs to Σ, the conjugate points of x are x and $-x$. It is also interesting to notice that the existence of conjugate points is related to the behavior of the function:

$$(\varphi_s^*\alpha)_{x_0} \wedge \varphi_s^*(d\alpha)_{x_0}/(\alpha \wedge d\alpha)_{x_0} \tag{2.16}$$

in particular when s goes to $\pm\infty$. Therefore, the asymptotic behavior of v, the fact that it expands or contracts volume is relevant to the conjugate points.

Before concluding this section, we introduce:

$$\Gamma = \{x \in_0 M \text{ s.t. } \lambda(s,x_0) > 1 \text{ for any } s \text{ such that } x_s \text{ is a coincidence point of } x_0\}, \tag{2.17}$$

and in the framework of (A1), the following hypotheses about v and the dynamics of α along v. We choose a distance on M, denoted d and a related norm for differentials, denoted $\| \ \|$:

(H2) v has a periodic orbit;
(H3) for a vector-field v_1, nonsingular and colinear to v, we have: $\exists k_1 > 0$ such that $\|D\varphi_s^1\| \leq k \ \forall s \in \mathbf{R}$, where φ_s^1 is the one-parameter group of v_1;
(H4) $\exists k_2$ and $k_3 > 0$ such that $\forall i \in \mathbf{Z}$, we have:

$$k_2 d(x,y) \leq d(f^i(x), f^i(y)) \leq k_3 d(x,y) \ \forall x,y \in M ;$$

H(5) $\exists k_4 > 0$ such that $|\mu_i(x) - \mu_i(y)| \leq k_4 d(x,y) \ \forall x,y \in M$;
(H6) $\exists \rho > 0$ such that $\forall x \in M$, the set $C_\rho(x) = \{f^i(x)/|\mu_i(x) - 1| < \rho;\ i \in \mathbf{Z}\}$ is finite.

(H2)-(H6) are heavy hypotheses, under which clean statements about critical points at infinity can be made. Nevertheless, these hypotheses are necessary only at the end of the analysis of the phenomenon; our feeling is that they can be considerably weakened (see [2] for further precisions).

3. TOPOLOGICAL ASPECTS OF THE VARIATIONAL PROBLEMS

3.1. The topology of $\mathcal{L}_\beta$ and C_β:

We limit our discussion to the case (A1) holds. We already pointed out Smale's theorem, which states that, if β is a contact form, the injection of $\mathcal{L}_\beta = \{x \in H^1(S^1;M) \text{ s.t. } \beta_x(\dot{x}) \equiv 0\}$ into $H^1(S^1;M)$ is a weak homotopy equivalence. Smale proves this theorem with showing that the projection:

$$\mathcal{L}_\beta \to M$$
$$x \to x(0) \tag{3.1}$$

is a Serre-fibration.

For this, he needs the hypothesis that β is a contact form and he gives an example where this is no more true, when β is no more a contact form. There is a slight improvement of Smale's theorem that we have completed. Introducing:

$$F = \{x \in M | (\beta \wedge d\beta)_x = 0\} \tag{3.2}$$

we have:

Proposition 6: Assume F is a codimension one submanifold of M and assume that β is transverse to F at any point x of F. $\mathcal{L}_\beta \hookrightarrow H^1(S^1;M)$ is then a weak homotopy equivalence.

Corollary: Generically on β, $\pi_0(\mathcal{L}_\beta) = \pi_1(M)$.

The proof of Proposition 5 follows the method of Stephen Smale; nevertheless, as β is no more a contact form, we need some other reduction of β in the neighborhood of a point of F (see [2]). The Corollary follows immediately Proposition 5, through a transversality argument.

Remark: When β is not transverse to F, the trace of β on F has singularities. Nearby the elliptic ones, β has a nice reduction (see D. Bennequin [4]) which I provided and which was shown by D. Bennequin, in a celebrated theorem, to provide exotic contact structures on $\mathbf{R}^3$.

In fact, the same paper by S. Smale gives some insight on the topology of C_β under (A1). Namely, S. Smale proves in [3] the following theorem, later developed in many directions:

Theorem (S. Smale [3]): Let N be a compact surface without boundary. Let $\mathrm{Imm}(S^1;N)$ be the space of immersions of S^1 in N and $\Omega(ST^*N)$ be the space of continuous closed curves in the cotangent S^1-bundle over N. Then the inclusion $\mathrm{Imm}(S^1,N) \hookrightarrow \Omega(ST^*N)$ is a weak homotopy equivalence.

This theorem is related to the topology of C_β as follows.

Consider α_0 the standard contact form on ST^*N and let α be a constant form obtained by multiplying α_0 by a C^∞ positive function λ. Let v be the vectorfield defining the natural fibration of ST^*N over N.

It is not difficult to see that, in this context, the form α_0, hence the form α, makes one (and only one) complete revolution along the fiber (in the sense of rotation introduced in Proposition 3; see Definitions 1, 2 and 3).

Suppose now $\beta = d\alpha(v,\cdot)$ is also a contact form, which means, from the Hamiltonian point of view, the convexity of the kinetic energy in the cotangent variable. As $\beta(v)$ is zero, β is also tangent to the fibers. Being transverse to α, β makes also one (and only one) complete revolution along the fiber.

Let x be an immersion of S^1 in N; $x(t_0)$ be a fixed point on the curve and $\dot{x}(t_0)$ be its tangent vector.

Let $p : ST^*N \to N$ be the natural projection.

As β turns once along the fiber, ξ, which belongs to the kernel of β, turns also once along this fiber. This implies the existence and the unicity of a point $(x(t_0),\theta)$, projecting on $x(t_0)$, in ST^*N such that:

$$dp_{(x(t_0),\theta)}(\xi_{(x(t_0),\theta)}) = \mu\dot{x}(t_0); \quad \mu > 0 , \tag{3.3}$$

where $dp_{(x(t_0),\theta)}$ is the differential of p at $(x(t_0),\theta)$ and $\xi_{(x(t_0),\theta)}$ is the value of ξ at $(x(t_0),\theta)$.

When t_0 runs in S^1, this device allows to lift the immersion x in a closed curve $y(\cdot) = (x(\cdot),\theta(\cdot))$ belonging to $\Omega(ST^*N)$. This curve y is defined by the following equation:

$$\left.\begin{aligned} &p(y(t)) = x(t) \\ &dp(\dot{y}(t)) = \dot{x}(t);\ dp_{y(t)}(\xi_{y(t)}) = \mu(t)\dot{x}(t); \quad \mu > 0 . \end{aligned}\right\} \tag{3.4}$$

Equation (3.4) yields that $\dot{y}(t) = \frac{1}{\mu(t)}\,\xi_{y(t)} + \lambda(t)v_{y(t)}$. Hence:

$$\dot{y}(t) = a_1\xi_{y(t)} + b_1 v; \quad a_1 > 0 . \tag{3.5}$$

As a_1 is continuous and positive, one can reparameterize the curve $y(\cdot)$; hence obtaining $z(\cdot)$ such that:

$$\dot{z}(t) = a\xi_{z(t)} + bv_{z(t)}; \quad a = \text{Cste.} > 0 . \tag{3.6}$$

We defined in this way a map h:

$$\text{Imm}(S^1,N) \xrightarrow{\ h\ } C_\beta . \tag{3.7}$$

Conversely, if y is a given curve in C_β, $p(y)$ belongs to $\text{Imm}(S^1,N)$. Indeed:

$$\dot{y} = a\xi + bv; \quad a = \text{positive constant} . \tag{3.8}$$

Hence

$$dp(\dot{y}) = \dot{\widehat{p(y)}} = a\,dp(\xi) . \tag{3.9}$$

As $dp(\xi)$ is not zero, $p(y)$ belongs to $Imm(S^1,N)$. We thus have another map:

$$\hat{h} : C_\beta \to Imm(S^1,N) \ . \tag{3.10}$$

One can check that $\hat{h} \circ h(x)$ is a reparameterization of $x(\cdot)$ while $h \circ \hat{h}(y)$ is a reparameterization of y.

This, together with some approximation argument appropriate to deal with the H^1-topology, proves that C_β has the same topology than $Imm(S^1,N)$; hence than $\Omega(ST^*N)$ by Smale's theorem.

The same argument works if M is a finite covering of ST^*N; $Imm(S^1,N)$ and $\Omega(ST^*N)$ are in this context to be replaced by $Imm_M(S^1,N)$ and $\Omega_M(ST^*N)$ where these indexed sets denote the results of curves which lift to M as closed curves. S. Smale theorem suggests the following result, which may be intuited through complicated variational arguments:

Conjecture: Assume β is a contact form. Under (A1), the injection of C_β in $H^1(S^1;M)$ is a weak homotopy equivalence.

3.2. The change of topology induced by J

We first introduce the following definitions; d is a distance on M.

Definition 4: Let v_1 be a nonsingular vectorfield colinear to v $(v_1 = \lambda v; \lambda > 0)$ and let φ_1^s be the one parameter group generated by v_1. v is said to be ξ-conservative if there exist v_1 and a constant K such that

$$\|D\varphi_s^1(\xi)\| \leq K \quad \forall s \in \mathbb{R} \ . \tag{3.11}$$

Here $\| \ \|$ stands for any norm of differentiable maps from M to M.

Definition 5: v is said to be α-nonresonant if there exists a nonsingular one differential form θ, transverse to α and tangent to v, such that:

$$\begin{aligned} &\text{there exists } \varepsilon_4 > 0 \text{ such that for any } x \text{ in } M \\ &\text{and } s_1 \text{ in } \mathbb{R} \text{ satisfying } d(\varphi_{s_1}(x),x) < \varepsilon_4, \\ &(\varphi^*_{s_1}\alpha)_x \text{ is transverse to } \theta_x \ . \end{aligned} \tag{3.12}$$

Definition 6: C_β is said to be branched on v if there exists $C > 0$ such that for any $\varepsilon > 0$, there exists x in C_β, with $\dot{x} = a\xi + bv$, $0 < a < \varepsilon$ and $\left|\int_0^1 b(t)dt\right| \geq C$.

Here are some comments about these definitions:

First, if v is ξ-conservative, there is no hyperbolic invariant set to v. Indeed, on such a set, Equation (3.11) cannot hold. Hence, v might be called "elliptic" in this situation.

Second, we consider a point x in M which is recurrent to v. Let f_x be the local Poincaré map on a section transverse to v at x.

As α rotates monotonically along v from x to $f_x(x)$, we can define:

$$\tilde{\varphi}(x) = \text{total rotation of } \alpha \text{ from } x \text{ to } f_x(x) \tag{3.13}$$

and

$$n_x = \left[\frac{\tilde{\varphi}(x)}{\pi}\right] . \tag{3.14}$$

If v is α-nonresonant, n_x is constant on the connected components of the v-recurrent set.

Third, C_β is branched on v if there exists curves of C_β close to recurrent orbits of v. We then have:

<u>Lemma 1</u>: If v has a periodic orbit and if (A_1) holds, C_β is branched on v.

Finally, we notice that in case v is the fiber vectorfield of an S^1-fibration over a surface N, then v is ξ-conservative, α-nonresonant and C_β is branched on v. With these definitions, we have the following Proposition:

<u>Proposition 7</u>: Assume (A_1) holds and v is ξ-conservative, α-nonresonant and C_β is branched on v. Then, for $\varepsilon > 0$ small enough, $\pi_0(J^\varepsilon)$ is infinite, where

$$J^\varepsilon = \{x \in C_\beta \text{ s.t. } J(x) < \varepsilon\} .$$

It is difficult to sum up the results of this section.

The conjecture we formulated may be proven through very complicated variational problems (one studies on $\mathcal{L}_\beta$ the functional $\left(\int_0^1 \alpha_x(\dot{x})^2 dt\right)^{1/2} - \int_0^1 \alpha_x(\dot{x})dt)$). The methods are nevertheless too complicated and another approach would be welcome, in order to build a clean proof.

Nevertheless, Smale's theorem shows the conjecture to hold for $(ST^*N, \lambda\alpha_0)$, $\lambda \in C^\infty(ST^*N, \mathbb{R}^{+*})$ and also for all finite coverings of ST^*N, (M,α) such that $q_*(\ker \alpha) = \ker \alpha_0$, where q is the projection $M \to ST^*N$. The simplest example of such coverings is, of course $(S^3, \lambda\alpha_0)$, α_0 being the standard contact form on S^3. In this case, with v defining a Hopf fibration of S^3 over S^2, $\pi_0(C_\beta) = 0$; while $\pi_0(J^\varepsilon)$ is infinite.

This difference of topology is drastic. Some thought shows that it cannot only be due to the periodic orbits of ξ.

Indeed, we may assume, by transversality arguments on λ, that given $a_0 > 0$, there are only finitely many of these orbits of length less than a_0 and that they all are nondegenerate.

On the other hand, the Morse index of J on C_β at these critical points increases to $+\infty$ with the length of these periodic orbits. Although the gradient is not Fredholm, a perturbation of J nearby these critical points, which gradient is Fredholm, induces then a finite difference of topology in the indices 0 and 1; therefore a contradiction. This shows that the only possibility is for the Palais-Smale condition to fail and that it is precisely this failure which will induce such a heavy difference of topology. Starting from here, we see that we are directly led to the idea of critical points at infinity, which have to explain such a drastic difference of topology. Therefore, these critical points at infinity cannot be mere sequences.

4. A COMPACTIFYING DEFORMATION LEMMA

We point out here that the compactifying deformation lemma we are presenting provides us with a control on each flow-line of the vector-fields we will introduce. These vector-fields, which depend on a parameter $\varepsilon > 0$, and which decrease the functional J, do not satisfy the Palais-Smale condition on sequences, unless we specify that these sequences belong to the same flow-line.

As we will see later on in this paper, there are intuitive reasons which imply that J does not satisfy the Palais-Smale condition or else any weaker compacity criterion. This is already hinted by the study of the topology of C_β and the differences of topology induced by J, which are by far too drastic to be due only to the ξ periodic orbits.

For the moment being, we can get a first understanding of the problem: along (pseudo)-gradient lines, one knows that $J(x)$ remains bounded; which provides a bound on the ξ-component of $\dot{x} = a\xi + bv$. (As x belong to C_β, $\dot{x}$ splits on (ξ, v) with a constant component on ξ.)

But we have no control on the v-component and it might happen that this v-component becomes infinite on a gradient line.

On the other hand, if $\partial J(x)$ goes to zero, we get: ($\partial J(x)$ is the gradient of J)

$$\partial J(x) \cdot z = -\int_0^1 b\eta \, dt \to 0 \quad (z = \lambda\xi + \mu v + \eta w) . \tag{4.1}$$

Hence

$$\left|\int_0^1 b\eta\, dt\right| < \varepsilon |z|_{H^1} . \tag{4.2}$$

The H^1-norm of z contains an L^2-norm of $\dot{\eta}$ and other terms also. Hence (4.2) will give a very weak information on b and anyway, no bound whatsoever (there are b-terms in $|z|_{H^1}$); hence the failure of the Palais-Smale condition.

In this situation, we construct a special deformation lemma, which will compactify the situation, by introduction of a certain "viscosity term" which allows to control the b-component. The deformation depends on a certain parameter $\varepsilon > 0$ and is induced by a vectorfield Z_ε such that the scalar product $\partial J(x) \cdot Z_\varepsilon(x)$ is positive. However, Z_ε can be zero while $\partial J(x)$ is not zero; which will bring the differential equation giving rise to critical points at infinity. Let

$$I(x) = \int_0^1 b^2 dt \quad (\dot{x} = a\xi + bv) . \tag{4.3}$$

I is a C^2 functional on C_β on which we wish to have control along deformation lines of the level sets of J.

For this purpose, we introduce the following vectorfield on C_β:

$$Z = |\partial J|\partial I + |\partial I|\partial J , \tag{4.4}$$

∂J and ∂I are the gradients of I and J and $|\partial I|$, $|\partial J|$ are their norms which respect to the H^1-metric of C_β (inherited for the H^1-metric on $H^1(S^1,M)$). We will note (,) the H^1 scalar product on the tangent space to C_β. The idea underlying the deformation lemma is to use Z selectively: where $\partial J(x)$ is small (in fact, some other quantity dominating $|\partial J(x)|$), we use $-Z$ to deform. Otherwise, we use $-\partial J(x)$.

Lemma 2: Z is a locally Lipschitz vectorfield on C_β. $(Z,\partial J)$ is positive as well as $(Z,\partial I)$. If one of these quantities is zero, then Z is zero.

Proof: The only delicate point is the fact that Z is Lipschitz. This is evident, as I and J are C^2, outside those points x of C_β where $\partial J(x)$ or $\partial I(x)$ vanishes. But, in those points, $|\partial J|$ and $|\partial I|$ are locally Lipschitz: hence the result.

Let $\varepsilon > 0$ be given. We consider:

$$\left.\begin{aligned} &\omega_\varepsilon : \mathbb{R} \to [0,1] \\ &\omega_\varepsilon(x) = 1 \quad \text{if} \quad x > \varepsilon \\ &\omega_\varepsilon(x) = 0 \quad \text{if} \quad x < \varepsilon/2 \\ &\omega_\varepsilon \in C^\infty , \end{aligned}\right\} \tag{4.5}$$

$$\left.\begin{aligned} &\ell_\varepsilon : C_\beta \to [0,1]; \quad \ell_\varepsilon \in C^\infty \\ &\ell_\varepsilon = 0 \quad \text{on an } \varepsilon/2\text{-neighborhood of the} \\ &\qquad \text{critical points of } J \\ &\ell_\varepsilon = 1 \quad \text{outside an } \varepsilon\text{-neighborhood of these points.} \end{aligned}\right\} \tag{4.6}$$

$$\left.\begin{aligned} &\varphi : C_\beta/\{\text{critical points of } J\} \to \mathbb{R} \\ &\varphi(x) = \frac{|\partial J|(x)}{|\partial I|(x)} I(x) \quad \text{if} \quad |\partial I|(x) \quad \text{is nonzero} \\ &\varphi(x) = +\infty \quad \text{is} \quad |\partial I|(x) = 0 . \end{aligned}\right\} \tag{4.7}$$

We introduce then the following vectorfield on C_β:

$$Z_\varepsilon(x) = \ell_\varepsilon(x)(\omega_\varepsilon(\varphi(x))\partial J(x) + Z(x)) . \tag{4.8}$$

This formula of Z_ε is a priori defined only on $C_\beta/\{$critical points of $J\}$ as φ is only defined on this set. But it clearly extends to all of C_β by setting $Z_\varepsilon = 0$ on a critical point of J. In fact, Z_ε is zero on an $\varepsilon/2$-neighborhood of these critical points, by definition of ℓ_ε.

We have the following proposition:

Proposition 8: Z_ε is locally Lipschitz on C_β.

Proof: The proof is reduced to the fact that $\omega_\varepsilon(\varphi)$ is locally Lipschitz on the set where ℓ_ε is nonzero.

If x is such that $\ell_\varepsilon(x)$ is nonzero, then $|\partial J|(x)$ is nonzero and $I(x)$ is nonzero (indeed $I(x) = 0$ implies $\partial J(x) = 0$). Hence $|\partial J|(y)$ and $I(y)$ are bounded from below by $a(x) > 0$ on a neighborhood of x.

If $|\partial I|(x)$ is nonzero, φ is continuous and even C^1 on this neighborhood; hence $\omega_\varepsilon(\varphi)$ is Lipschitz.

If $|\partial I|(x) = 0$, then $|\partial I|(y)$ is small on this neighborhood. Hence $\varphi(y)$ is larger than ε (if we restrain the neighborhood). Hence $\omega_\varepsilon(\varphi)$ is equal to one on this neighborhood and therefore is Lipschitz.

The vectorfield Z_ε will provide us with the compactifying deformation lemma that we state now:

We consider the differential equation on C_β

$$\left.\begin{aligned} &\frac{\partial x}{\partial s} = -Z_\varepsilon(x) \\ &x(0) \quad \text{given} . \end{aligned}\right\} \tag{4.9}$$

<u>Compactifying Deformation Lemma</u>:

1. On an integral curve of (4.8) $x(s)$, $J(x(s))$ decreases and $J(x(s))$ remains bounded by a constant depending of ε, $J(x(0))$ and $I(x(0))$, for all time $s > 0$ such that $J(x(s))$ is positive.

2. Let $0 < a_0 < a_1$. Suppose $J^{a_0} = \{x \in C_\beta | J(x) < a_0\}$ is not retract by deformation of $J^{a_1} = \{x \in C_\beta | J(x) < a_1\}$.

Then there exists $\varepsilon_0 > 0$ such that for any $0 < \varepsilon < \varepsilon_0$, there is a point x_ε in C_β with:

$$a_0 < J(x_\varepsilon) < a_1 \quad \text{and either} \quad \partial J(x_\varepsilon) = 0 \quad \text{or}$$

$$Z(x_\varepsilon) = 0; \qquad \varphi(x_\varepsilon) < \varepsilon . \tag{4.10}$$

<u>Proof</u>: Let $f(x) = I(x(s))$, $g(s) = J(x(s))$

$$g'(s) = -(\partial J, Z) = -\ell_\varepsilon(x(s))(\omega_\varepsilon|\partial J|^2 + (Z,\partial J)) .$$

We know that $(Z,\partial J)$ is positive. Hence:

$$\begin{aligned} g(s) - g(0) &= - \int_0^s \ell_\varepsilon(x(\tau))[\omega_\varepsilon|\partial J|^2 + (z,\partial J)](x(\tau)d\tau \\ &< - \int_0^1 \ell_\varepsilon\omega_\varepsilon|\partial J|^2 d\tau . \end{aligned} \tag{4.11}$$

Hence, for s positive and as long as $g(s)$ remains positive, we have:

$$\int_0^s \omega_\varepsilon\ell_\varepsilon(x(\tau))|\partial J|^2 d\tau < g(0) = J(x(0)) . \tag{4.12}$$

We first notice that $g'(s)$ is negative; hence $J(x(s))$ decreases.

On the other hand, $f'(s) = -(\partial I, Z_\varepsilon) = -\ell_\varepsilon(\omega_\varepsilon(\partial I,\partial J) + (Z,\partial I))$. As $(Z,\partial I)$ is positive, we have:

$$\begin{aligned} f'(s) &< -\ell_\varepsilon\omega_\varepsilon(\partial I,\partial J) < \omega_\varepsilon\ell_\varepsilon|\partial I||\partial J| \\ &= \ell_\varepsilon\omega_\varepsilon(\varphi)\frac{|\partial I|}{|\partial J|} \cdot \frac{1}{I}(I|\partial J|^2) . \end{aligned} \tag{4.13}$$

Hence

$$f(s) < \ell_\varepsilon \frac{\omega_\varepsilon(\varphi)}{\varphi} I|\partial J|^2 . \tag{4.14}$$

But, by the very definition of ω_ε, we have:

$$\omega_\varepsilon(\varphi) < 2/\varepsilon\varphi . \tag{4.15}$$

Hence:

$$f'(s) < (2\ell_\varepsilon/\varepsilon)\omega_\varepsilon(\varphi)|\partial J|^2 I = (2\ell_\varepsilon/\varepsilon\omega_\varepsilon(\varphi)|\partial J|^2)f(s) . \tag{4.16}$$

Hence, using (4.12)

$$f(s) < f(0)e^{2/\varepsilon \int_0^1 \omega_\varepsilon(\varphi)|\partial J|^2 \ell_\varepsilon d\tau} < f(0)e^{2/\varepsilon g(0)} . \tag{4.17}$$

The first statement of the lemma is then proven.

The proof of the second statement requires the following two lemmas whose proof is straightforward.

Lemma 3: Let (x_n) be a sequence in C_β such that $0 < a_0 < J(x_n) < a_1$ and $(I(x_n))$ is bounded. If $\partial J(x_n)$ goes to zero, there is a subsequence converging weakly to x in C_β with $0 < a_0 < J(x) < a_1$ and $\partial J(x) = 0$.

Lemma 4: Let (x_m) be a sequence in C_β such that $0 < a_0 < J(x_m) < a_1$ and $(I(x_m))$ is bounded. If $(\partial J(x_m),Z(x_m)) \to 0$ and if (x_m) converges weakly to x in C_β, with $\partial J(x)$ nonzero, there is a strongly convergent subsequence to x.

Remark: If $(\partial J(x_m),Z(x_m)) \to 0$, then $|Z(x_m)| \to 0$; and if (x_m) converges weakly to x such that $\partial J(x)$ is nonzero, with $J(x_m)$ and $I(x_m)$ bounded, then $|\partial J(x_m)|$ is bounded away from zero. Therefore $\partial I(x_m) \to 0$; $I(x_m)$ and $J(x_m)$ are bounded. It is then easy to prove Lemma 4.

Proof of the second statement: Arguing by contradiction, we may assume there is no critical point of J in the set $\{x|a_0 < J(x) < a_1\}$. Hence, for $0 < \varepsilon < \varepsilon_0$, ℓ_ε is equal to 1 on this set.

Let now x_0 be such that $a_0 < J(x_0) < a_1$.

We denote by $x(s,x_0)$ the solution of (4.9) having x_0 as initial data. The situation divides in two cases.

1st case: $\forall s > 0$, $J(x(s,x_0)) > a_0 > 0$.
Then $I(x(s,x_0))$ and $J(x(s,x_0))$ are uniformly bounded on $[0,+\infty[$ and the solution of (4.8) exists for all positive s.

Reminding that $g(s) = J(x(s,x_0))$, we have

$$g'(s) = -\ell_\varepsilon(x(s,x_0))(\omega_\varepsilon|\partial J|^2 + (Z,\partial J)) \ . \tag{4.18}$$

Hence, as $\ell_\varepsilon(x(s,s_0)) = 1$,

$$g'(s) = -\omega_\varepsilon|\partial J|^2 - (Z,\partial J) \ . \tag{4.19}$$

From the boundedness of $g(s)$, we deduce

$$\int_0^{+\infty} [\omega_\varepsilon|\partial J|^2 + (Z,\partial J)]d\tau < +\infty \ . \tag{4.20}$$

As $(Z,\partial J)$ is positive, (4.20) yields the existence of a sequence (x_n) such that:

$$a_0 < J(x_n) < a_1 \ , \tag{4.21}$$

$$\omega_\varepsilon(\varphi)(x_n)|\partial J|^2(x_n) + (\partial J,z)(x_n) \to 0 \ , \tag{4.22}$$

$$I(x_n) < I(x_0)e^{2/\varepsilon \, a_1} \quad \text{(see (4.17))} \ . \tag{4.23}$$

As $(J(x_n))$ and $(I(x_n))$ are bounded, (x_n) is H^1-bounded and we can extract from (x_n) a weakly convergent subsequence to x belonging to C_β, with $a_0 < J(x) < a_1$. We will call this subsequence (x_n) again.

Our hypothesis is that x is not critical.

Hence (x_n) is a sequence such that $a_0 < J(x_n) < a_1$; $(I(x_n))$ is bounded and, by (4.22) $(\partial J,Z)(x_n)$ goes to zero. Applying Lemma 4, we derive that (x_n) converges in fact strongly to x; (4.22) then implies:

$$\omega_\varepsilon(\varphi)(x)|\partial J|^2 + (\partial J,Z)(x) = 0 \ . \tag{4.24}$$

Hence, as $\partial J(x)$ is nonzero,

$$Z(x) = 0 \quad \text{and} \quad \varphi < \varepsilon \tag{4.25}$$

which proves the second statement of the lemma is this case.

<u>2nd case</u>: For any x_0 belonging to $\{x | a_0 < J(x) < a_1\}$, there exists a positive s such that $J(x(s,x_0)) = a_0$.

Let then $s(x_0)$ be the first time s such that $J(x(s(x_0),x_0)) = a_0$. Let

$$z_0 = x(s(x_0),x_0) \ . \tag{4.26}$$

If $(\partial J(z_0),Z(z_0))$ is not zero, the function $s(\cdot)$ is continuous at x_0.

Hence, if

$$(\partial J(x), Z_\varepsilon(x)) > 0 \ \forall x \quad \text{such that} \quad a_0 \leq J(x) \leq a_1 \tag{4.27}$$

then $s(\cdot)$ is globally continuous and the map:

$$J^{a_1} \times [0,1] \to J^{a_1}$$

$$(x_0,t) \to \begin{cases} x(ts(x_0),x_0) & \text{if} \quad J(x_0) \geq a_0 \\ x_0 & \text{if} \quad J(x_0) \leq a_0 \end{cases}$$

defines a retraction by deformation of J^{a_1} on J^{a_0}.

This is excluded from our hypothesis; and (4.27) is consequently impossible; which yields the existence of z_0 such that:

$$J(z_0) = a_0; \ (\partial J(z_0), z_\varepsilon(z_0)) = 0 \quad \text{i.e.}$$

$$\ell_\varepsilon(\omega_\varepsilon(\varphi)|\partial J|^2 + (Z,\partial J))(z_0) = 0 . \tag{4.28}$$

But $\ell_\varepsilon(z_0)$ is equal to 1 $(J(z_0) = a_0)$ and $|\partial J|(z_0)$ is nonzero. Hence:

$$(\partial J(z_0), Z_\varepsilon(z_0)) = 0 \quad \text{and} \quad \omega_\varepsilon(\varphi)(z_0) = 0 \tag{4.29}$$

which implies

$$Z_\varepsilon(z_0); \qquad \varphi(z_0) \leq \varepsilon. \tag{4.30}$$

This ends the proof of the compactifying deformation lemma.

5. ANALYTICAL ASPECTS OF THE CRITICAL POINTS AT INFINITY

5.1. The equation of critical points at infinity

In this section, we study these sequences, given by the compactifying lemma, which satisfy:

$$Z(x_\varepsilon) = 0; \quad \varphi(x_\varepsilon) \leq \varepsilon; \quad a_0 \leq J(x_\varepsilon) \leq a_1 . \tag{5.1}$$

For the moment being we are interested in making explicit the equation satisfied by these sequences. For this purpose, we introduce:

$$\dot{x}_\varepsilon = a\xi + bv; \qquad \omega = \frac{|\partial I|(x_\varepsilon)}{|\partial J|(x_\varepsilon)} . \tag{5.2}$$

We will drop for sake of simplicity the subscripts ε in the variables we will use. a is a constant. b is an $L^2(S^1)$-function. We will assume:

$$\int_0^1 b^2 dt \to +\infty \quad \text{when} \quad \varepsilon \to 0. \tag{5.3}$$

This is indeed the interesting case, when there is no compacity.

We then have the following proposition which gives the equation satisfied by b.

Proposition 9: Under (5.3), (5.1) is equivalent to:

$$\left.\begin{aligned} &\ddot{b} + b\left(-\frac{\omega a}{2} + \frac{b^2}{2} - \int_0^1 \frac{b^2}{2}\right) + a^2 b\tau - ab^2\dot{\bar{\mu}}_\xi + b\dot{b}\bar{\mu} = 0 \\ &b(0) = b(1); \qquad \dot{b}(0) = \dot{b}(1) \\ &\int_0^1 \frac{b^2}{\omega} \to 0 \,. \end{aligned}\right\} \tag{5.4}$$

5.2. Geometric interpretation of the equation of critical points at infinity

We write down the equation satisfied by the critical points at infinity in a matricial form. Let

$$\left.\begin{aligned} A_1^* &= 1 - \frac{b^2}{\omega a} + \int_0^1 \frac{b^2 dt}{\omega a}, \\ B_1^* &= -\frac{2b}{\omega}, \\ C_1^* &= -\frac{2\dot{b}}{\omega a}. \end{aligned}\right\} \tag{5.5}$$

Then we have:

$$\left.\begin{aligned} \dot{A}_1^* &= -\frac{2b\dot{b}}{\omega a} = -bC_1^*, \\ \dot{B}_1^* &= -\frac{2\dot{b}}{\omega}, \\ \dot{C}_1^* &= -\frac{2\ddot{b}}{\omega a} = bA_1^* - \frac{2ab\tau}{\omega} - b\bar{\mu}_\xi B_1^* - b\bar{\mu}C_1^* \end{aligned}\right\} \tag{5.6}$$

or else

$$\dot{Z}_1^* = \begin{bmatrix} \dot{A}_1^* \\ \dot{B}_1^* \\ \dot{C}_1^* \end{bmatrix} = b \begin{bmatrix} 0 & 0 & -1 \\ 0 & 0 & 0 \\ 1 & -\bar{\mu}_\xi & -\bar{\mu} \end{bmatrix} \begin{bmatrix} A_1^* \\ B_1^* \\ C_1^* \end{bmatrix} + \begin{bmatrix} 0 \\ -\frac{2\dot{b}}{\omega} \\ -\frac{2ab}{\omega} \end{bmatrix}$$

$$= b \begin{bmatrix} 0 & 0 & -1 \\ 0 & 0 & 0 \\ 1 & -\bar{\mu}_\xi & -\bar{\mu} \end{bmatrix} Z_1^* + \begin{bmatrix} 0 \\ -\frac{2\dot{b}}{\omega} \\ \frac{2ab}{\omega} \end{bmatrix}. \tag{5.7}$$

The matrix $\begin{bmatrix} 0 & 0 & -1 \\ 0 & \underline{0} & \underline{0} \\ 1 & -\mu_\xi & -\mu \end{bmatrix}$ is ${}^t\Gamma$ where Γ is defined in Proposition 4. Hence (5.7) is also:

$$\dot{Z}_1^* = -b\,{}^t\Gamma Z_1^* + \begin{bmatrix} 0 \\ -\frac{2\dot{b}}{\omega} \\ -\frac{2ab\tau}{\omega} \end{bmatrix}. \tag{5.8}$$

Equation (5.8) and Equation (D2) which gives the dynamics of α along v, have to be thought together.

Indeed these equations are very close:

In Equation (D2), $\dot{Z}^*$ is the derivative of $Z^*(s)$ with respect to s or either to v, as s represents the time on the v-orbit.

In Equation (5.8) $\dot{Z}_1^*$ is the derivate of Z_1^* with respect to time t along $x(t)$ or either with respect to $\dot{x} = a\xi + bv$.

Let us rewrite (5.8) with respect to:

$$\frac{\partial}{\partial s} = \frac{1}{b}\frac{\partial}{\partial t} = \frac{a}{b}\,\xi + v\,. \tag{5.9}$$

We then find:

$$\frac{\partial Z_1^*}{\partial s} = -{}^t\Gamma Z_1^* + \begin{bmatrix} 0 \\ -\frac{2\dot{b}}{\omega b} \\ -\frac{2a\tau}{\omega} \end{bmatrix}. \tag{5.10}$$

Now, when b is very large, $\frac{a}{b}$ is small; hence $\frac{\partial}{\partial s}$ looks like v. As ω goes to infinity, $\frac{a\tau}{\omega}$ goes to zero.

Hence (5.10) is very close to the equation governing the dynamics of α along v, provided b is very large and $\frac{\dot{b}}{\omega b}$ is small.

Under these conditions, which will amount later on to the fact that b is large, (5.10) acquires a geometric significance: it is very close to the transport equations of forms along v.

This is a key point.

6. THE CONVERGENCE THEOREM; THE GEOMETRICAL CURVES

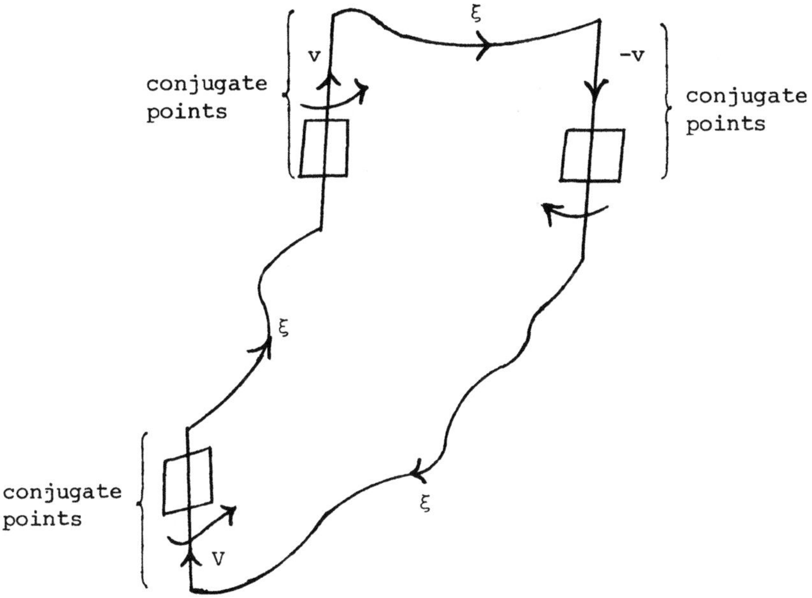

(6.1)

(6.1) is a geometrical description of the critical points at infinity. To understand this description, we set in (5.4), $\omega = \frac{1}{\varepsilon}$. x is a curve in C_β with:

$$\dot{x} = a\xi + bv; \quad a \text{ being a constant}; \quad b \in L^2(S^1,\mathbf{R}) \tag{6.2}$$

b and a satisfying (5.4).

To understand qualitatively what is going on, we analyze the convergence process.

Due to (5.4), in particular to the fact that $\int_0^1 b^2 dt/\omega \to 0$, we are able (see [2]) to distinguish on the curves $x(\omega)$ or x_ε, two types of pieces:

1. <u>The pieces rather tangent to v</u>: there, $\int_0^1 b^2/\omega$ is rather large.

2. <u>The pieces rather tangent to ξ</u>: there, $\int_0^1 b^2/\omega$ is rather small.

On the first kind of pieces, the geometric interpretation of the critical points at infinity holds. The curves are then close to a $\pm v_-$ orbit. Writing down, as in (D2), the equation function of the time s, satisfied by the form α_{x_0} in the transported frame along $v, ((D\varphi^*_{-s}\alpha)_{x_0}, (D\varphi^*_{-s}\beta)_{x_0})$, we interpretate (5.4) as the transport equation of α along v. The condition $\int_0^1 b^2/\omega \to 0$ tells us, when $\omega \to +\infty$ (or $\varepsilon \to 0$) that these pieces run from one point to one of its conjugate; the error being of the order $o(\Delta t)$, where Δt is the time spent on such pieces.

For $\varepsilon \to 0$, the curve x_ε approximating the object (6.1) on the deformation line forms a <u>small bubble</u> in section to v; i.e. if one projects a neighborhood of this piece rather tangent to v on a section to v, one finds:

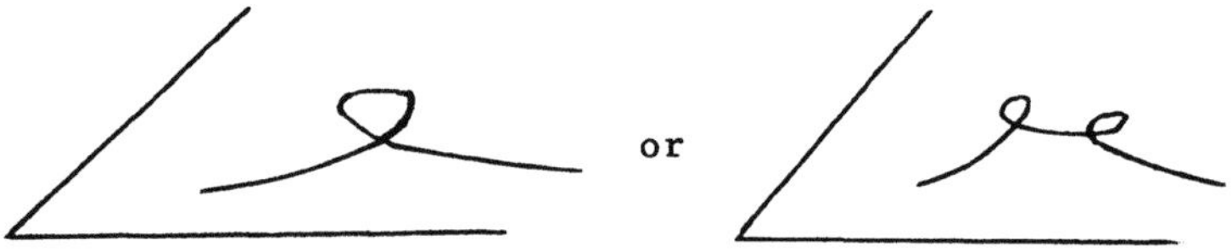

(6.3)

or more bubbles.

These are thus points where the tangent vector to x_ε, when projected, completes rapidly an integer number of rotations, possibly growing when $\varepsilon \to 0$ through the following process:

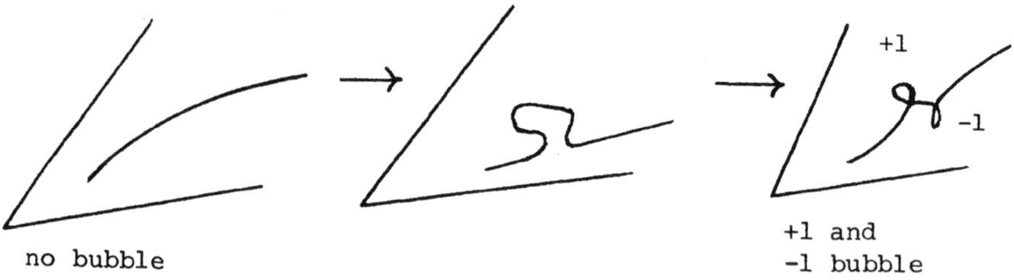

(6.4)

However, the resulting movement is very particular: the bubble as deployed along v will go from one point to a conjugate of this point.

Therefore, generically, these bubbles build up at <u>precise</u> locations in M. We will see later on that these singularities have further more a <u>very precise and restricted</u> normal form. To see the phenomenon, we could draw v-orbits passing through each point x of M and distinguish on these orbits the <u>coincidence points to x</u>:

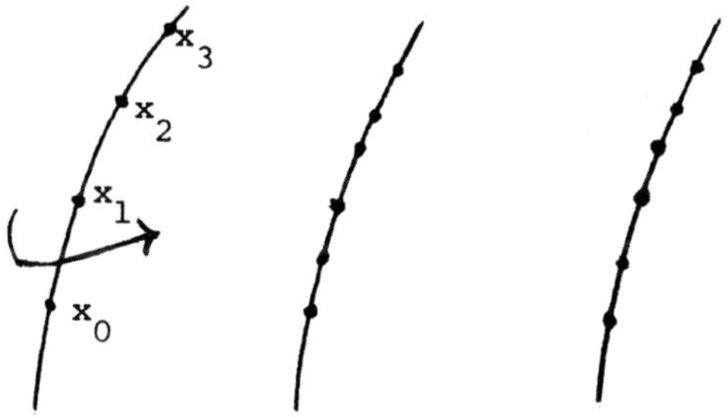

(6.5)

We thus have a **Z**-structure along M related to α along v. For some distinguished points, we have conjugate points:

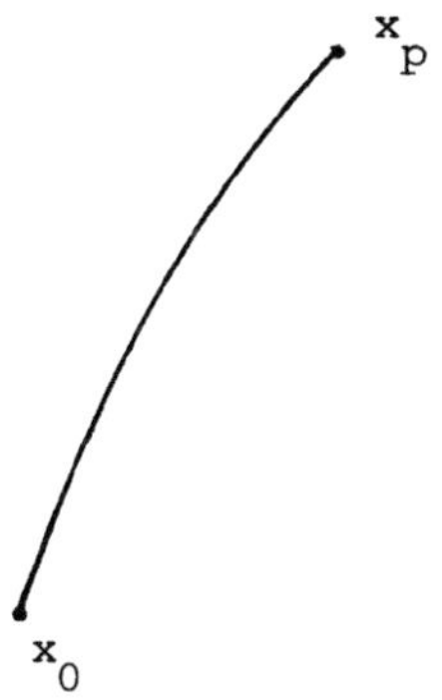

$x_0 \in$ hypersurface of M. (6.6)

If x_p corresponds to the time s_1 along the v-orbit starting at x_0, we have:

$$D\varphi_{s_1}^* \alpha = \alpha \tag{6.7}$$

and we may compute the <u>second</u> variation of α

$$\delta(D\varphi_s^* \alpha - \alpha)(s) \ . \tag{6.8}$$

This gives rise, as we will see later on, to a quadratic form on tangent vectors to M at x_0, q_0; and an associated quadratic form on tangent vectors to M at $x_{s_1} = x_p, q_1$.

Thus, these conjugate points come out with:

1 - a <u>precise</u> location;
2 - a <u>precise</u> normal form to the singularity;
3 - an <u>integer</u> (the rotation of α from x_0 to $x_{s_1} = x_0$);
4 - <u>two quadratic forms</u> q_0 and q_1;
5 - a way to approach them by curves which <u>project on local sections on bubbles</u>.

2 - the ξ-pieces

These are pieces where the curve is tangent to the Reeb vector-field; thus the curve is tangent to ξ (in this **Z**-structure we introduced) until it hits a point admitting a conjugate point. Then, under certain conditions stated in [5], it jumps to the conjugate point.

> The ξ-pieces come also with a quadratic form q_3 defined by the second variation of J along them with fixed ends. This quadratic form is related to a rotation of v along the ξ-piece (see [2]). (6.9)

The Reeb vector-field ξ is, in the case of the cotangent unit sphere bundle of a Riemannian manifold Σ, such that its periodic orbits project on geodesics of Σ. In that case, there is no other conjugate point for a point x_0 than itself and the Dirac masses describe a complete circle S^1 over a given point in Σ in $ST^*\Sigma$. In other simple, but more complicated cases, this is what happens:

Take the case of S^3 fibering over S^2 with the Hopf fibration

$$p : S^3 \to S^2 . \tag{6.10}$$

Consider $\alpha = \lambda\alpha_0$, λ a positive function on S^3 and α_0 the standard contact form of S^3. Let v be the vector-field of the fibers over S^2. In this case, the Reeb vector-field ξ, when describing a fiber S^1 over a point x_0 of S^2, describes in the tangent plane to S^2 at x_0 the following:

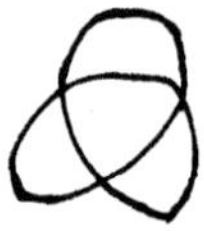

i.e. two circles . (6.11)

We thus have two choices of length on S^2, hence two notions of geodesics. Then, (6.1) projects as:

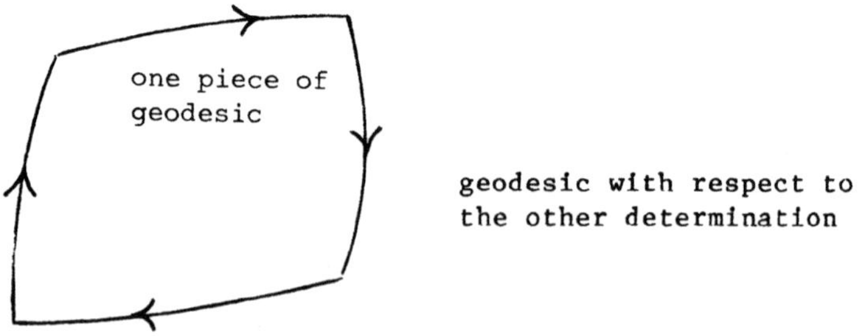

(6.12)

The location of the corners is restricted and there is a Morse index related to q_0, q_1, q_3.

This is a general picture of what happens.

We reproduce here the theorem we announced in [5]:

Assume:

(H_1) α turns well along v;
(H_2) v has a periodic orbit,
(H_3) for one vector-field v_1, nonsingular and colinear to v, we have: $\exists k_1 > 0$ such that $\|D\theta_s^1\| < k_1$ $\forall s \in \mathbf{R}$, where θ_s^1 is the one-parameter group of v_1;
(H_4) $\exists k_2$ and $k_3 > 0$ such that $\forall i \in \mathbf{Z}$, we have:

$$k_2 d(x,y) < d(f^i(x),f^i(y)) < k_3 d(x,y) \quad \forall x, y \in M ;$$

(H_5) $\exists k_4 > 0$ such that $|\mu_i(x) - \mu_i(y)| < k_4 d(x,y)$ $\forall x,y \in M$;
(H_6) $\exists \rho > 0$ such that for any $x \in M$, the set $C_\rho(x) = \{f^i(x)/|\mu_i(x) - 1| < \rho;\ i \in \mathbf{Z}\}$ is finite.

Then, under these hypotheses which can be considerably weakened (see [2]), we have:

Theorem: The critical points at infinity of the variational problem are continuous and closed curves made up with pieces $[x_{2i},x_{2i+1}]$ tangent to ξ and pieces $[x_{2i+1},x_{2i+2}]$ tangent to v. x_{2i+2} is conjugate to x_{2i+1}. If the Betti numbers of the loop space are unbounded, there are infinitely many of these curves. Furthermore, if n is the number of v-pieces of one of these curves, we have:

$$n < Ca \tag{6.13}$$

where a is the length of the curve along ξ and C is a universal constant.

7. EXPANSION OF J NEARBY INFINITY. THE INDEX OF A CRITICAL POINT AT INFINITY

7.1. The parameterization normal form

We are then left with these geometric curves made up of ξ-pieces and v-pieces. The v-pieces have been seen to run from a point to a conjugate.

On such a piece, as seen from Section 5 on, the function $b(t)$ is very particular.

Indeed, as stated in (5.5), the vector:

$$\left.\begin{aligned} A_1^* &= 1 - \frac{b^2}{\omega a} + \frac{\int_0^1 b^2 dt}{\omega a} \\ B_1^* &= -\frac{2b}{\omega} \\ C_1^* &= -\frac{2\dot{b}}{\omega a} \end{aligned}\right\} \tag{7.1}$$

nearly satisfies the transport equation of the forms.

Furthermore, if we are looking at a v-piece between x_{2i} and x_{2i+1}, then nearby x_{2i}, $\begin{bmatrix} A_1^* \\ B_1^* \\ C_1^* \end{bmatrix}$ is nearly $\begin{bmatrix} 1 \\ 0 \\ 0 \end{bmatrix}$.

Consequently b has in fact a <u>first normal form</u> on a critical point at infinity: Namely, we introduce the function φ_i on the v-piece between x_{2i} and x_{2i+1} satisfying:

$$\left.\begin{aligned} &\frac{\partial^2 \varphi_i}{\partial s^2} + \varphi_i + \bar{\mu}\,\frac{\partial \varphi_i}{\partial s} = 0 \\ &\varphi_i(0) = \varphi_i(s_i) = 1 \\ &\frac{\partial \varphi_i}{\partial s}(0) = \frac{\partial \varphi_i}{\partial s}(s_i) = 0\,. \end{aligned}\right\} \tag{7.2}$$

Here s is the time parameter along the v-orbit from x_{2i} to x_{2i+1}. φ_i exists and is uniquely defined by (7.2) as x_{2i} and x_{2i+1} are conjugate. We thus have:

$$\frac{b(t)}{\sqrt{\omega a}} \sim \pm\sqrt{1 - \varphi_i(s(t))}\,. \tag{7.3}$$

We state this in:

<u>Proposition 10</u>: Along a nearly tangent to v-piece between two nearby conjugate points, $\frac{b(t)}{\sqrt{\omega a}}$ is equivalent to $\pm\sqrt{1-\varphi_i(s(t))}$ where φ_i satisfies:

$$\left.\begin{aligned} &\frac{\partial^2 \varphi_i}{\partial s^2} + \varphi_i + \bar{\mu}\,\frac{\partial \varphi_i}{\partial s} = 0 \\ &\varphi_i(0) = \varphi_i(s_i) = 1 \\ &\frac{\partial \varphi_i}{\partial s}(0) = \frac{\partial \varphi_i}{\partial s}(s_i) = 0 \\ &s \in [0,s_i];\ \text{time on the } (\pm) \text{ v-orbit from } x_{2i} \text{ to} \\ &x_{2i+1} \text{ which are the conjugate points.} \end{aligned}\right\} \tag{7.4}$$

Thus φ_i satisfies:

$$\left.\begin{aligned} &\frac{\partial^2 \varphi_i}{\partial s^2} + \varphi_i + \bar{\mu}\,\frac{\partial \varphi_i}{\partial s} = 0 \\ &\varphi_i(0) = 1; \quad \frac{\partial \varphi_i}{\partial s}(0) = 0\,. \end{aligned}\right\} \tag{7.5}$$

φ_i is extremal only at the coincidence points of x_{2i} and the only possibility for b is to accomplish a piece of v-orbit from x_{2i} to x_{2i+1}, then come back from x_{2i+1} to x_{2i} etc., (see [2]).

If we consider a deformation line of (4.9) going to a critical point at infinity, these oscillations are in finite number (upperbounded). Otherwise, we leave an L^∞-neighborhood of this critical point at infinity and one constructs a deformation lemma to move all such curves away from infinity.

As we are dealing with an actual jump, this number is odd.

Thus, we are left, as a model, with only one jump and a definite sign for b on such a piece.

As we wish to present general ideas rather than justify all the technical details, we will assume for sake of simplicity that whenever a jumps occurs, a single oscillation is associated with it.

So that a critical point at infinity is this geometric curve, together with a parameter $\sqrt{\omega a} \to +\infty$, the v-pieces being described with $b(t)/\sqrt{\omega a} \sim \pm\sqrt{1 - \varphi_i(s(t))}$, where $\pm$ is fixed by the orientation along v of $[x_{2i}, x_{2i+1}]$.

7.2. The variations along a critical point at infinity inwards C_β

There are two kinds of variations along such a geometric curve with this limit parameterization we pointed out in 7.1.

The first kind, we will present here consists in opening up the oscillations in order to see if we are dealing with an actual critical point at infinity. This will be made clear later on.

These variations are inwards C_β.

We want to know if a sequence $(\varepsilon \to 0)$ of flow lines of (4.9) does arrive to the limit object.

In order to discriminate between these two possibilities, we need a first expansion of J along inwards C_β-variation. An inward variation has to bring the length along ξ to be a strictly positive constant which is nearly a, a being the length along ξ of the limit object (the curve x).

We are thus led to introduce along a v-piece $[x_{2i}, x_{2i+1}]$ of x, which we will assume for sake of simplicity to be oriented by $+v$, the differential equation (see [2] for further details):

$$\left.\begin{aligned} &\overset{\bullet}{\overbrace{\lambda + \bar{\mu}\eta}} - \eta = \frac{a}{\sqrt{\omega a}\,\sqrt{1 - \varphi_i(s)}} \qquad s \in [0, s_i] \,. \\ &\dot{\eta} = -\lambda \end{aligned}\right\} \tag{7.6}$$

In (7.6), $\cdot$ is a derivation by $\frac{\partial}{\partial s} = v$; ω is a large positive parameter and a is the length along ξ of the curve as already stated.

Another way to see (7.6) is to set:

$$\frac{\partial}{\partial t} = b\,\frac{\partial}{\partial s} = \sqrt{\omega a}\,\sqrt{1 - \varphi_i(s)}\,\frac{\partial}{\partial s} \tag{7.7}$$

and we then have:

$$\left.\begin{aligned} &\frac{\partial}{\partial t}(\lambda + \bar{\mu}\eta) - b\eta = a \\ &\frac{\partial \eta}{\partial t} = -\lambda b \end{aligned}\right\} \quad z = \lambda\xi + \mu v + \eta w \,. \tag{7.8}$$

The homogeneous equation:

$$\left.\begin{aligned} &\dot{\overbrace{\lambda + \bar{\mu}\eta}} - \eta = 0 \\ &\dot{\eta} = -\lambda \end{aligned}\right. \quad s \in [0, s_i] \Bigg\} \tag{7.9}$$

has solutions satisfying:

$$(\lambda + \bar{\mu}\eta)(s_i) = (\lambda + \bar{\mu}\eta)(0) \qquad (7.9)\text{bis}$$

as x_{2i} and x_{2i+1} are conjugate points. Indeed (7.9) expresses the relations which have to be satisfied by a transported vector along a v-piece.

There is thus an indeterminacy in (7.6) which we will discuss later on, when we will introduce the index of a critical point at infinity.

Notice that the parameterization introduced by (7.8), with $b = \sqrt{\omega a}\,\sqrt{1 - \varphi_i(s)}$ corresponds to the first normal form we pointed out in Proposition 10.

In (7.6) there is a problem:

Indeed, φ_i satisfies on $[x_{2i}, x_{2i+1}]$ parameterized by v:

$$\left.\begin{aligned} &\frac{\partial^2}{\partial s^2}\varphi_i + \varphi_i + \bar{\mu}\frac{\partial \varphi_i}{\partial s} = 0 \\ &\varphi_i(0) = \varphi_i(s_i) = 1 \\ &\frac{\partial \varphi_i}{\partial s}(0) = \frac{\partial \varphi_i}{\partial s}(1) = 0 \,. \end{aligned}\right\} \tag{7.10}$$

Thus, $1 - \varphi_i$ has a zero of second order at 0 and s_i. Nearby this point, we have:

$$1 - \varphi_i(s) \sim C_i s^2 \ \text{ at } \ 0;\ 1 - \varphi_i(s) \sim C_i'(s - s_i)^2 \ \text{ at } \ s_i \,. \tag{7.11}$$

Thus

$$\int_0^{s_i} \frac{1}{\sqrt{1 - \varphi_i(s)}}\, ds \ \text{ diverges logarithmically at both ends} \,. \tag{7.12}$$

This implies, by integration of the first equation in (7.6), that $(\lambda + \bar{\mu}\eta, \eta)$ cannot possibly be L^∞, a fortiori L^∞-small.

We analyze here what is going on in (7.6).

Lemma 5: Consider a solution of (7.6), $(\lambda,\eta)(s)$, and a solution of (7.9), $(\lambda_1,\eta_1)(s)$ taking the same value at a point τ_i in $[0,s_i]$. Then:

$$|\eta(s) - \eta_1(s)| < \frac{C}{\sqrt{\omega}}$$

$$\int_0^{s_i} |\lambda(s) - \lambda_1(s)|^2 ds < \frac{C}{\omega}$$

$$\int_0^{s_i} |\dot{\eta}(s) - \dot{\eta}_1(s)|^2 ds < \frac{C}{\omega} .$$

Furthermore:

$\lambda + \bar{\mu}\eta(s)$ is equivalent to $-\frac{1}{\sqrt{C_i}}\sqrt{\frac{a}{\omega}}\log s$ near $s = 0$

$\lambda + \bar{\mu}\eta(s)$ is equivalent to $-\frac{1}{\sqrt{C_i'}}\sqrt{\frac{a}{\omega}}\log(s_i - s)$ near $s = s_i$.

Lastly,

$$\Delta_i = \lim_{\substack{\varepsilon\to 0\\ \varepsilon'\to 0}} \{\int_\varepsilon^{s_i-\varepsilon'} \frac{ds}{\sqrt{1-\varphi_i(s)}} + \sqrt{\frac{\omega}{a}}\,[(\lambda+\bar{\mu}\eta)(\varepsilon) - (\lambda+\bar{\mu}\eta)(s_i-\varepsilon')]\} \tag{7.13}$$

exists and is independent of the solution of (7.6) considered as well as on ω and a. This quantity if thus attached to $[x_{2i},x_{2i+1}]$ and only to it.

Proof of Lemma 5: $(\lambda - \lambda_1, \eta - \eta_1)$ satisfy (7.6) with zero conditions at τ_i:

$$\overset{\bullet}{\overbrace{(\lambda - \lambda_1) + \bar{\mu}(\eta - \eta_1)}} - (\eta - \eta_1) = \frac{a}{\sqrt{\omega a}\sqrt{1-\varphi_i(s)}}$$

$$\overset{\bullet}{\overbrace{\eta - \eta_1}} = -(\lambda - \lambda_1) \quad s \in [0,s_i],\ (\lambda - \lambda_i)(\tau_i) = (\eta - \eta_1)(\tau_i) = 0 . \tag{7.14}$$

We thus have:

$$(\eta - \eta_1)(s) = -\int_{\tau_i}^{s} [\lambda - \lambda_1 + \bar{\mu}(\eta - \eta_1)]d\tau$$

$$+ \int_{\tau_i}^{s} \bar{\mu}(\eta - \eta_1)d\tau$$

$$= -\int_{\tau_i}^{s} \Big[(\eta - \eta_1) + \int_{\tau_i}^{\tau} \frac{a}{\sqrt{\omega a}\,\sqrt{1 - \varphi_i(x)}}\,dx\Big]d\tau$$

$$+ \int_{\tau_i}^{s} \bar{\mu}(\eta - \eta_1)d\tau \ . \tag{7.15}$$

Thus:

$$(\eta - \eta_1)(s) = \int_{\tau_i}^{s} (\bar{\mu} - 1)(\eta - \eta_1)d\tau$$

$$+ \sqrt{\frac{a}{\omega}} \int_{\tau_i}^{s} \Big(\int_{\tau}^{\tau_i} \frac{1}{\sqrt{1 - \varphi_i(x)}}\,dx\Big)d\tau \ . \tag{7.16}$$

As by (7.10), we know that:

$$\sqrt{1 - \varphi_i(s)} \sim C_i s \quad \text{at} \quad 0; \quad \sqrt{1 - \varphi_i(s)} \sim C_i'(s_i - s) \quad \text{at} \quad s_i \ , \tag{7.17}$$

we have:

$$\int_{\tau_i}^{s} \int_{\tau}^{\tau_i} \frac{1}{\sqrt{1 - \varphi_i(x)}}\,dx \quad d\tau < C_1 \quad \forall s \ . \tag{7.18}$$

Thus:

$$\Big|(\eta - \eta_1)(s) - \int_{\tau_i}^{s} (\bar{\mu} - 1)(\eta - \eta_1)d\tau\Big| < C_1\sqrt{\frac{a}{\omega}} \ . \tag{7.19}$$

This, together with the vanishing of $\eta - \eta_1$ at τ_i, implies the existence of C such that:

$$|\eta(s) - \eta_1(s)| < \frac{C}{\sqrt{\omega}} \ . \tag{7.20}$$

The other inequalities of Lemma 5 follow easily by integration of

(7.14), use of (7.20) and use of the vanishing of $\lambda - \lambda_1$ and $\eta - \eta_1$ at τ_i.

We have now to show (7.13):

The existence of a limit to:

$$\int_{\varepsilon}^{s_i-\varepsilon'} \frac{ds}{\sqrt{1-\varphi_i(s)}} + \sqrt{\frac{\omega}{a}}\,[(\lambda+\bar{\mu}\eta)(\varepsilon) - (\lambda+\bar{\mu}\eta)(s_i-\varepsilon')] \tag{7.21}$$

follows from the fact that this expression is by integration of the first equation in (7.6):

$$-\sqrt{\frac{\omega}{a}}\int_{\varepsilon}^{s_i-\varepsilon'} \eta(s)ds . \tag{7.22}$$

Now, η_1 being a solution of (7.9) is bounded; and $|\eta - \eta_1|$ is also bounded, as we just proved, by $\frac{C}{\sqrt{\omega}}$. Thus η is bounded on $[0,s_i]$ and (7.22), hence (7.21) has a limit, which we call Δ_i. Δ_i does not depend on the particular solution of (7.6) chosen. Indeed, if we consider another solution, (λ',η'), the difference $(\lambda - \lambda',\eta - \eta')$ is a solution of (7.9). Thus, by (7.9)bis, $[(\lambda - \lambda') + \bar{\mu}(\eta - \eta')](s_i) = [(\lambda - \lambda') + \bar{\mu}(\eta - \eta')](0)$. Hence:

$$[(\lambda+\bar{\mu}\eta)(\varepsilon) - (\lambda+\bar{\mu}\eta)(s_i-\varepsilon')] - [(\lambda'+\bar{\mu}\eta')(\varepsilon) - (\lambda'+\bar{\mu}\eta')(s_i-\varepsilon')] \tag{7.23}$$

goes to zero with ε and ε'.

Hence the result.

Δ_i is also independent of ω and a, as, in fact, it can be computed on a solution of:

$$\left.\begin{aligned} \dot{\overbrace{\gamma_i + \bar{\mu}\eta_i}} - \eta_i &= \frac{1}{\sqrt{1-\varphi_i(s)}} \\ \dot{\eta}_i &= -\gamma_i \end{aligned}\right\} \tag{7.24}$$

as equal to:

$$\Delta_i = \lim_{\substack{\varepsilon\to 0\\ \varepsilon'\to 0}} \left\{\int_{\varepsilon}^{s_i-\varepsilon'} \frac{ds}{\sqrt{1-\varphi_i(s)}} + [(\gamma_i+\bar{\mu}\eta_i)(\varepsilon) - (\gamma_i+\bar{\mu}\eta_i)(s_i-\varepsilon')]\right\} . \tag{7.25}$$

The proof of Lemma 5 is thereby complete.

A variation governed by (7.6) comes now with two problems: the first one, which is not very serious, is due to the indeterminacy in it; one can add to such a variation any variation subject to (7.9). In order to fix once for all the variation we are looking at, we will impose:

$$\lambda(s_i/2) = \eta(s_i/2) = 0 . \tag{7.26}$$

The influence of the solutions of (7.9) is analyzed separately in 7.3. The solution of (7.6)-(7.26) is denoted:

$$(\tilde{\lambda},\tilde{\eta}) . \tag{7.27}$$

Notice that, by Lemma 5, and the fact that $(\tilde{\lambda}_1,\tilde{\eta}_1) \equiv (0,0)$, where $(\tilde{\lambda}_1,\tilde{\eta}_1)$ is the solution of (7.9) taking the same value (i.e. (0,0)) then $(\tilde{\lambda},\tilde{\eta})$ at $\tau_i = s_i/2$, $(\tilde{\lambda},\tilde{\eta})$ satisfies:

$$\left.\begin{aligned} &|\tilde{\eta}(s)| < \frac{C}{\sqrt{\omega}} ; \int_0^{s_i} |\tilde{\lambda}(s)|^2 ds < \frac{C}{\omega} ; \int_0^{s} |\dot{\tilde{\eta}}|^2 ds < \frac{C}{\omega} ; \\ &(\tilde{\lambda} + \bar{\mu}\tilde{\eta})(s) \sim - \frac{1}{\sqrt{C_i}} \sqrt{\frac{a}{\omega}} \log s \quad \text{at} \quad 0 \end{aligned}\right\} \tag{7.28}$$

$$\sim - \frac{1}{\sqrt{C_i'}} \sqrt{\frac{a}{\omega}} \log(s_i - s) \quad \text{near} \quad s = s_i .$$

The second problem is more serious:

By (7.28) such a variation cannot be made to be L^∞, a fortiori not L^∞ or H^1 small.

Indeed $\tilde{\lambda} + \bar{\mu}\tilde{\eta}(s)$ diverges logarithmically at $s = 0$ and $s = s_i$. $\tilde{\eta}$ remains meanwhile bounded, and even, by (7.28), going to zero with ω. Thus, at 0 and s_i, the variation is infinite along ξ. Nevertheless, we can try to extend it along the ξ-pieces, subject to the differential equation:

$$\left.\begin{aligned} &\dot{\overbrace{\lambda + \bar{\mu}\eta}} = \varphi(t) \\ & \qquad\qquad\qquad \text{here} \quad \cdot = \xi = \text{derivation along} \quad \xi . \\ &\dot{\eta} = \mu a \end{aligned}\right\} \tag{7.29}$$

We will state further on what are the conditions on φ nearby the points x_{2i+1}. For the moment being, let us try to understand what is going on the v-pieces. The equation (7.6) may be rewritten, in a more intrinsic form: setting:

$$z(s) = (\tilde{\lambda}\xi + \tilde{\eta}w) \tag{7.30}$$

we have:

$$\left[\frac{\partial}{\partial s} + v; z(s)\right] = \frac{a}{\sqrt{\omega a}} \frac{1}{\sqrt{1 - \varphi_i(s)}} \xi + \frac{a}{\sqrt{\omega a}} \theta v;$$

$$\theta \in C^\infty(]0, s_i[) . \tag{7.31}$$

Indeed, (7.6) and (7.31) are equivalent as can be checked by applying α and β to (7.31). As we are dealing with a variation along a v-piece and as the functional $J(x) = \int_0^1 \alpha_x(\dot{x})dt$ (this is rephrasing of our functional which makes sense not only on C_β or $\mathcal{L}_\beta$, but even on unparameterized, however oriented, curves) is invariant under reparameterization, the θ-term is not relevant. What matters is the variation transverse to $\dot{x}$, i.e. to v along these pieces. Clearly:

$$z = \sqrt{\frac{a}{\omega}}\, z_0 , \tag{7.32}$$

z_0 satisfying:

$$\left[\frac{\partial}{\partial s} + v, z_0\right] = \frac{1}{\sqrt{1 - \varphi_i(s)}} \xi + \theta v \tag{7.33}$$

$z_0(s_i/2) = 0$; z_0 has no component on v .

If we draw the variation, we have:

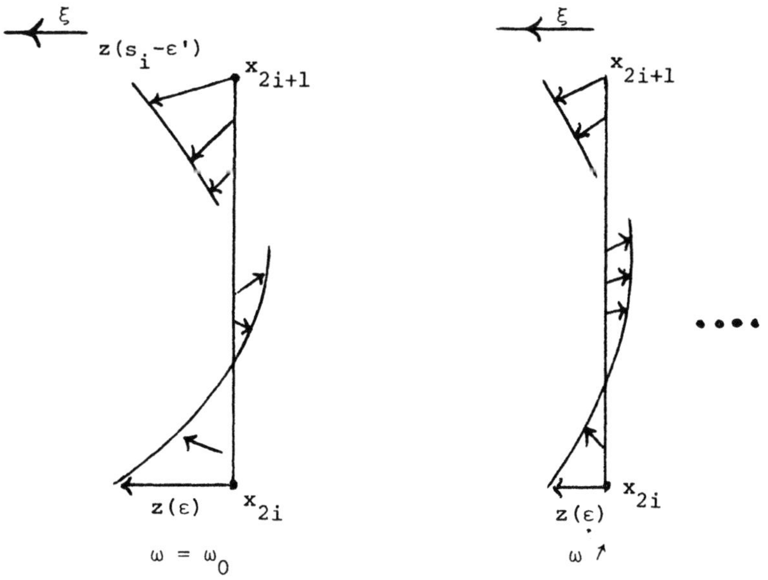

(7.34)

Remark: The linking as shown in (7.34) actually occurs.

There is thus, as $\omega \to +\infty$, more and more control on the variation z which is of the order of $\frac{C}{\sqrt{\omega}}$ on a fixed (when $\omega \nearrow +\infty$) compact set in $]0,s_i[$. Nevertheless, at the ends, 0 and s_i, we always have a logarithmic divergence along ξ. If we extend now to the ξ-pieces subject to (7.29), we have:

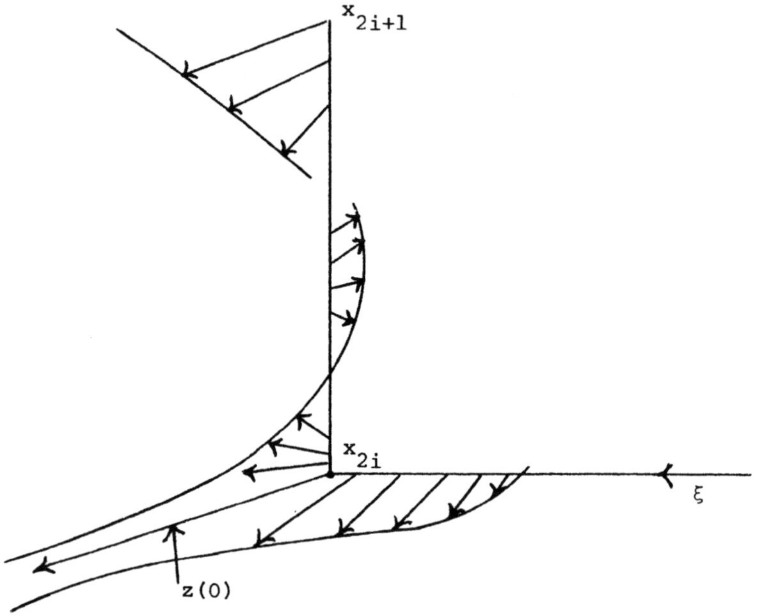

(7.35)

Following a choice of φ in (7.29) behaving as:

$$\left.\begin{array}{l} \varphi(t) \sim \sqrt{\frac{a}{\omega}}\, \dfrac{C''_i}{t - t_i} \quad \text{nearby} \quad t = t_i; \\ t_i \quad \text{corresponding to} \quad x_{2i} \quad \text{on the } \xi\text{-piece}; \\ C''_i \quad \text{bounded constant}; \end{array}\right\} \tag{7.36}$$

one can take care of the logarithmic divergence of $\tilde{\lambda} + \widetilde{\overline{\mu\eta}}$ at x_{2i} (also x_{2i+1} on the other ξ-piece) and control in $\sqrt{\frac{a}{\omega}}$ the H_1-norm of the variation on each compact subset of the $]x_{2i-1},x_{2i}[$.

This will be made precise later on.

The only problem is thus at the corners, i.e. at the points x_{2i}, x_{2i+1}.

There, in fact, it is natural that we find some problems, as $\dot{x}_\varepsilon$ changes very rapidly. On the other hand, the curves x_ε stay nevertheless in an L^∞-neighborhood of x.

We wish thus to cut out the variation on a neighborhood of the corners (a neighborhood which we can take to be smaller and smaller as ω goes to $+\infty$) in order to have an L^∞ control on it (going to zero with ω); and then to make a first expansion of J.

Let us first derive what kind of first expansion we may expect, considering the variation as taking place in $\mathcal{L}_\beta$ with the functional $J(y) = \int_0^1 \alpha_y(\dot{y})dt$ (which coincides with J on C_β and makes sense on unparameterized oriented curves of $\mathcal{L}_\beta$).

We have:

Lemma 6: Consider x a curve of $\mathcal{L}_\beta$, unparameterized. Let $J : \mathcal{L}_\beta \to \mathbf{R}$

$$y \to \int_0^1 \alpha_y(\dot{y})dt .$$

Consider the variation z defined by (7.30), (7.29), cut out at the corners so that all the (7.28) estimates hold. Call z_1 this new variation. Then,

$$\partial J(x) \cdot z_1 = \sqrt{\frac{a}{\omega}}\left(\sum_i \Delta_i\right) + o\left(\frac{1}{\sqrt{\omega}}\right) .$$

Proof: We have: $\partial J(x) \cdot x_1 = \int \left[\frac{\partial}{\partial s} \alpha(z_1) - d\alpha(\dot{x}, z_1)\right]ds$ if we parameterize the curve by s $(\dot{x} = \frac{\partial x}{\partial s})$; which is as well:

$$\partial J(x) \cdot z_1 = \sum_i \int_{[x_{2i}, x_{2i+1}]} [\dot{\overbrace{\tilde{\lambda}_1 + \bar{\mu}\tilde{\eta}_1}} - \tilde{\eta}_1]ds$$

$$+ \sum_i \int_{[x_{2i+1}, x_{2i+2}]} \dot{\overbrace{\tilde{\lambda}_1 + \bar{\mu}\tilde{\eta}_1}}\, ds \tag{7.37}$$

where $\tilde{\lambda}_1$ and $\tilde{\eta}_1$ are the components of z_1 and the parameterization is the one of (7.30) and (7.29) on the corresponding v and ξ-pieces. Thus:

$$\partial J(x) \cdot z_1 = - \sum_i \int_{[x_{2i}, x_{2i+1}]} \tilde{\eta}_1 ds \tag{7.38}$$

the other terms indeed cancel in (7.37). Now:

$$- \sum_i \int_{[x_{2i}, x_{2i+1}]} \tilde{\eta}_1 ds = - \sum_i \int_{\varepsilon}^{s_i - \varepsilon'} \tilde{\eta}_1 ds + O\left(\frac{1}{\sqrt{\omega}}\right) \tag{7.39}$$

as z_1 satisfies (7.28). But $z = z_1$ outside a small neighborhood of the corners. Thus:

$$-\int_{\varepsilon}^{s_i-\varepsilon'} \tilde{\eta}_1 ds = -\int_{\varepsilon}^{s_i-\varepsilon'} \tilde{\eta}\, ds =$$

$$= \sqrt{\frac{a}{\omega}}\Big[\int_{\varepsilon}^{s_i-\varepsilon'} \frac{ds}{\sqrt{1-\varphi_i(s)}}$$

$$+ \sqrt{\frac{\omega}{a}}\,((\tilde{\lambda} + \bar{\mu}\tilde{\eta})(\varepsilon) - (\tilde{\lambda} + \bar{\mu}\eta)(s_i - \varepsilon'))\Big] . \qquad (7.40)$$

By Lemma 5, this has a limit, when ε and ε' go to zero, equal to $\sqrt{\frac{a}{\omega}}$. Thus:

$$\partial J(x) \cdot z_1 = -\sum_i \int_{[x_{2i},x_{2i+1}]} \tilde{\eta}_1 ds = \sqrt{\frac{a}{\omega}}\,(\sum_i \Delta_i) + o\Big(\frac{1}{\sqrt{\omega}}\Big). \qquad (7.41)$$

Hence Lemma 6.

Corollary: Such a curve x is a critical point at infinity (i.e. an end for a flow-line) if $\sum_i \Delta_i > 0$. If $\sum_i \Delta_i < 0$, x does not define a critical point at infinity. The case $\sum_i \Delta_i = 0$ is left open.

7.3. The variations tangent to the border-line. The index of a critical point at infinity

We are now left with indeterminacies.

Namely, we are considering variations such that on the v-pieces, we have:

$$\left.\begin{aligned} &\dot{\overline{\lambda + \bar{\mu}\eta}} - \eta = 0 \\ &\dot{\eta} = -\lambda \ . \end{aligned}\quad \cdot = \frac{\partial}{\partial s} = \pm v;\ \mu \text{ arbitrary}, \quad s \in [0,s_i]\right\} \qquad (8.1)$$

On the ξ-pieces we have:

$$\left.\begin{aligned} &\dot{\overline{\lambda + \bar{\mu}\eta}} = \varphi \\ &\dot{\eta} = \mu a \end{aligned}\right\} \qquad (8.2)$$

$\cdot = \frac{\partial}{\partial s} = \xi$; φL^{∞}-small.

(8.1) defines, up to the μ-indeterminacy, the equation of a transported vector (by v) along the v-piece $[x_{2i}, x_{2i+1}]$. This can be easily checked by applyiing α and β to the transport equation:

$$\left[\frac{\partial}{\partial s} + v,\ \lambda\xi + \mu v + \eta w\right] = \psi(s)v,\ \psi \text{ arbitrary} . \tag{8.3}$$

Thus, calling:

$$z_i(s) \text{ the variation subject to (8.1) on } [0, s_i] , \tag{8.4}$$

we have:

$$z_i(s_i) = D\varphi_{s_i}(z_i(0)) + \delta s_i v; \quad \delta s_i \in \mathbf{R} . \tag{8.5}$$

Consequently, in order to compute $J(x + z)$, we may always see $x + z$ as being geometrically realized as follows:

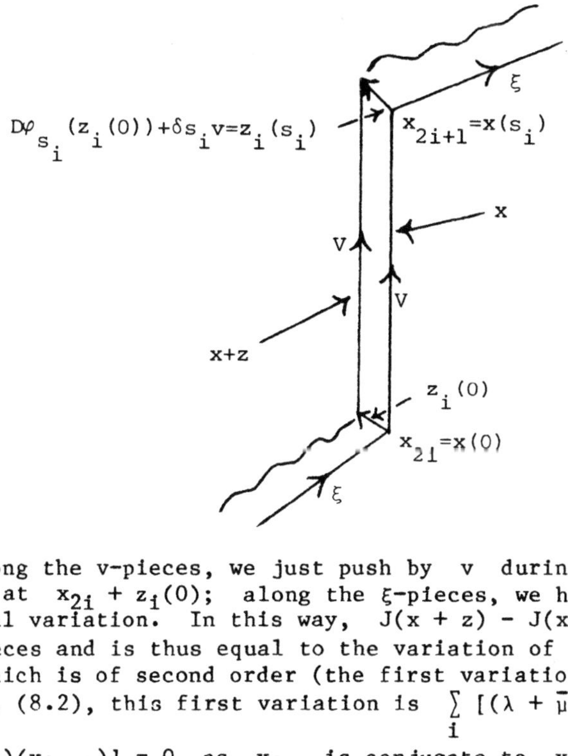

(8.6)

i.e., along the v-pieces, we just push by v during a time $s_i + \delta s_i$ starting at $x_{2i} + z_i(0)$; along the ξ-pieces, we have as usual a tangential variation. In this way, $J(x + z) - J(x)$ comes only from the ξ-pieces and is thus equal to the variation of $J(x)$ along these pieces which is of second order (the first variation in zero. Indeed, by (8.1), (8.2), this first variation is $\sum_i [(\lambda + \bar{\mu}\eta)(x_{2i+2}) - (\lambda + \bar{\mu}\eta)(x_{2i+1})] = 0$ as x_{2i} is conjugate to x_{2i+1}). The expansion of $J(x + z) - J(x)$ is:

$$J(x + z) - J(x) = z \cdot (\partial J(x) \cdot z) =$$

$$= z \cdot \sum_i \{[(\lambda + \bar{\mu}\eta)(x_{2i+2}) - (\lambda + \bar{\mu}\eta)(x_{2i+1})]$$

$$- (\int_{x_{2i+1}}^{x_{2i+2}} b\eta)\}$$

$$= z \cdot (\sum_i [(\lambda + \bar{\mu}\eta)(x_{2i}) - (\lambda + \bar{\mu}\eta)(x_{2i+1})]$$

$$- \int_{x_{2i+1}}^{x_{2i+2}} b\eta) . \tag{8.7}$$

We first compute:

$$z \cdot (\int_{x_{2i+1}}^{x_{2i+2}} b\eta) \tag{8.8}$$

b is as noted previously equal to $d\alpha(\dot{x},w) = \gamma(\dot{x})$. Thus, the first variation of b along z is:

$$\frac{d}{dt}\gamma(z) - d\gamma(\dot{x},z) = \dot{\mu} + a\eta\tau - b\eta\bar{\mu}_\xi . \tag{8.9}$$

On a ξ-piece, b is zero. Thus this variation is:

$$\dot{\mu} + a\eta\tau . \tag{8.10}$$

Thus, using again the fact that b is zero on such a piece (thus $bz \cdot \eta = 0$ and $b\eta(x_{2i+2}) = b\eta(x_{2i+1}) = 0$), we derive:

$$z \cdot [- \int_{x_{2i+1}}^{x_{2i+2}} b\eta] = - \int_{x_{2i+1}}^{x_{2i+2}} (\dot{\mu} + a\eta\,\tau)\eta . \tag{8.11}$$

We are thus left with:

$$z \cdot [(\lambda + \bar{\mu}\eta)(x_{2i}) - (\lambda + \bar{\mu}\eta)(x_{2i+1})] \tag{8.12}$$

which has the following simple interpretation:

Consider the differential equations:

$$\left\{\begin{array}{l} \dot{\overbrace{\lambda + \bar{\mu}\eta}} = \eta \\ \dot{\eta} = -(\lambda + \bar{\mu}\eta) + \bar{\mu}\eta \text{ . initial data given by } z_i(0) \\ \cdot = \frac{\partial}{\partial s} = \pm v \text{ on the } \pm v\text{-piece } \underline{\text{from}} \ \underline{x_{2i} + z_i(0)} \ \underline{\text{to}} \\ \underline{x_{2i+1} + z_i(s_i)} \text{ during the time } [0, s_i + \delta s_i] \end{array}\right. \tag{8.13}$$

$$\left\{\begin{array}{l} \dot{\overbrace{\lambda + \bar{\mu}\eta}} = \eta \\ \dot{\eta} = -(\lambda + \bar{\mu}\eta) + \bar{\mu}\eta \text{ . initial data given by } z_i(0) \\ \cdot = \frac{\partial}{\partial s} = \pm v \text{ on the } \pm v\text{-piece } \underline{\text{from}} \ \underline{x_{2i}} \ \underline{\text{to}} \ \underline{x_{2i+1}} \\ \text{during the time } [0, s_i + \delta s_i] \text{ .} \end{array}\right. \tag{8.13$_{bis}$}$$

Then, with evident notations:

$$\begin{aligned} z \cdot [(\lambda + \bar{\mu}\eta)(x_{2i}) - (\lambda + \bar{\mu}\eta)(x_{2i+1})] \\ = (\lambda + \bar{\mu}\eta)(x_{2i} + z_i(0)) - (\lambda + \bar{\mu}\eta))(x_{2i+1} + z_i(s_i)) \\ - (\lambda + \bar{\mu}\eta)(x_{2i}) + (\lambda + \bar{\mu}\eta)(x_{2i+1}) \end{aligned} \tag{8.14}$$

at first order.

To give an intrinsic form to this expression, we come back to Proposition 4.

A transported vector $Z = \begin{bmatrix} A \\ B \\ C \end{bmatrix}(s)$ along v satisfies the differential equation:

$$\dot{Z} = \Gamma Z \text{ in the basis } (\xi, v, [\xi, v]) \text{ .} \tag{8.15}$$

Let

$$\left.\begin{array}{l} V(s) = \begin{bmatrix} a_0(s) & b_0(s) & c_0(s) \\ a_1(s) & b_1(s) & c_1(s) \\ a_2(s) & b_2(s) & c_2(s) \end{bmatrix} \text{ be the resolvant matrix of (8.15)} \\ V(0) = \text{Id .} \end{array}\right\} \tag{8.16}$$

Then:

$$\alpha_{x_s}(D\varphi_s(Z(0))) = [1,0,0]V(s)\begin{bmatrix} A(0) \\ B(0) \\ C(0) \end{bmatrix} = a_0(s)A(0)$$
$$+ b_0(s)B(0) + c_0(s)C(0) \tag{8.17}$$

and

$$\alpha_{x_s}(D\varphi_s(Z(0))) - \alpha_{x_0}(Z(0)) = (a_0(s) - 1)A(0)$$
$$+ b_0(s)B(0) + c_0(s)C(0) . \tag{8.18}$$

Setting:

$$z(s) = \lambda\xi + \mu v + \eta w , \tag{8.19}$$

we thus have:

$$(\lambda + \bar{\mu}\eta)(s) - (\lambda + \bar{\mu}\eta)(0) = (a_0(s) - 1)\lambda(0)$$
$$+ b_0(s)\mu(0) + c_0(s)\eta(0) . \tag{8.20}$$

In fact V depends also on the starting point of the transport differential equations, which we will denote y:

$$V(s) = V(s,y) . \tag{8.21}$$

Thus

$$(\lambda + \bar{\mu}\eta)(s) - (\lambda + \bar{\mu}\eta)(0) = (a_0(s,y) - 1)\lambda(0)$$
$$+ b_0(s,y)\mu(0) + c_0(s,y)\eta(0) . \tag{8.22}$$

With $y = x_{2i}$ and $s = s_i$, we have:

$$a_0(s_i,x_{2i}) = 1; \quad b_0(s_i,x_{2i}) = c_0(s_i,x_{2i}) = 0 \tag{8.23}$$

as x_{2i+1} is conjugate to x_{2i}.

Now, by (8.5), the variation z we are considering, under (8.1) and (8.2) is defined by:

$$\left.\begin{array}{l} \lambda_i,\mu_i,\eta_i\text{: coordinates of } z_i(0) \text{ at } x_{2i} \\ \text{and } \delta s_i \text{ such that } z_i(s_i) = D\varphi_{s_i}(z_i(0)) + \delta s_i v \end{array}\right\} \tag{8.24}$$

and we may view the variation along this v-piece as pushing along the transport vector of $z_i(0)$ along this v-piece during the time $s_i + \delta s_i$. The variation (8.14), in view of (8.22), (8.23) and (8.24) is thus:

$$z \cdot [(\lambda + \bar{\mu}\eta)(x_{2i}) - (\lambda + \bar{\mu}\eta)(x_{2i+1})]$$

$$= -\{[\frac{\partial a_0}{\partial s}(s_i,x_{2i})\lambda_i + \frac{\partial b_0}{\partial s}(s_i,x_{2i})\mu_i + \frac{\partial c_0}{\partial s}(s_i,x_{2i})\eta_i]\delta s_i$$

$$+(\frac{\partial a_0}{\partial y}\Big|_{(s_i,x_{2i})} \cdot z_i(0))\lambda_i + (\frac{\partial b_0}{\partial y}\Big|_{(s_i,x_{2i})} \cdot z_i(0))\mu_i$$

$$+ (\frac{\partial c_0}{\partial y}\Big|_{(s_i,x_{2i})} \cdot z_i(0))\eta_i\} . \tag{8.25}$$

We are now ready to define the index of a critical point at infinity. Consider:

$$\text{a } \pm v\text{-piece } [x_{2i},x_{2i+1}] . \tag{8.26}$$

Let

$V(s,y)$ be the resolvant matrix of the transport equation for the vectors starting at y along the v-orbit, in the basis $(\xi,v,[\xi,v])$, (8.27)

i.e. $\frac{\partial V}{\partial s} = \Gamma V;\ \Gamma = \begin{bmatrix} 0 & 0 & -1 \\ 0 & 0 & \bar{\mu}_\xi(s,y) \\ 1 & 0 & \bar{\mu}(s,y) \end{bmatrix}$. Let

$$V(s,y) = \begin{bmatrix} a_0(s,y) & b_0(s,y) & c_0(s,y) \\ a_1(s,y) & b_1(s,y) & c_1(s,y) \\ a_2(s,y) & b_2(s,y) & c_2(s,y) \end{bmatrix} . \tag{8.28}$$

Let z be a variation of the critical point at infinity x satisfying:

$$\left.\begin{aligned} &\dot{\overbrace{\lambda + \bar{\mu}\eta}} - \eta = 0 \\ &\dot{\eta} = -\lambda \\ &\cdot = \frac{\partial}{\partial s} = \pm v;\ \mu \text{ arbitrary, } s \in [0,s_i] \text{ on the} \\ &\qquad \text{piece } [x_{2i},x_{2i+1}] ; \end{aligned}\right\} \tag{8.29}$$

$$\left.\begin{aligned}&\widehat{\lambda + \bar{\mu}\eta}^{\,\bullet} = \varphi(s)\\&\dot{\eta} = \mu a\\&\bullet = \frac{\partial}{\partial s} = \xi \quad \text{on a } \xi\text{-piece;} \quad \varphi \text{ is } L^\infty\text{-small .}\end{aligned}\right\} \tag{8.30}$$

Let

$z_i(0) = \lambda_i\xi + \mu_i v + \eta_i[\xi,v]$ be the value of this variation at x_{2i} ; (8.31)

$$z_i(s_i) = \text{value of } z \text{ at } x_{2i+1} = D\varphi_{s_i}(z_i(0)) + \delta s_i v; \tag{8.32}$$

$$\hat{\mu}_i = a_1(s_i,x_{2i})\lambda_i + b_1(s_i,x_{2i})\mu_i + c_1(s_i,x_{2i})\eta_i \ ; \tag{8.33}$$

$$\hat{\eta}_i = a_2(s_i,x_{2i})\lambda_i + b_2(s_i,x_{2i})_i + c_2(s_i,x_{2i})\eta_i \ ; \tag{8.34}$$

$$|z|^2_{H^1(x_{2i+1},x_{2i+2})} = \int_{[x_{2i+1},x_{2i+2}]}(\dot{\lambda}^2 + \dot{\mu}^2 + \dot{\eta}^2)ds \ . \tag{8.35}$$

We then have:

<u>Proposition 11:</u>

$$\begin{aligned}J(x + z) = a + \sum_i \Big[(\hat{\mu}_i + \delta s_i)\hat{\eta}_i - \mu_i\eta_i - \delta s_i\Big[\frac{\partial a_0}{\partial s}(s_i,x_{2i})\lambda_i +\\ + \frac{\partial b_0}{\partial s}(s_i,x_{2i})\mu_i + \frac{\partial c_0}{\partial s}(s_i,x_{2i})\eta_i\Big] -\\ - \Big(\frac{\partial a_0}{\partial y}\Big|_{(s_i,x_{2i})} \cdot z_i(0)\Big)\lambda_i - \Big(\frac{\partial b_0}{\partial y}\Big|_{(s_i,x_{2i})} \cdot z_i(0)\Big)\mu_i -\\ - \Big(\frac{\partial c_0}{\partial s}\Big|_{s_i,x_{2i})} \cdot z_i(0)\Big)\eta_i + \frac{1}{a}\int_{[x_{2i},x_{2i+2}]}(\dot{\eta}^2 - a^2\eta^2\tau)ds\Big]\\ + O(|z|^2) \ ,\end{aligned}$$

where η is an H^1-arbitrary function equal to η_i at x_{2i} and $\hat{\eta}_i$ at x_{2i+1}. This formula gives the index of the critical point at infinity. <u>Notice that it does not depend on</u> φ. Here

$$|z| = \sum_i |\delta s_i| + \sum_i [|\mu_i|+|\eta_i|+|\lambda_i|] + \sum_i |z|_{H^1(x_{2i+1},x_{2i+2})} \ .$$

<u>Proof:</u> It follows from (8.7), (8.32) and (8.11).

The second variation is the sum of the expressions in (8.32) and (8.11). In (8.11), we have $\dot{\eta} = \mu a$. We thus integrate by part:

$$-\int_{[x_{2i+1},x_{2i+2}]}(\dot{\mu}+a\eta\tau) = \mu(x_{2i+1})\eta(x_{2i+1}) - \mu(x_{2i+2})\eta(x_{2i+2}) + \int_{[x_{2i+1},x_{2i+2}]} a\mu^2 - a\eta^2\tau$$

$$= \mu(x_{2i+1})\eta(x_{2i+1}) - \mu(x_{2i+2})\eta(x_{2i+2}) + \frac{1}{a}\int_{[x_{2i+1},x_{2i+2}]}[\dot{\eta}^2-a\eta^2\tau]. \quad (8.36)$$

Now

$$\mu(x_{2i+2}) = \mu_{i+1};\ \eta(x_{2i+2}) = \eta_{i+1} \text{ by (8.31)} \quad (8.37)$$

and by (8.32), (8.27) and (8.33), (8.34):

$$\mu(x_{2i+1}) = \hat{\mu}_i + \delta s_i;\ \eta(x_{2i+1}) = \hat{\eta}_i\ . \quad (8.38)$$

This yields the result.

The fact that the remainder term is $o(|z|^2)$ with the definition stated of $|z|$, retaining the H^1-norm only on the ξ-pieces is just due to the invariance of J under reparameterization, in particular along the v-pieces where the variation is L^∞ in λ and η which are governed by (8.1), hence are L^∞-bounded by $\sum_i |\lambda_i| + |\eta_i|$; while the μ-variation along these pieces can be absorbed through a reparameterization, yielding only a $\sum_i |\delta s_i|$ term. Hence the proof of Proposition 11.

CONCLUSION

As one can clearly see throughout this description, the critical points at infinity, as a concept, are neither sequences, nor normal forms. They are geometrical objects together with a parameterization, involving quantities going to $+\infty$ and Morse indices to relate them to the difference of topology they produce in the level sets.

These are ends of orbits and in this sense, are ordinary objects in dynamical systems which happen to have very nice representations when there is more structure in the problem, as geometry and variational framework.

In fact, it is very much likely that such a conception is useful to solve the so-called "non-compactness"; out of it, one either loses the global variations (i.e. when one restricts to sequences) and therefore cannot derive general existence theorems; or one loses, developing only the topological aspects, the geometrical objects as such, which are seen in some other problems to be closely related to the critical points themselves.

REFERENCES

[1] A. Weinstein, 'On the hypotheses of Rabinowitz' periodic orbit theorems', J. Diff. Equ. **33** (1979), 353-358.
[2] A. Bahri, 'Pseudo-orbits of contact forms', to appear.
[3] S. Smale, 'Regular curves on Riemannian manifolds', Trans. Amer. Math. Soc. **87** (1958), 492-512.
[4] D. Bennequin, 'Quelques remarques sur la rigidité symplectique', Seminaire Sud-Rhodanian de Geometrie III. Geometrie symplectique et de contact, 1-150.
[5] A. Bahri, 'Un probleme variationnel sans compacité en geometrie de contact', Comptes. Rendus de l'Academie des Sciences Paris, **t299**, Serie I, no. 15 (1984).

SOME APPLICATIONS OF THE MORSE-CONLEY THEORY TO THE STUDY OF PERIODIC SOLUTIONS OF SECOND ORDER CONSERVATIVE SYSTEMS

V. Benci
Istituto di Matematiche Applicate
Universitá di Pisa
Pisa, Italy

INTRODUCTION

The aim of this paper is to show how the Morse-Conley theory can be applied to the study of periodic solutions of second-order Hamiltonian systems.

Also, in [BZ], the Morse-Conley theory has been successfully applied to this type of problem. However, the approach of [BZ] is different from ours and applies to a more narrow class of problems (i.e. problems with asymptotically quadratic Hamiltonians).

The reason is that in [BZ] the authors reduce the problem to a finite dimensional one, while we use an infinite-dimensional version of the Morse-Conley theory as developed in [B1] and [B2] and this fact allows to treat also superlinear problems.

A quite different approach to the study of periodic solutions via the Morse theory has been developed by Ekeland [E] using the dual action principle.

In this paper we restrict ourselves to the study of second order conservative systems, since the study of general Hamiltonian systems involve much more technicalities (cf. [BL]).

1. THE GENERALIZED MORSE-CONLEY INDEX FOR VARIATIONAL SYSTEMS

Let M be a Hilbert manifold and let $f \in C^2(M)$. We denote by $\eta(t,x)$ the flow relative to the differential equation

$$\frac{dx}{dt} = - \frac{f'(x)}{1 + \|f'(x)\|} .$$

When no ambiguity is possible we shall write $x \cdot t$ instead of $\eta(t,x)$. If U is an open set in M we set

$$G^T(U) = \{x \in \bar{U} | x \cdot [-T,T] \subset \bar{U}\} ,$$

P. H. Rabinowitz et al. (eds.), Periodic Solutions of Hamiltonian Systems and Related Topics, 57–78.

$$\Gamma^T(U) = \{x \in G^T(U) | x \cdot [0,T] \cap \partial U \neq \emptyset\}$$

where ∂U denotes the boundary of U.

Also we set

$$\Sigma = \{U \subset M | U \text{ is open and } \exists T > 0 \text{ such that } G^T(U) \subset U\} .$$

S will denote the set of formal power series (in t) with nonnegative coefficients (or to be more precise with coefficients which are cardinal numbers).

The generalized Morse-Conley Index (GIM) is a map

$$i : \Sigma \to S$$

defined as follows

$$i_t(U) = \lim_{T \to +\infty} \sum_{k=0}^{\infty} \dim[\bar{H}^k(G^T(U), \Gamma^T(U))] t^k \tag{1.1}$$

where $\bar{H}^*(\cdot,\cdot)$ denotes the Alexander-Spanier [Sp] cohomology with coefficients in some field, which in this paper will be $\mathbf{Q}$. The limit in (1.1) exists in a trivial sense; in fact in [Bl], it is proved that, for T large enough, $\bar{H}^*(G^T(U), \Gamma^T(U))$ does not depend on T.

When no ambiguity is possible we shall write $i(U)$ instead of $i_t(U)$.

Now we shall list some of the properties of the GIM which have been proved in [Bl].

Theorem 1.1. The GIM satisfies the following properties

(i) if $U \in \Sigma$ then $G^T(U) \in \Sigma$ and $i(G^T(U)) = i(U)$ $\forall T > 0$;

(ii) if $U \in \Sigma$ then $\eta(T,U) \in \Sigma$ and $i(\eta(T,U)) = i(U)$ $\forall T > 0$;

(iii) if $U,V \in \Sigma$ and $\exists T > 0$ such that $G^T(U) \subset V$ and $G^T(V) \subset U$, then $i(U) = i(V)$;

(iv) if $x \in U$ and for every $x \in U$ $\exists t > 0$ such that $x \cdot t \notin U$, then $i(U) = 0$;

(v) if $U \in \Sigma$ is contractible and positively invariant then $i(U) = 1$;

(vi) if $U,V \in \Sigma$ and $\bar{U} \cap \bar{V} = \emptyset$, then $i(U \cup V) = i(U) + i(V)$;

(vii) if η_i is a flow in $M_i (i = 1,2)$, then a flow $\eta_1 \times \eta_2$ is defined on $M_1 \times M_2$; in this case if $U_i \in \Sigma(\eta_i)$ $(i = 1,2)$, then $U_1 \times U_2 \in \Sigma(\eta_1 \times \eta_2)$ and

$$i(U_1 \times U_2, \eta_1 \times \eta_2) = i(U_1,\eta_1) \cdot i(U_2,\eta_2) .$$

Definition 1.2. Let $U_1, U_2 \in \Sigma$ with $U_1 \cap U_2 = \emptyset$. We say that U_2 is over U_1 if there exists $T > 0$ such that $U_1 \cap G^T(U_1,U_2)$ is positively invariant with respect to $G^T(U_1 \cup U_2)$.

Definition 1.3. Let $U \in \Sigma$. A family of sets $\{U_k\}_{k \leq N}$ is called a Morse decomposition of U if

(i) $\bar{U} = \bigcup_{k=1}^{N} \bar{U}_k$,

(ii) $U_k \in \Sigma$ for $k = 1,\ldots,N$,

(iii) $U_k \cap U_k = \emptyset$ for $k \neq h$

(iv) U_{h+1} is over $\bigcup_{j=1}^{h} U_i$ for $h = 1,\ldots,N-1$.

Example. Let f be a Liapunov function for (M,η) (i.e. a function strictly decreasing on non-stationary trajectories), and let $c_1 < c_2 < \cdots < c_{N-1}$ be a sequence of regular values for f (i.e. $f(x) = c_i \Longrightarrow f'(x) \neq 0$, $i = 1,\ldots,N-1$). Now set $c_0 = -\infty$ and $c_N = +\infty$ and

$$U_k = \{x \in U | c_{k-1} < f(x) < c_k\},\ k = 0,\ldots,N;\ U \in \Sigma .$$

It is easy to check that $\{U_k\}$ is a Morse decomposition of U.

Theorem 1.4. If $\{U_k\}_{k \leq N}$ is a Morse decomposition of U, then there exists $Q \in S$ such that

$$\sum_{k=1}^{N} i(U_k) = i(U) + (1 + t)Q(t) .$$

Now let $\Gamma \subset \Sigma$ be a family of sets which satisfy the following property,

$$\text{if } U \in \Gamma \text{ then any sequence } \{x_n\} \subset U \text{ such that } f'(x_n) \to 0 \text{ has a converging subsequence .} \tag{1.2}$$

The property (1.2) is related to the well-known condition of Palais-Smale.

Definition 1.5. We say that $f \in C^1(M)$ satisfies P.S. if any bounded sequence $x_n \in M$ such that $f(x_n)$ is bounded and $f'(x_n) \to 0$ has a converging subsequence.

Then if f satisfies P.S. it follows that

$$\Gamma = \{U \in \Sigma | f|_U \text{ is bounded}\} . \tag{1.3}$$

The couple $\{\eta,\Gamma\}$ is called variational system. In [B1] and [B2] there is a detailed study of variational systems. As we will see the sets $U \in \Gamma$ are sets for which the main properties of the Morse-Conley theory which are true in finite dimension are preserved.

Before recalling these properties some notation is necessary,

$$f_a^b = \{x \in M | a < f(x) < b\} ,$$

$$f_a = f_a^{+\infty}; \qquad f^b = f_{-\infty}^b ,$$

$$K(U) = \{x \in U : f'(x) = 0\} \qquad U \subset M ,$$

$$\mathcal{K} = \{K \subset M | K = K(U) \text{ for some } U \in \Gamma, \text{ and } K \text{ is connected}\}.$$

For $x \in M$, $f''(x)$ can be regarded as a bounded selfadjoint operator on the tangent space of M at x. We assume that the nonpositive part of the spectrum of $f''(x)$ consists of isolated eigenvalues of finite multiplicity.

Then, for $x \in K(M)$, we set

$$\left.\begin{aligned} m(x) &= \text{dimension of the space spanned by the eigenvectors of } f \text{ corresponding to negative eigenvalues} \\ n(x) &= \dim[\ker f(x)] \\ m^*(x) &= m(x) + n(x) . \end{aligned}\right\} \qquad (1.4)$$

$m(x)$ is called the Morse index of x and $n(x)$ the nullity of x. If $n(x) = 0$ then x is called "non-degenerate". For $k \subset K(M)$ we set

$$m(K) = \min_{x \in K} m(x) ,$$

$$m^*(K) = \max_{x \in K} m^*(x) .$$

<u>Proposition 1.6.</u>

(i) If f satisfies P.S. and $f|_U$ is bounded below $(U \in \Sigma)$, then $i(U) = 0$ implies $K(U) = \emptyset$.

(ii) If f satisfies P.S. and $a,b \in \mathbb{R}$ are regular values of f, then $f_a^b \in \Sigma$ and $i(f_a^b) = \sum_{q=0}^{\infty} \dim[H_q(f^b,f^a)]t^q$ where H_* denotes the singular homology with coefficients in $\mathbb{Q}$.

(iii) If $U \in \Gamma$ then $i(U)$ is finite (i.e. $i_t(U)$ is a polynomial in t with nonnegative coefficients).

(iv) Let $K \in \mathcal{K}$ and let $U,V \in \Gamma$ with $K(U) = K(V) = K$. Then $i(U) = i(V)$.

Proposition 1.6 (iv) suggests the following Definition.

<u>Definition 1.7.</u> If $K \in \mathcal{K}$ then we set

$$i(K) = i(U)$$

where U is a sufficiently small neighborhood of K such that

$K(U) = K$. In particular the index of an isolated critical point x_0 is defined (identifying x_0 with $\{x_0\}$). Moreover, we have

Proposition 1.8. If x_0 is a nondegenerate critical point then we have

$$i(x_0) = t^{m(x_0)} .$$

If x_0 is degenerate, we get some information from the following proposition.

Proposition 1.9. If $U \in \Gamma$ and $K = K(U)$ we have

$$i(U) = \sum_{q=m(K)}^{m^*(K)} a_q t^q$$

where the a_q's are nonnegative numbers.

In particular if $K \in \mathcal{K}$ we have

$$i(K) = \sum_{q=m(K)}^{m^*(K)} a_q t^q .$$

Definition 1.10. If $K \in \mathcal{K}$ we define the multiplicity of K the integer number $i_1(K)$.

If $i_1(K) = 1$ we say that K is topologically nondegenerate. If a point x_0 is nondegenerate, then $\{x_0\} \in \mathcal{K}$ and by Proposition 1.8 x_0 is topologically nondegenerate.

The Definition 1.10 is justified by the following proposition.

Proposition 1.11. Let $K \in \mathcal{K}$ with $i(K) = \sum_{q=m(K)}^{m^*(K)} a_q t^q$, and let U be a sufficiently small neighborhood of K.

Then every sufficiently C^2-small perturbation g which satisfies P.S. and whose critical points in U are nondegenerate, has at least

$$\sum_{q=m(x)}^{m^*(x)} a_q$$

critical points. Moreover, at least a_q of them have Morse index q (for $q = m(x), m(x) + 1, \ldots, m^*(x)$).

Notice that a generic perturbation of f has all nondegenerate critical points. Therefore the conclusion of Proposition 1.11 holds for a generic perturbation of f which satisfies P.S.

Now we can write the "Morse relations" for variational systems as defined above.

Definition 1.12. Let $X \in \Gamma$ and let $K = K(X)$. A family of sets $\{u_j\}_{j\in I}$ is called ε-Morse covering of K if

(i) $\bar{U}_j$ is connected for $j \in I$,

(ii) $K \subset \bigcup_{j=1}^{N} U_j \subset N_\varepsilon(K)$,

(iii) $U_j \in \Gamma$ and $\sum_{j\in I} i(U_j) = i(X) + (1 + t)Q(t) \quad Q \in \mathcal{S}$.

The above definition is justified by the following theorem:

Theorem 1.13. If $X \in \Gamma$, then for every $\varepsilon > 0$ there exists a finite ε-Morse covering of $K(X)$.

From Definition 1.7 and the above theorem we get the following corollary:

Corollary 1.14. If $U \in \Gamma$ and $K(U)$ consists of a finite number of connected components $K_1,\ldots,K_N$, then

$$\sum_{j=1}^{N} i(K_j) = i(U) + (1 + t)Q(t) .$$

From Corollary 1.14, the classic Morse relations follow:

Corollary 1.15. Let $U \in \Gamma$ and suppose that $K(U)$ contains only nondegenerate critical points. Then they are a finite number. Moreover, if $a(q)$ denotes the number of critical points having Morse index q, we have

$$\sum_{q=m(K(U))}^{m^*(K(U)} a(q)t^q = i(U) + (1 + t)Q \qquad Q \in \mathcal{S} .$$

The next theorem generalizes the Morse relations to a set where f is not bounded above.

Theorem 1.16. Let f be a function which satisfies P.S. and let $K = K(f_c)$.

Then, for every $\varepsilon > 0$ there exists an ε-Morse covering of K. Notice that, in Theorem 1.16, the series $\sum_{j\in I} i(U_j)$ and $Q(t)$ (which appears in (iii) of Definition 1.2) may have some coefficient equal to $+\infty$.

Proof. Let $c_n > c$ be an increasing sequence of regular values of f diverging to $+\infty$. By Theorem 1.4 we have, for every $n \in \mathbf{N}$,

$$i(f_c^{c_n}) + i(f_{c_n}^{\infty}) = i(f_c) + (1 + t)Q_n^1(t), \quad Q_n^1 \in \mathcal{S} . \tag{1.5}$$

By Theorem 1.13 (with $X = f_c^{c_n}$) we have

$$\sum_{j=1}^{k_n} i(U_j) = i(f_c^{c_n}) + (1 + t)Q_n^2(t), \quad Q_n^2 \in S \ .$$

Comparing the above formula with (1.5) we get

$$\sum_{j=1}^{k_n} i(U_j) + i(f_{c_n}) = i(f_c) + (1+t)Q_n, \quad Q_n = Q_n^1 + Q_2^n. \tag{1.6}$$

Now if $p = \sum_{\ell=0}^{\infty} a_\ell t^\ell \in S$, we set $\{p\}_\ell = a_\ell$. Then (1.6) reads

$$\{\sum_{j=1}^{k_n} i(U_j)\}_\ell + \{i(f_{c_n})\}_\ell = \{i(f_c)\}_\ell + \{(1+t)Q_n(t)\}_\ell \ . \tag{1.7}$$

The theorem is proved if we can take the limit in (1.7) for every $\ell \in \mathbf{N}$.

We consider two cases

(a) $\{i(f_{c_n})\}_\ell = 0$ for n large enough,

(b) $\{i(f_{c_n})\}_\ell \neq 0$ for a subsequence $c_n' \uparrow +\infty$.

If (a) holds we have done, since we can take the limit in (1.7) (notice that the sequence $\sum_{j=1}^{k_n} i(U_j)$ is monotonically increasing as $n \to +\infty$). If (b) holds, then, by Proposition 1.6 (iv), we have that

$$H_\ell(M, f^{c_n'}) \neq 0 \quad \text{for the subsequence } c_n' \ .$$

Let $[\alpha]$ denote the support of a nontrivial homology class $\alpha \in H_\ell(M, f^{c_n'})$ and let $c_m' > \max_{x\in[\alpha]} f(x)$.

Consider the exact homology sequence:

$$\cdots \to H_\ell(f^{c_m'}, f^{c_n'}) \xrightarrow{i_\ell} H_\ell(M, f^{c_n'}) \xrightarrow{j_\ell} H_\ell(M, f^{c'_m}) \xrightarrow{\partial_\ell} \cdots \ .$$

By our choice of c_m', $j_\ell(\alpha) = 0$; then by the exactness of the sequence $\exists \beta \in H_\ell(f^{c_m'}, f^{c_n'})$ s.t. $j_\ell(\beta) = \alpha$.

This fact shows that

$$\{i(f_{c_n'}^{c_m'})\}_\ell \neq 0$$

and by Theorem 1.3 there exists $U_m \subset f_{c'_n}^{c'_m}$ such that $U \in \Gamma$ and

$$\{i(U_m)\}_\ell \neq 0 .$$

Since this is true for all the terms of the subsequence c'_n defined by (b), it follows that taking the limit in (1.7)

$$\{\sum_{j=1}^{k_n} i(U_j)\}_\ell$$

diverges to $+\infty$.

Thus the equality (iii) of Definition 1.17 is satisfied also in this case. □

2. THE MASLOV INDEX AND THE ROTATION NUMBER

For $\sigma \in S^1 = \{z \in \mathbf{C} \mid |z| = 1\}$ we set

$$L^2_{\sigma,T} = \{x \in L^2_{loc}(\mathbf{R},\mathbf{C}^N) \mid x(t + T) = \sigma x(t)\}$$

where $L^2_{loc}(\mathbf{R},\mathbf{C}^N)$ is the set of function $x : \mathbf{R} \to \mathbf{C}^N$ which are measurable and whose square is locally integrable.

$L^2_{\sigma,T}$ is a Hilbert space if it is equipped with the following scalar product,

$$(x,y)_{L^2_{\sigma,T}} = \frac{1}{T}\int_0^T (x(t),y(t))_{\mathbf{C}^N} dt . \tag{2.1}$$

Now let $A(t)$ be a family of real symmetric $N \times N$ matrices depending continuously on t and periodic of period τ and consider the following ordinary differential equation

$$\ddot{y} + A(t)y = -\lambda y, \qquad y \in \mathbf{C}^N, \quad \lambda \in \mathbf{R} , \tag{2.2}$$

with the condition

$$y(t + T) = \sigma y(t), \qquad \sigma \in S^1, \quad T = k\tau, \quad k \in \mathbf{N} . \tag{2.3}$$

Now let $W^2_{loc}(\mathbf{R},\mathbf{C}^N)$ denote the space of function having two square locally integrable derivatives.

If $L_{\sigma,T}$ is the extension to $W^2_{loc}(\mathbf{R},\mathbf{C}^N) \cap L^2_{\sigma,T}$ of the operator

$$-\ddot{y} - A(t)y \tag{2.3'}$$

then it is well-known that $L_{\sigma,T}$ is a selfadjoint unbounded operator on $L^2_{\sigma,T}$.

Then the eigenvalue problem (2.2), (2.3) becomes

$$L_{\sigma,T}y = \lambda y, \ y \in D(L_{\sigma,T}) = L^2_{\sigma,T} \cap W^2_{loc}(\mathbf{R},\mathbf{C}^N) \ . \tag{2.4}$$

It is easy to check from elementary facts of spectral theory that $L_{\sigma,T}$ has discrete spectrum with only a finite number of negative eigenvalues.

This fact allows us to define a function

$$\mathfrak{J}(T,\cdot) : S^1 \to \mathbf{N}$$

as follows

$$\mathfrak{J}(T,\sigma) = \{\text{number of negative eigenvalues of } L_{\sigma,T} \text{ counted with their multiplicity}\} \ .$$

We shall call the function $\mathfrak{J}(T,1)$ the Maslov index relative to the equation $\ddot{y} + A(t)y = 0$ in the interval $[0,T]$.

Now let $W(t)$ be the Wronskian matrix relative to the equation

$$\ddot{y} + A(t)y = 0$$

i.e. the matrix which sends the initial data $\begin{bmatrix} u_0 \\ v_0 \end{bmatrix}$ to $\begin{bmatrix} y(t) \\ \dot{y}(t) \end{bmatrix}$.

The map $W(T) : \mathbf{C}^n \to \mathbf{C}^n$ (where T is the period of $A(t)$) is called the Poincaré map or the monodromy map. The eigenvalues of $W(T)$ are usually called Floquet multipliers (relative to the interval $(0,T)$).

Proposition 2.1. The function $\mathfrak{J}(T,\sigma)$ satisfies the following properties:

(i) $\mathfrak{J}(T,\bar{\sigma}) = \mathfrak{J}(T,\sigma)$ where $\bar{\sigma}$ is the complex conjugate of σ.

(ii) If $\mathfrak{J}(1,\sigma)$ is discontinuous at the point σ^* then σ^* is a Floquet multiplier.

(iii) $|\mathfrak{J}(T,\sigma_2) - \mathfrak{J}(T,\sigma_1)| \le \ell$, $\forall \sigma_1,\sigma_2 \in S^1 - \{+1,-1\}$ where 2ℓ is the number of nonreal Floquet multiplier on S^1 counted with their multiplicity. Thus, in particular $|\mathfrak{J}(T,\sigma_2) - \mathfrak{J}(T,\sigma_1)| \le N$ for every $\sigma_1,\sigma_2 \in S^1 - \{+1,-1\}$.

(iv) $\mathfrak{J}(kT,\theta) = \sum_{\sigma^k=\theta} \mathfrak{J}(T,\sigma)$.

Proof.

(i) If $y(t)$ is an eigenfunction of $L_{\sigma,T}$ the complex conjugate function $\overline{y(t)}$ is an eigenfunction of $L_{\bar{\sigma},T}$ corresponding to the same eigenvalue. Therefore $L_{\sigma,T}$ and $L_{\bar{\sigma},T}$ have the same number of negative eigenvalues.

(ii) The eigenvalues of $L_{\sigma,T}$ depend continuously on σ. Since $L_{\sigma,T}$ is selfadjoint, they are all real. Therefore the number of the nonpositive eigenvalues $\mathring{J}(T,\sigma)$ can change only for those σ^* such that 0 is an eigenvalue of $L_{\sigma^*,T}$. This means that if σ^* is a discontinuity of $\mathring{J}(\sigma,T)$, the following problem

$$\ddot{y} + A(t)y = 0 \tag{2.5}$$

$$y(t + T) = \sigma^* y(t) \tag{2.6}$$

has a nontrivial solution $y(t)$. If $W(t)$ is the Wronskian matrix of (2.5), then $\begin{bmatrix} y(t) \\ \dot{y}(t) \end{bmatrix} = W(t)\begin{bmatrix} x \\ v \end{bmatrix}$ for some $x,v \in \mathbf{C}^N$.

Then the condition (2.6) for $t = 0$ reads

$$W(T)\begin{bmatrix} x \\ v \end{bmatrix} = \sigma^*\begin{bmatrix} x \\ v \end{bmatrix} .$$

Therefore σ^* is an eigenvalue of $W(T)$.

(iii) $W(T)$ is a symplectic matrix; then if λ is an eigenvalue of $W(T)$ also $\bar{\lambda},\lambda^{-1},\bar{\lambda}^{-1}$ are eigenvalues of $W(T)$. Therefore

(a) the number of eigenvalues of $W(T)$ different from ± 1 is even.

(b) the sum of the multiplicity of the eigenvalues $+1$ and -1 is even (if ± 1 are not eigenvalue then their multiplicity has to be assumed 0 in order to make sense of the above statement).

In particular the eigenvalues of $W(T)$ on $S^1 - \{+1,-1\}$ is an even number 2ℓ. We can assume that all the eigenvalues are simple (otherwise use a perturbation argument).

Therefore $\mathring{J}(T,\sigma)$ has at most 2ℓ points of discontinuity and at each of them the jump of $\mathring{J}(T,\sigma)$ is ± 1 since we have assumed that all the eigenvalues are simple.

Now

$$|\mathring{J}(T,\sigma_2) - \mathring{J}(T,\sigma_1)| = |\mathring{J}(T,e^{iw_1}) - \mathring{J}(T,e^{iw_2})|$$

$$\text{with } w \in (-\pi,\pi) - \{0\} .$$

By (i) we have that $\mathring{J}(T,e^{iw}) = \mathring{J}(T,e^{-iw})$. Then we can assume that $w \in (0,\pi)$. But the function $\mathring{J}(T,e^{iw})$ has at most ℓ jumps and this proves the statement.

(iv) If $v \in L^2_{\sigma,T}$ then v has the following series expansion:

$$v(t) = e^{iwt}\sum_{n=-\infty}^{+\infty} c_n e^{2\pi in/T} \text{ with } w \in [0,2\pi) \text{ with } e^{iwt} = \sigma.$$

Using the above formula we have that $v \in \bigoplus_{\sigma^k=\theta} L^2_{T,\sigma}$ has the following expansion:

$$v(t) = \sum_{\ell=0}^{k-1} e^{iw_\ell t} \sum_{n=-\infty}^{+\infty} c_{\ell,n} e^{2\pi i n/T} \quad \text{with} \quad e^{iw_\ell T} = \sigma_\ell ,$$

$$\sigma_\ell \in \sqrt[k]{\theta} , \qquad \ell = 0,\dots,k-1 .$$

But we have $w_\ell = w_0 + 2\pi\ell/kT$, $\ell = 1,\dots, k-1$. Then

$$v(t) = e^{iw_0 T} \sum_{k=0}^{k-1} \sum_{n=-\infty}^{+\infty} c_{n,\ell} e^{2\pi i(kn+\ell)/kT}$$

and rearranging the terms we have

$$v(t) = e^{iw_0 T} \sum_{m=-\infty}^{+\infty} a_m e^{2\pi i m/kT} .$$

The above formula shows that

$$L^2_{kT,\theta} = \bigoplus_{\sigma^k=\theta} L^2_{T,\sigma} .$$

Now the operator $L_{\theta,kT}$ leaves invariant the spaces $L^2_{T,\sigma}$ (if $\sigma^k = \theta$). Now if $\sigma^-(L)$ denotes the negative spectrum of L we have:

$$\sigma^-(L_{\theta,kT}) = \bigcup_{\sigma^k=\theta} \sigma^-\left(L_{\theta,kT}\big|_{L^2_{\sigma,T}}\right) = \bigcup_{\sigma^k=\theta} \sigma^-(L_{\sigma,T}) .$$

From the above formula the conclusion follows. □

Now we can define the rotation number as follows:

$$\rho = \frac{1}{2\pi\tau} \int_{S^1} \mathring{J}(\tau,\sigma)d\sigma = \frac{1}{2\pi\tau} \int_0^{2\pi} \mathring{J}(\tau,e^{iw})dw .$$

<u>Proposition 2.2.</u> The rotation number satisfies the following properties:

(i) $\rho = \lim\limits_{\substack{T\to-\infty \\ T=k\tau}} \frac{1}{T} \mathring{J}(T,1)$,

(ii) $\rho = \frac{1}{2\pi T} \int_{S^1} \mathring{J}(T,\sigma)d\sigma, \qquad T = k\tau$,

(iii) $|T\rho - \mathring{J}(T,\sigma)| < \ell$ for every $\sigma \in S^1 - \{+1,-1\}$, $T = k\tau$. 2ℓ is the number of Floquet multipliers on $S^1 - \{+1,-1\}$.

(iv) For every $\sigma \in S^1$ we have $\lim_{\substack{T\to+\infty \\ T=k\tau}} \frac{1}{T}\mathring{J}(T,\sigma) = \rho$.

<u>Proof</u>.

(i) By Proposition (2.1) (iv) we have

$$\lim_{k\to+\infty} \frac{1}{k\tau}\mathring{J}(k\tau,1) = \lim_{k\to+\infty} \frac{1}{k\tau}\sum_{\ell=0}^{k-1} \mathring{J}(\tau,e^{2\pi i\ell/k}) . \tag{2.7}$$

By the definition of the Cauchy integral we have

$$\lim_{k\to+\infty} \frac{2\pi}{k}\sum_{\ell=0}^{k-1} \mathring{J}(\tau,e^{2\pi i\ell/k}) = \int_0^{2\pi} \mathring{J}(\tau,e^{iw})dw = 2\pi\tau\rho . \tag{2.8}$$

Then by (2.7) and (2.8) we have

$$\lim_{\substack{T\to+\infty \\ T=k\tau}} \frac{1}{T}\mathring{J}(k\tau,1) = \frac{1}{2\pi\tau}\lim_{k\to+\infty}\frac{2\pi}{k}\sum_{\ell=0}^{k-1} \mathring{J}(\tau,e^{2\pi\ell/k}) = \rho .$$

(ii) From (i) it follows that ρ is independent on $T = k\tau$.

(iii) $|T\rho - \mathring{J}(T,\sigma) = \frac{1}{2\pi}\left|\int_{S^1} \mathring{J}(T,\theta)d\theta - \int_{S^1}\mathring{J}(T,\sigma)d\theta\right| <$

$< \frac{1}{2\pi}\int_{S^1-\{+1,-1\}} |\mathring{J}(T,\theta) - \mathring{J}(T,\sigma)|d\theta < \ell$ by Proposition 2.1(iv).

(iv) It follows from (i) and (iv).

<u>Example</u>. Consider the equation

$$\ddot{y} + Ay = 0 ,$$

$$y(0) = y(T)$$

where A is a time independent real symmetric matrix with ℓ positive eigenvalues $w_1^2,\dots,w_\ell^2$ and $N - \ell$ negative eigenvalues.

Then the negative eigenvalues of $-\ddot{y} - Ay$ on $L^2_{1,T}$ are $\lambda_{n,j} = (\frac{2\pi}{T})n^2 - w_j^2$ with $n \in \mathbf{N}$, $\ell = 2,\dots,k-1$ and $n < \frac{w_jT}{2\pi}$.
Notice that for $n > 1$ they have double multiplicity. Therefore

$$\mathring{J}(T,1) = \ell + 2\cdot \#\{(n,j)|n < \frac{w_jT}{2\pi}\} = \ell + 2\sum_{j=1}^{\ell}\left[\frac{w_jT}{2\pi}\right] .$$

Then by Proposition 2.2(i) we have

$$\rho = \lim_{T\to+\infty} \frac{1}{T} \dot{J}(T,1) = \lim_{T\to+\infty} \frac{\ell}{T} + \frac{2}{T} \sum_{j=1}^{\ell} \left[\frac{w_j T}{2\pi}\right] = \frac{1}{\pi} \sum_{j=1}^{\ell} w_j \ .$$

3. THE GENERALIZED MORSE-CONLEY INDEX FOR PERIODIC SOLUTIONS OF SECOND ORDER CONSERVATIVE SYSTEMS

In this section we consider the following system of ordinary differential equations

$$\ddot{x} + V'(t,x) = 0, \qquad x \in \mathbb{R}^N \tag{3.1}$$

with $V \in C^2(\mathbb{R} \times \mathbb{R}^N)$. We suppose that $V(t,\cdot)$ is τ-periodic.

We set

$$W^T = \{x \in W^2_{loc}(\mathbb{R},\mathbb{R}^N) \,|\, x \text{ is T-periodic}\} \ .$$

W^T is an Hilbert space if it is equipped with the following scalar product:

$$(x,y)_{W^T} = \frac{1}{T} \int_0^T (\dot{x} \cdot \dot{y} + x \cdot y)dt$$

where "$\cdot$" denotes the scalar product in $\mathbb{R}^N$.

The equations (3.1) are the Euler-Lagrange equations corresponding to the functional

$$f(x) = \frac{1}{T} \int_0^T \{\frac{1}{2} |\dot{x}|^2 + V(t,x)\}dt, \qquad x \in W^T \ . \tag{3.2}$$

It is well-known that $f(x)$ is a functional of class C^2 on W^T. Therefore, any T-periodic solution of (3.1) can be interpreted as a critical point of the functional (3.2).

If we apply the theory of Section 1, we can define a Morse index for every T-periodic solution of (2.9) (cf. Definition (1.4)) which we shall denote by $m(x,T)$ to emphasize the fact that the Morse index is computed in the space W^T.

Of course, we can also define the nullity $n(\bar{x},T)$ and the number $m^*(\bar{x},T) = m(\bar{x},T) + n(\bar{x},T)$ as in Definition (1.3). Now let us consider the linearization of the equation (3.1) at x:

$$\ddot{y} + V''(t,\bar{x}(t))y = 0 \ . \tag{3.3}$$

It is easy to check that $m(x,T)$ is the number of negative

eigenvalues of the self-adjoint operator

$$y \longmapsto -\ddot{y} - V''(t,\bar{x}(t))y \tag{3.3'}$$

in $L^2((0,T],\mathbb{R}^n)$.

$n(\bar{x},T)$ is the multiplicity of the eigenvalue 0 of (3.3') and hence it is the number of independent solutions of equation (3.3).

A T-periodic solution $\bar{x}$ of (3.1) is called nondegenerate if it is nondegenerate as critical point of the functional (3.2) i.e. if $n(x,T) = 0$. Clearly $\bar{x}$ is nondegenerate if and only if the linear system (3.3) does not have any nontrivial T-periodic solution, or, if you like, if 1 is not a Floquet multiplier of the equation (3.3) relative to the interval $(0,T)$. We recall that a number $\alpha \in \mathbf{C}$ is called a Floquet exponent if e^{α} is a Floquet multiplier.

Definition 3.1. Let $\bar{x}$ be a T-periodic solution of the equation (3.1) and let $2\pi i w_j$ $(j = 1,\dots,\ell < N)$ be the purely imaginary Floquet exponent of the linearized equation (3.3). Then if $w_j \notin \mathbf{Q}$ for $j = 1,\dots,\ell$ we say that $\bar{x}$ is nonresonant.

It is easy to check that if $\bar{x}$ is a nonresonant τ-periodic solution, then $\bar{x}$ is T-nondegenerate for every $T = k\tau$, $k \in \mathbf{N}$.

If $\bar{x}$ is a T-degenerate solution of (3.1) then the Definition 1.10 can be applied to define the multiplicity of $\bar{x}$.

We can associate to the equation (3.3) a Maslov index $\dot{J}_{\bar{x}}(T,\sigma)$ as in Section 2 where $A(t) = V''(t,\bar{x}(t))$ and consequently a rotation number $\rho(\bar{x})$.

Proposition 3.2. If $\bar{x}$ is a T-periodic solution of (3.1) $(T = kT$, $k \in \mathbf{N})$ then

(i) $m(\bar{x},T) = \dot{J}_{\bar{x}}(T,1)$.

Moreover if $\bar{x}$ is not degenerate

(ii) $T \cdot \rho(\bar{x}) - N \leq m(\bar{x},T) \leq T\rho(\bar{x}) + N$.

Proof.

(i) Is a trivial consequence of the definitions.

(ii) Since 1 is not a Floquet multiplier, then for σ_1 very close to 1 $(\sigma \in S^1)$ σ_1 is not a Floquet multiplier and $m(T,\bar{x}) = \dot{J}_{\bar{x}}(T,\sigma)$ by Proposition 2.1(ii).

Then the conclusion follows from Proposition 2.2(iii). □

Now let Γ^T be the family of subsets of W^T defined as in (1.2).

Now we want to examine the relationship between the index of a set U $(U \in \Gamma^T)$ and the rotation number of the solution of (3.1) contained in U.

Proposition 3.3. Let $U \in \Gamma^T$ and let $i(U) = \sum_{\ell=m_1}^{m_2} a_\ell t^\ell$ with $a_m \neq 0$ $(m_1 \leq m \leq m_2)$. Then $\exists x \in U$ such that

$$\frac{m - N}{T} \leq \rho(x) \leq \frac{m + N}{T} .$$

Proof. Since $U \in \Gamma^T$ we can apply Proposition 1.11.

Then for every $\varepsilon > 0$ there exists $g_\varepsilon > 0$ such that $i(U)$ relative to g_ε is the same than the index relative to f and all the critical points of g_ε in U are nondegenerate.

Then, since $a_m \neq 0$, there exists x_ε critical point of g_ε such that

$$\frac{1}{T}(m - N) \leq \rho(x_\varepsilon) \leq \frac{1}{T}(m + N)$$

(we have used Proposition 3.2(ii)).

Now, letting $\varepsilon \to 0$, $x_\varepsilon \to x$ and $\rho(x_\varepsilon) \to \rho(x)$ and this proves our assumption. □

Corollary 3.4. Let x_0 be a degenerate critical point whose index is

$$i(x_0) = \sum_{\ell=m_1}^{m_2} a_\ell t^\ell \cdot (a_{m_1} \neq 0,\ a_{m_2} \neq 0,\ m_1 \leq m_2) .$$

Then

$$\frac{1}{T}(m_2 - N) \leq \rho(x) \leq \frac{1}{T}(m_1 + N) .$$

Proof. Apply Proof 3.4. □

Next we shall examine some facts which occur in the autonomous case i.e. we consider the equation

$$\ddot{x} + V'(x) = 0, \qquad x(t) \in \mathbb{R}^N . \tag{3.4}$$

In this situation every critical point $x \in \mathbb{R}^N$ of V is a constant solution of (3.4).

Proposition 3.5. Let $U \in W^T(U \in \Gamma)$ be a set which does not contain constant solutions. Then there exists a polynomial $P(t)$ with integer (but not necessarily positive) coefficients such that

$$i(U) = (1 + t)P(t) .$$

Proof. See [B2] Proposition 4.8. □

4. SOME APPLICATIONS IN THE NON-AUTONOMOUS CASE

In this section we try to get some information on the structure of the periodic solutions of the equation (3.1).

We suppose that $V(t,x)$ satisfies the following asymptotic conditions:

$$\text{there exists } R > 0 \text{ and } p > 2 \text{ such that}$$
$$0 < V(t,x) \leq \frac{1}{p} V_x(t,x) \cdot x, \ \forall t \in \mathbf{R}, \ \forall x \text{ with } |x| > R. \quad (4.1)$$

Condition (4.1) implies that $V(t,x)$ grows more than $|x|^2$ as $|x| \to +\infty$. Moreover, this condition implies the following facts:

Lemma 4.1. Suppose that V satisfies (4.1). Then the functional (3.2) satisfies P.S.

Proof. See e.g. [R]. □

Lemma 4.2. Let

$$f_c = \{x \in W^T | f(x) > c\} .$$

Then there exists $c_0 \in \mathbf{R}$ such that

$$f_c \in \Sigma \text{ and } i(f_c) = 0 \text{ for every } c \leq c_0.$$

Proof. See [B2] lemma 3.7. □

Theorem 4.3. Suppose that V satisfies (4.1) and let x_0 be a nonresonant τ-periodic solution of 3.1. Then, for every $\varepsilon > 0$ there exists a T-periodic solution $x \neq x_0$ (with $T = k\tau$; $T \leq \tau + \frac{2N+1}{\varepsilon}$) such that

$$|\rho(x) - \rho(x_0)| \leq \varepsilon .$$

Proof. Take $T = kT$ with $\frac{2N+1}{\varepsilon} \leq T < \tau + \frac{2N+1}{\varepsilon}$. Since x_0 is nonresonant, there is a neighborhood $N_\delta(x_0)$ in W^T which does not contain periodic solutions of (4.1). Now take a δ-Morse covering $\{U_\ell\}$ of f_c (where f_c is as in Lemma 4.2, $c \leq c_0$). Then, by Theorem 1.16

$$i(x_0) + \sum_{\ell \cdot I} i(U_\ell) = (1 + t)Q(t) .$$

By the above formula there exists $\ell \in I$ such that either

$$i(U_\ell) = t^{m+1} \quad (4.2)$$

or

$$i(U_\ell) = t^{m-1} .$$

where m is the Morse index of x_0. We consider the first possibility (if the second one holds we argue in the same way).

By Proposition 3.2(ii) we have

$$i(x_0) = t^m \quad \text{with} \quad \rho(x_0)T - N \leq m \leq \rho(x_0) \cdot T + N . \tag{4.3}$$

By Proposition 3.3 and (4.2), there exists $x \in U_\ell$ such that

$$\frac{1}{T}(m + 1 - N) \leq \rho(x) \leq \frac{1}{T}(m + 1 + N) . \tag{4.4}$$

Comparing (4.3) and (4.4) we get

$$|\rho(x) - \rho(x_0)| \leq \frac{1}{T}(2N + 1) \leq \varepsilon . \qquad \square$$

The next theorem we are going to prove has stronger assumptions and gives better information about the T-periodic solution of equation (3.1).

<u>Theorem 4.4.</u> Suppose that V satisfies (4.1). Let $T = p\tau$ with p prime number, and suppose that all the T-periodic solutions of (3.1) are isolated (as points in W^T). Let $x_1, x_2, \ldots, x_n, \ldots$ be the τ-periodic solutions of equation (3.1). We suppose that they are T-nondegenerate and ordered by increasing rotation number

$$\rho(x_1) \leq \rho(x_2) \leq \rho(x_3) \leq \ldots .$$

Then for every number $\rho \in [\rho(x_{2n-1}), \rho(x_{2n})]$ $(2n < p)$ there is a T-periodic solution $\bar{x}$ such that

$$|\rho(\bar{x}) - \rho| \leq \frac{N + 1}{T} .$$

<u>Proof.</u> By the Theorem 1.16 relative to the space W^T we have

$$\sum_{j \in J} i(x_j) + \sum_{j \in I} i(U_j) = (1+t)Q(t) \text{ with } Q(t) = \sum_\ell q_\ell t^\ell \tag{4.5}$$

where $\{U_j\}_{j \in I}$ is an ε-Morse covering of the T-periodic solutions of (3.1) which are not τ-periodic and $\{x_j\}_{j \in J}$ is the set of τ-periodic solutions.

Now fix $\rho \in [\rho(x_{2n-1}) + T \cdot (N + 1), \rho(x_{2n}) - T \cdot (N + 1)]$ and take

$$m = \{\text{integer part of } \rho \cdot T\} .$$

Consider only the terms of (4.5) of order less or equal to m:

$$\sum_{\ell=1}^{m} a_\ell t^\ell + \sum_{\ell=0}^{m} b_\ell t^\ell = (1+t)\sum_{\ell=0}^{m-1} q_\ell t^\ell + q_m t^m \tag{4.6}$$

where

$$\sum_{\ell=1}^{m} a_\ell t^\ell = \sum_{j=1}^{2n-1} i(x_j) \tag{4.7}$$

and the term $\sum_{\ell=0}^{m} b_\ell t^\ell$ comes from the ε-Morse covering relative to the solutions which are not τ-periodic.

Since we have supposed that such solutions are isolated, by Proposition 4.1 of [B2] we have that

$$b_\ell = q\beta_\ell \quad \text{for some} \quad \beta_\ell \in \mathbf{N} .$$

Then rewriting (4.6) for $t = -1$, we get

$$\sum_{\ell=1}^{m} (-1)^\ell a_\ell + p \sum_{\ell=0}^{m} (-1)^\ell \beta_\ell = (-1)^m q_m . \tag{4.8}$$

By (4.7), the first term of (4.8) is an odd number less or equal to $2n - 1$, and by our assumption less than p.

Thus, the sum of the two terms of the left-hand side of (4.8) is different from 0. Thus, $q_m \neq 0$. Then, by (4.5), there exists U_j such that

$$i(U_j) = t^m + \text{possible other terms} .$$

Proposition 3.4 implies that there exists $\bar{x} \in U_j$ such that

$$\frac{1}{T}(m - N) < \rho(\bar{x}) < \frac{1}{T}(m + N)$$

and by the definition of m we have that

$$\rho - \frac{N+1}{T} < \rho(\bar{x}) < \rho + \frac{N+1}{T} .$$

Thus, the theorem is proved for $\rho \in [\rho(x_{2n-1}) + T(N + 1), \rho(x_{2n}) - T(N + 1)]$. Considering also the solutions x_{2n-1} and x_{2n} the theorem is proved for every $\rho \in [\rho(x_{2n-1}), \rho(x_{2n})]$. □

We conclude this section with a theorem which is the analogous of Theorem 4.3 in the asymptotically quadratic case.

We say that $V(t,x)$ is asymptotically quadratic if there exists a matrix $A_\infty(t)$ such that

$$V_x(t,x) = A_\infty(t)x + O(|x|) \quad \text{as} \quad |x| \to +\infty . \tag{4.9}$$

If V is asymptotically quadratic we can consider the linearized system at ∞

$$\ddot{y} + A_\infty(t)y = 0 \tag{4.10}$$

and associate to (4.10) a rotation number ρ_∞. Then we have the following result:

Theorem 4.5. Suppose that V satisfies (4.9) and suppose that (4.10) has no T-periodic solution different from 0.

Let x_0 be a nondegenerate T-periodic solution of (3.1) with rotation number $\rho(x_0)$ such that

$$|\rho(x_0) - \rho_\infty| > \frac{2N}{T} . \tag{4.11}$$

Then the system (3.1) has a T-periodic solution $\bar{x}$ such that

$$|\rho(\bar{x}) - \rho(x_0)| < \frac{2N + 1}{T} .$$

Sketch of the proof. If we take a ball in W^T of sufficiently large radius R, arguing as in [B2], we have that

$$B_R \in \Gamma^T \quad \text{and} \quad i(B_R) = t^{m(\infty)} .$$

It is easy to check that

$$T \cdot \rho_\infty - N \leq m(\infty) \leq T \cdot \rho_\infty + N . \tag{4.12}$$

Then the Morse relation take the form

$$i(x_0) + \sum_{e \in I} i(U_e) = t^{m(\infty)} + (1 + t)Q(t) .$$

Let $i(x_0) = t^m$. Then, by (4.11) and (4.12)

$$|m - m(\infty)| \neq 0 .$$

Therefore we have that $Q(t) \neq 0$.

From now on we can argue as at the end of the Theorem 4.3. □

5. ONE APPLICATION TO THE AUTONOMOUS CASE

Now we consider the autonomous equation (3.4). We restrict ourselves to the superlinear case i.e. we still assume that V satisfies (4.1).

In this case the Theorems 4.3 and 4.4 do not apply since every solution of equation (3.4) is degenerate. In fact if x is a T-periodic solution of (3.4), $y = \dot{x}$ is a T-periodic solution of the linearized equation

$$\ddot{y} + V''(x(t))y = 0 \ .$$

Let $\rho_0 = \max\{\rho(x) | x$ is a constant solution of (3.4).$\}$

Theorem 5.1. For every $\rho > \rho_0$ there is a T-periodic solution $\bar{x}$ such that

$$|\rho - \rho(\bar{x})| < \frac{N+1}{T} \ .$$

Proof. For the sake of simplicity we will suppose that the constant solutions $x_1,\ldots,x_n$ of (3.4), i.e. the critical points of $V(x)$ are T nondegenerate solutions.

The general case can be treated using a perturbation argument of the type used in [B2] theorem 4.13.

By Theorem 1.16 relative to the space W^T we have

$$\sum_{j=1}^{n} i(x) + \sum_{j\in I} i(U_j) = (1+t)Q(t) \quad \text{with}$$

$$Q(t) = \sum_{\ell} q_\ell t^\ell, \qquad q_\ell > 0 \tag{5.1}$$

where $\{U_j\} \cup \{ \bigcup_{j=1}^{n} B_\varepsilon(x_j)\}$ is an ε-Morse covering.

Now we claim that n is an odd number. In fact, the critical points of V (which we are supposed nondegenerate) satisfy the following Morse relations:

$$\sum_{\ell} a_\ell t^\ell = i(\mathbb{R}^n) + (1+t)Q(t) \ .$$

Since $i(\mathbb{R}^n) = 1$ by our assumption on the potential V, taking the above relation with $t = 1$ we get

$$\text{number of critical points of } V = 1 + 2 \cdot Q(1)$$

which proves our claim. By Proposition 3.5

$$i(U_j) = (1+t)P_j(t) \ .$$

Then equation (5.1) can be written as follows

$$\sum_{j=1}^{n} i(x_j) + (1+t)\sum_{\ell} b_\ell t^\ell = (1+t)\sum_{\ell} q_\ell t^\ell$$

$$\text{where} \quad \sum_{\ell} b_\ell t^\ell = \sum_{j} P_j(t) \ . \tag{5.2}$$

Now take $\rho > \bar{\rho} + \frac{N+1}{T}$ and let

$$m = \text{integer part of } \frac{\rho T}{\pi} .$$

The equation (5.2) up to the order m reads

$$\sum_{j=1}^{n} i(x_j) + (1+t) \sum_{\ell=1}^{m-1} b_\ell t^\ell + b_m t^m = \sum_{\ell=1}^{m-1} q_\ell t^\ell + q_m t^m . \quad (5.3)$$

Now taking $t = -1$, from the above equation we get

$$\sum_{j=1}^{n} i_1(x) + (-1)^m b_m = (-1)^m q_m .$$

Since $\sum_{j=1}^{n} i_1(x)$ is an odd number, it follows that b_m (or q_m) is different from zero.

In either case, from equation (5.1) it follows that there exists U_j such that

$$i(U_j) = t^m + \text{other possible terms} .$$

Then by Proposition 3.4, there exists $\bar{x}$ such that

$$\frac{m-N}{T} \leq \rho(\bar{x}) \leq \frac{m+N}{T} .$$

The conclusion follows from the definition of m. □

REFERENCES

[B1] Benci, V., 'A new approach to the Morse-Conley theory', to appear in Recent Advances in Hamiltonian Systems, G. F. Dell'Antonio, ed.

[B2] Benci, V., 'Some applications of the generalized Morse-Conley index', Mathematics Research Center Technical Summary Report (1986) and to appear in Conferences del Seminario di Matematica dell'Universita di Bari.

[BL] Benci V. and Longo, D., in preparation.

[CZ] Conley, C. and Zehnder, E., 'Morse type index theory for flows and periodic solutions for Hamiltonian equations', Comm. Pure Appl. Math. **37** (1984), 207-253.

[E] Ekeland, I., 'Une théorie de Morse pour le systemes Hamiltoniens convexes', Ann. Inst. Henri Poincaré **1** (1984), 19-78.

[R] Rabinowitz, P. H., 'Minimax methods in critical point theory with applications to differential equations', CBMS Regional Conf. Series in Math. **65**, A.M.S., Providence, RI (1986).

A "BIRKHOFF-LEWIS" TYPE RESULT FOR NONAUTONOMOUS HAMILTONIAN SYSTEMS*

Vieri Benci
Istituto di Matematiche
Applicate Università
Pisa, Italy

Donato Fortunato
Dipartimento di Matematica
Università
70125 Bari, Italy

ABSTRACT. This paper contains results concerning the existence of long periodic solutions of nonautonomous Hamiltonian systems near the origin.

1. STATEMENT OF THE RESULTS

Consider the following Hamiltonian system of differential equations

$$-J\dot{z} = H_z(t,z) \tag{1.1}$$

where $z = (p,q) \in \mathbf{R}^{2N}$, J denotes the symplectic structure in $\mathbf{R}^{2N}$ and H_z denotes the gradient of the Hamiltonian function H with respect to z. We assume that:

H_1) H is C^2 near $z = 0$ and $H(t,0) = 0$, $H_z(t,z) = 0$,

H_2) H is 1-periodic in the t-variable.

In this paper we are concerned with the existence of subharmonic solutions for (1.1), i.e. we look for k periodic solutions $(k \in \mathbf{N})$ of (1.1).

More precisely we search for subharmonic solutions near the origin having arbitrarily long minimal periods.

The existence of long periodic solutions of (1.1) near a periodic solution has been the object of a considerable amount of study (cf.

---

*Sponsored by M.P.I. (fondi 60% "Problemi diff. nonlineari e teoria dei punti critici"; fondi 40% "Eq. diff. e calcolo delle variazioni").

P. H. Rabinowitz et al. (eds.), Periodic Solutions of Hamiltonian Systems and Related Topics, 79–84.

[6,7] and references). The first results of this type are due to Birkhoff and Lewis [3].

In order to state the results we have obtained, we need to introduce some notations and to recall some well-known facts.

By H_1) we can write

$$H(t,z) = \frac{1}{2}(Az|z) + \hat{H}(t,z)$$

where $A = A(t)$ denotes the Hessian matrix of H at $z = 0$, and $\hat{H}(t,z) = o(|z|^2)$. Consider now the linearized equation at $z = 0$

$$\dot{z} = JAz \tag{1.2}$$

and let $X(t)$ the solution of $\dot{X}(t) = JAX(t)$ and $X(0) = Id$. The eigenvalues of the symplectic matrix $X(1)$ are called the Floquet multipliers of $z = 0$.

We shall make the following assumptions:

H_3) $z = 0$ possesses at least one simple Floquet multiplier $\sigma = e^{iw}(w \in \mathbf{R})$ on the unit circle (i.e. $\sigma = e^{iw}$ is a simple eigenvalue of $X(1)$).

H_4) 1 is not a Floquet multiplier for (1.2).

H_5) There exist constants $r > 0$ and $p > 2$ such that

$$(\hat{H}_z(t,z)|z) \geq p\hat{H}(t,z) \geq \text{const.}\ |z|^p$$

for any $z \in \mathbf{R}^{2N}$, $|z| \leq r$ and any $t \in \mathbf{R}$.

H_6) If $z = z(t)$ is a periodic function with minimal period q, q rational, and $\hat{H}_z(t,z(t))$ is periodic with minimal period q, then q is an integer.

The following theorem holds

Theorem 1.1. Suppose that $H_1),\ldots,H_6)$ hold. Then there exists an integer m such that for any k prime, $k \geq m$, there exists a subharmonic solution z_k of (1.1) with minimal period k. Moreover z_k converges to zero as $k \to \infty$ uniformly in $C^1(\mathbf{R},\mathbf{R}^{2N})$.

Remark 1.2. Assumption H_4) is a nonresonance assumption; it requires that the linearized system (1.2) admits no nontrivial periodic solution with period 1.

Assumption H_6) means that $\hat{H}$ depend on t in an "essential way". This assumption is satisfied, for example, if $\hat{H}(t,z) = g(t)H^*(z)$ and g is a periodic function with minimal period 1.

Remark 1.3. If $A = 0$ a result related to Theorem 1.1 has been proved in [6].

If $A \neq 0$ the existence of long periodic solutions near the origin has been proved in [3-5] without assumption H_5). On the other hand in those papers stronger nonresonance assumptions than H_4) are made.

2. SKETCH OF THE PROOF OF THEOREM 1.1

The proof of Theorem 1.1 will be carried out in various steps.

Step 1. The local nature of the result "permits" to modify the Hamiltonian function outside a neighborhood of the origin by using a suitable cut-off function (cf. [6]). Then we can assume, without loosing generalities, that $H_5)$ holds globally (i.e. for any $z \in \mathbb{R}^{2N}$) and that $\hat{H}(t,z) = \text{const.}\ |z|^p$ outside a neighborhood of the origin.

Step 2. Let $k \in \mathbb{N}$ and denote by L_k the self-adjoint realization in $L^2(0,k)$ of the operator $z \to -J\dot{z} - Az$ with periodic conditions.

It is not difficult to show that the spectrum $\sigma(L_k)$ of L_k consists of isolated eigenvalues of finite multiplicity. Moreover, $\sigma(L_k)$ is unbounded both from below and from above. By virtue of $H_3)$ and by using Floquet theory and the symplectic nature of (1.2) (cf. [0]) it is possible to prove that for any $k \in \mathbb{N}$ there exists $\lambda_k \in \sigma(L_k)$ such that

$$0 < \lambda_k < \frac{c}{k} \tag{2.1}$$

where c is a constant independent on k.

Step 3. Let $k \in \mathbb{N}$. The k-periodic solutions of (1.1) are the critical points of the action functional

$$f_k(z) = \frac{1}{2}\langle L_k z, z\rangle - \int_0^k \hat{H}(t,z(t))dt, \quad z \in H^{1/2}(0,k) \tag{2.2}$$

where $\langle \cdot,\cdot \rangle$ denotes the canonical pairing between $H^{1/2}(0,k)$ and its dual, and $H^{1/2}(0,k)$ is the fractional order Sobolev space of the k-periodic $\mathbb{R}^{2N}$-valued functions.

By using min-max methods we shall construct a suitable sequence $\{z_k\}$ of critical points of (2.2). Set

$$H_+ = \overline{\text{span}\{\psi_j^k : \lambda_j^k > 0\}} \qquad H_- = H_+^{\perp}$$

where ψ_j^k denote the eigenfunctions of L_k corresponding to the eigenvalues λ_j^k and the closure is taken in $H^{1/2}(0,k)$. Since $\hat{H}(t,z) = o(|z|^2)$ at $z = 0$, there exist $\rho, \beta > 0$ such that

$$f_k(z) \geq \beta \quad \text{for} \quad z \in H_+ \quad \|z\| = \rho . \tag{2.3}$$

Let d_1, d_2 be two positive numbers with $d_1 > \rho$, and consider a normalized eigenfunction φ_k of L_k corresponding to a positive eigenvalue λ_k satisfying (2.1). Set

$$Q_k \equiv \{v + s\varphi_k \mid v \in H_-,\ \|v\| \leq d_2,\ \ 0 \leq s \leq d_1\} .$$

Since we can assume (see Step 1) that $\hat{H}(t,z) \equiv \text{const.}\ |\dot{x}|^p$ $(p > 2)$

for $|z|$ large, we deduce that, if d_1 and d_2 are sufficiently large,

$$f_k(z) \leq 0 \quad \text{for any} \quad z \in \partial Q_k \tag{2.4}$$

where ∂Q_k denotes the boundary of Q_k.

Standard calculations show that f_k satisfies the Palais-Smale condition; moreover ∂Q_k and $S = \{u \in H_+ | \, \|u\| = \rho\}$ "intersect" with respect to a suitable class of homeomorphisms (see, for example, lemma 3.5 of [2]).

Then, by (2.3) and (2.4) and using a generalized version of the mountain pass lemma (see e.g. [1]), it can be proved that there exists a k periodic solution z_k of (1.1) such that

$$0 < f_k(z_k) \leq \sup f_k(Q_k) \, . \tag{2.5}$$

<u>Step 4</u>. Let z_k be the k periodic solution of (1.1) found in Step 3. We show that

$$0 < f_k(z_k) \leq \frac{c}{k^\gamma} \, , \qquad \gamma = \frac{2}{p-2} \tag{2.6}$$

where c is a constant independent on k.

Set $|\cdot|_q$ the norm in $L^q(0,k)$ and let

$$z \in Q_k, \qquad z = v + s\varphi_k, \qquad v \in H_-, \qquad s \in [0,d_1]$$

then, by H_5) (see also Step 1):

$$\begin{aligned} f_k(z) &\leq \frac{1}{2}\lambda_k s^2 - \int_0^k \hat{H}(t,z)dt \leq \\ &\leq \frac{1}{2}\lambda_k s^2 - c_1|v + s\varphi_k|_p^p \leq \\ &\leq \frac{1}{2}\lambda_k s^2 - c_1(k)^{(2-p)/2} \cdot |v + s\varphi_k|_2^p \leq \\ &\leq \frac{1}{2}\lambda_k s^2 - c_2(k)^{(2-p)/2} s^p \leq c_3(\lambda_k)^{p/(p-2)} \cdot k \end{aligned} \tag{2.7}$$

where c_1, c_2, c_3 denote positive constants independent on k. Then by (2.1) and (2.7) we deduce that there exists $c_4 > 0$, independent on k, such that:

$$\forall z \in Q_k, \qquad f_k(z) \leq \frac{c_4}{k^\gamma} \, , \qquad \gamma = \frac{2}{p-2} \, . \tag{2.8}$$

Then (2.6) follows from (2.5) and (2.8).

Let us now prove that

$$|z_k|_p \to 0 \quad \text{as} \quad k \to \infty . \tag{2.9}$$

In fact

$$\frac{1}{2}(L_k z_k | z_k) - \int_0^k \hat{H}(t,z_k)dt = f_k(z_k)$$

$$-\frac{1}{2}(L_k z_k | z_k) + \frac{1}{2}\int_0^k (\hat{H}_z(t,z_k)|z_k)dt = 0 .$$

Then, by H_5) (see Step 1), we get

$$f_k(z_k) \geq \left(\frac{1}{2} - \frac{1}{p}\right) \int_0^k (\hat{H}_z(t,z_k)|z_k)dt \geq c_5 |z_k|_p^p . \tag{2.10}$$

Then, by (2.8) and (2.10), we derive (2.9).

Moreover, observe that (2.10) and (2.8) imply that

$$\int_0^k (\hat{H}_z(t,z_k)|z_k)dt \to 0 \quad \text{as} \quad k \to \infty . \tag{2.11}$$

Step 5. By using (2.9) and (2.11) and since z_k solve (1.1), it is not difficult to obtain an L^∞ bound for the sequence $\{z_k\}$. Then, since z_k solve (1.1), we deduce that also $\{\dot{z}_k\}$ is bounded in L^∞ norm. Then, by using again (2.9), we deduce that $z_k \to 0$ in L^∞ norm. Moreover, using again the equation (1.1), we deduce that

$$z_k \to 0 \quad \text{uniformly in} \quad C^1(\mathbf{R},\mathbf{R}^{2N}) . \tag{2.12}$$

Step 6. By the nonresonance assumption H_4) we derive that 0 is an isolated (in the $H^{1/2}$ norm) 1-periodic solution of (1.1). Now, arguing by contradiction, suppose that there are infinitely many k prime such that z_k has not minimal period k. Then, by assumption H_6), we deduce that z_k has minimal period 1. This and (2.12) contradict the fact that 0 is an isolated 1-periodic solution of (1.1).

REFERENCES

[0] V. Benci and D. Fortunato, in preparation.

[1] V. Benci and P. H. Rabinowitz, 'Critical point theorems for indefinite functionals', Inv. Math. **52** (1979), 336-352.

[2] V. Benci, A. Capozzi and D. Fortunato, 'Periodic solutions of Hamiltonian systems with superquadratic potential', Ann. Mat. Pura Appl. **143** (1986), 1-46.

[3] G. D. Birkhoff and D. C. Lewis, 'On the periodic motions near a given periodic motion of a dynamical system', Ann. Mat. Pura Appl. **12** (1933), 117-133.

[4] T. C. Harris, 'Periodic solutions of arbitrarily long periods in Hamiltonian systems', J. Diff. Eq. **4** (1968), 131-141.

[5] J. Moser, 'Proof of a generalized form of a fixed point theorem due to G. D. Birkhoff', Springer Lecture Notes in Math. No. 597 (1977), 464-494.

[6] P. H. Rabinowitz, 'On subharmonic solutions of Hamiltonian systems', Comm. Pure Appl. Math. **33** (1980), 609-633.

[7] P. H. Rabinowitz, 'Periodic solutions of Hamiltonian systems: a survey', SIAM J. Math. Anal. **13** (1982), 343-352.

A REMARK ON A PRIORI BOUNDS AND EXISTENCE FOR PERIODIC SOLUTIONS OF HAMILTONIAN SYSTEMS

Vieri Benci
Istituto di Matematiche
Applicate Università
Pisa, Italy

Helmut Hofer
Department of Mathematics
Rutgers University
New Brunswick, NJ 08903

Paul H. Rabinowitz
Department of Mathematics
University of Wisconsin-Madison
Madison, WI 53706

During the past few years, several papers have been written which give sufficient conditions for a Hamiltonian so that the Hamiltonian system

$$\dot{p} = -H_q(p,q) \ , \tag{1}$$

$$\dot{q} = H_p(p,q) \tag{2}$$

has periodic solutions of prescribed energy. See e.g. [1-8]. To normalize matters suppose the energy is one and $S \equiv H^{-1}(1)$. Thus we seek solutions of (1) on S. In [5], Weinstein observed that the earlier papers [1-4] possessed a common differential geometric feature and conjectured that whenever this feature was present, S contains a periodic orbit of (1). Recently Viterbo [9] proved a generalized version of Weinstein's conjecture. Still more recently, at this conference, Hofer and Zehnder [10] simplified Viterbo's argument and extended his result. They proved that if S bounds a compact neighborhood of 0 in $\mathbb{R}^{2n}$, and $(H_p, H_q) \neq 0$ on S, then given any $\delta > 0$, there is an ε with $|\varepsilon| < \delta$ and such that (1-2) has a periodic solution on $S_\varepsilon \equiv H^{-1}(1 - \varepsilon)$.

Naturally this result leads one to try to find solutions of (1-2) on S by taking a sequence $\delta_m \to 0$ and trying to show the corresponding sequence of T_m periodic solutions, $z_m \equiv (p_m, q_m)$, on S_{ε_m} converge to a periodic limit on S. Indeed in [10] it was shown that this process could be carried out whenever there are estimates for all T_ε periodic solutions z_ε on S_ε, for all small ε, of the form

P. H. Rabinowitz et al. (eds.), Periodic Solutions of Hamiltonian Systems and Related Topics, 85–88.

$$\alpha < T_\varepsilon^{-1} A(z_\varepsilon) < \beta \tag{3}$$

where

$$A(z_\varepsilon) = \int_0^{T_\varepsilon} p_\varepsilon \cdot \dot{q}_\varepsilon dt$$

and α, β are independent of ε. Such a priori bounds have been obtained by Benci and Rabinowitz [11] for certain classes of Hamiltonian systems including most of those studied in [1-8]. The purpose of this note is to give a sufficient condition for the bounds (3) which both slightly generalizes the case treated in [11] and encompasses all of the cases treated in [1-8].

We will always assume S is a connected set which bounds a compact neighborhood of 0 in $\mathbb{R}^{2n}$ and $H_z \equiv (H_p, H_q) \neq 0$ on S. To begin a couple of simple observations are in order. Taking the inner product of (1) with $-q$ or (2) with p where these are T periodic functions and integrating yields:

$$A(z) = \int_0^T p \cdot H_p dt = \int_0^T q \cdot H_q dt . \tag{4}$$

Let U be compact neighborhood of S in which $H_z = (H_p, H_q) \neq 0$. Suppose $K \in C^1(U, \mathbb{R})$. If z is T periodic, clearly

$$\int_0^T K_z(z) \cdot \dot{z} dt = 0 . \tag{5}$$

With these observations we have the following

<u>Criterion for a priori bounds</u>: Suppose there is a function K as above and constants $a, b \geq 0$ with $a + b > 0$ and

$$ap \cdot H_p(z) + bq \cdot H_q(z) + K_z(z) \cdot JH_z(z) > 0 \tag{6}$$

for all $z \in U$ where $J = \begin{pmatrix} 0 & -id \\ id & 0 \end{pmatrix}$ and id denotes the $n \times n$ identity matrix. Then the a priori bounds (3) obtain.

Indeed by (6), there is a constant $\gamma > 0$ such that for all $z \in U$,

$$ap \cdot H_p(z) + bq \cdot H_q(z) + K_z(z) \cdot JH_z(z) \geq \gamma . \tag{7}$$

Hence letting $z(t) = (p(t), q(t))$ be a T periodic solution of (1-2) in (2) and integrating over $[0,T]$ gives

$$(a + b)A(z) \geq \gamma T . \tag{8}$$

On the other hand, we trivially have

$$A(z) \leq \gamma_1 T \tag{9}$$

where

$$\gamma_1 = \min(\max_{z \in U}|p \cdot H_p|, \max_{z \in U}|q \cdot H_q|)$$

and (3) follows immediately from (8)-(9).

It remains to give interesting examples of when (6) holds. We will give two.

Example 10: S bounds a starshaped neighborhood of 0 in $\mathbf{R}^{2n}$, i.e. $z \cdot H_z(z) > 0$ for all $z \in S$. Then taking $a = b = 1$ and $K \equiv 0$ gives (6). This example contains the cases treated in [2] and [3].

Example 11: Suppose for all $z \in S$,

$$p \cdot H_p(z) > 0 \tag{12}$$

if $p \neq 0$. This is a slight generalization of the case treated in [11] and contains in particular cases treated in [1,2,4,6-8]. We will use the argument of [11] to verify (6). Note first that (12) implies $H_p(0,q) = 0$ for $(0,q) \in S$. Hence $H_q(0,q) \neq 0$ via our standing assumption on H_z on S. It is not difficult - see e.g. [11] or [12-13] - to construct a map $W \in C^1(\mathbf{R}^{2n},\mathbf{R}^n)$ such that

$$W(0,q) \cdot H_q(0,q) > 0 \tag{13}$$

for all $(0,q) \in U$, a compact neighborhood of S. Set $K(z) \equiv -W(z) \cdot p$. Then $K \in C^1(\mathbf{R}^{2n},\mathbf{R})$ and

$$K_z(z) \cdot JH_z(z) = -(W_z(z)JH_z(z)) \cdot p + W(z) \cdot H_q(z) \,. \tag{14}$$

Let

$$M = \max_{z \in U}|K_z(z) \cdot JH_z(z)| \,.$$

From (13), it easily follows that there are constants $\sigma,\gamma > 0$ such that

$$W(z) \cdot H_q(z) \geq 2\gamma \tag{15}$$

and

$$|(W_z(z)JH_z(z)) \cdot p| \leq \gamma \tag{16}$$

for all $z \in U$ and $|p| \leq \sigma$. Set

$$a = \frac{M + \gamma}{\inf\limits_{z \in U, |p| \geq \sigma} p \cdot H_p(z)}, \qquad b = 0 .$$

Then if $|p| \leq \sigma$

$$ap \cdot H_p + K_z(z) \cdot JH_z(z) \geq -\gamma + 2\gamma \geq \gamma$$

while if $|p| \geq \sigma$

$$ap \cdot H_p + K_z(z) \cdot JH_z(z) \geq M + \gamma - M = \gamma .$$

Hence (6) holds and we have bounds for this case.

REFERENCES

[1] Seifert, H., 'Periodische Bewegungen mechanischen systeme', Math. Z. **51** (1948), 197-216.
[2] Weinstein, A., 'Periodic orbits for convex Hamiltonian systems', Ann. Math. **108** (1978), 507-518.
[3] Rabinowitz, P. H., 'Periodic solutions of Hamiltonian systems', Comm. Pure Appl. Math. **31** (1978), 157-184.
[4] Rabinowitz, P. H., 'Periodic solutions of a Hamiltonian system on a prescribed energy surface', J. Diff. Eq. **33** (1979), 336-352.
[5] Weinstein, A., 'On the hypotheses of Rabinowitz's periodic orbit theorems', J. Diff. Eq. **33** (1979), 353-358.
[6] Gluck, H. and W. Ziller, 'Existence of periodic solutions of conservative systems', Seminar on Minimal Submanifolds, Princeton University Press (1983), 65-98.
[7] Hayashi, K., 'Periodic solutions of classical Hamiltonian systems', Tokyo Univ. J. Math. **6** (1983), 473-486.
[8] Benci, V., 'Closed geodesics for the Jacobi metric and periodic solutions of prescribed energy of natural Hamiltonian systems', to appear Ann. Inst. H. Poincaré, Analyse Nonlineaire.
[9] Viterbo, C., 'A proof of the Weinstein conjecture in $\mathbf{R}^{2n}$', preprint, Sept. 1986.
[10] Hofer, H. and E. Zehnder, 'Periodic solutions on hypersurfaces and a result by C. Viterbo', to appear in Inv. Math.
[11] Benci, V. and P. H. Rabinowitz, 'A priori bounds for periodic solutions of Hamiltonian systems', to appear in Ergodic Theory and Dynamical Systems.
[12] Palais, R. S., 'Critical point theory and the minimax principle', Proc. Sym. Pure Math. **15** (1970), Amer. Math. Soc., Providence, RI, 185-212.
[13] Clark, D. C. 'A variant of the Ljusternik-Schnirelman theory', Ind. Univ. Math. J. **22** (1972), 65-74.

ON A CLASS OF NONLINEAR PROBLEMS WITH LACK OF COMPACTNESS

A. Capozzi
Dipartimento di Matematica
Campus Universitario
Bari 70125, Italy

ABSTRACT. In this note there are a brief illustration of the problem of finding critical points of functionals, which don't verify "globally" the Palais-Smale condition and the announce of some results concerning with the research of periodic solutions of Hamiltonian systems in presence of resonance at infinity.

Many problems in mathematical physics can be reduced to the study of the following equation

$$Au = f(u) \qquad u \in H^{\alpha}(\Omega, \mathbf{R}^k) \tag{1}$$

where $\Omega \subset \mathbf{R}^n$ is a bounded open set, $\alpha > 0$, $H^{\alpha}(\Omega,\mathbf{R}^k)$ is the usual Sobolev space, A is a continuous self-adjoint operator in H, f is the Nemytskii operator associated with $\nabla F:\mathbf{R}^k \to \mathbf{R}^k$ ($F \in C^1(\mathbf{R}^k,\mathbf{R})$).

Under suitable assumptions f is a continuous potential operator on Hilbert space $H = H^{\alpha}(\Omega,\mathbf{R}^k)$, the potential Φ being given by

$$\Phi(u) = \int_{\Omega} F(u(x))dx \qquad u \in H , \tag{2}$$

then the solutions of (1) are the critical points of the functional

$$\varphi(u) = 1/2(Au,u)_H - \Phi(u) \qquad u \in H . \tag{3}$$

Many tools have been developed for the research of critical points of the functional (3) under the assumption that φ verifies the Palais-Smale condition.

Recently many authors have studied the problem of searching for critical points of (3) when the Palais-Smale condition is not "globally" satisfied (problems with "lack of compactness"). This happens, for example, in the searching for nontrivial solutions to

P. H. Rabinowitz et al. (eds.), Periodic Solutions of Hamiltonian Systems and Related Topics, 89–91.

nonlinear problems in presence of the "critical Sobolev exponent" (cf. [2]) or periodic solutions of Hamiltonian systems when the potential is bounded (cf. [4] and its references) or in presence of "resonance" (cf. [1,6] and their references).

In [4] the problem of finding periodic solutions of Hamiltonian systems is reconduced to the problem of finding critical points of the "action" functional restricted to a suitable subspace H of the Hilbert space, in which one usually works. In such a subspace the functional verifies the Palais-Smale condition.

In [1] and [6] it is recognized the strip, in which the Palais-Smale condition is verified, and the critical points of the functional are found via Luisternik-Schnirelman theory or Morse type arguments.

In this note we will announce some results obtained in [3] and in [5]. In [5] the problem is studied by using direct methods and some results contained in [8]. In [3] the problem is studied by using some results contained in [8] and a variant of some results contained in [7], which concern with the estimate of Morse-index of particular critical points.

Consider the problem of finding T-periodic solutions $(T > 0$ given) of the second order system of n ordinary differential equations

$$-\ddot{x} - \lambda_k x = \nabla V(x,t) \tag{4}$$

where $x \in \mathbf{R}^n$, $t \in R$, $\ddot{x} = d^2x/dt^2$, $V \in C^1(\mathbf{R}^n \times \mathbf{R},\mathbf{R})$ is T-periodic in t, ∇V is the gradient of V respect to variable x and $\lambda_k \in \sigma(\mathcal{L})$ ($\mathcal{L}$ is the self-adjoint realization in $L^2(|0,T|;\mathbf{R}^n)$ of the operator $x \to -\ddot{x}$). We suppose that

$$\left.\begin{array}{l} V(x,t) \to 0 \\ \nabla V(x,t) \to 0 \end{array}\right\} \text{ as } |x| \to \infty \text{ uniformly in } t \in \mathbf{R} \tag{V_1}$$

and that

$$\nabla V(0,t) = 0 \quad \text{for any} \quad t \in \mathbf{R}\,. \tag{V_2}$$

In [3] it has been proved that

<u>Theorem 1.</u> <u>If</u> $V \in C^2(\mathbf{R}^n \times \mathbf{R},\mathbf{R})$ <u>satisfies</u> (V_1), (V_2) <u>and</u>

$$V(0)V_{xx}(0,t) \ \underline{\text{is positive definite}}\ , \tag{V_3}$$

<u>where</u> $V_{xx}(0,t)$ <u>is the Hessian matrix of</u> V <u>at the origin, then</u> (4) <u>has a nontrivial T-periodic solution.</u>

In [5] it has been proved that

Theorem 2. *If* V *satisfies* (V_1), (V_2) *and*

$$\frac{V(x,t)}{|x|^2} \to +\infty \quad \text{as} \quad |x| \to 0 \quad \text{uniformly in} \quad t \in R \qquad (V_4)$$

then (4) *has a nontrivial T-periodic solution.*

For other results we refer to [3] and [5].

REFERENCES

[1] A. Ambrosetti and V. Coti Zelati, 'Critical points with lack of compactness and singular dynamical systems', preprint.

[2] H. Brezis and L. Nirenberg, 'Positive solutions of nonlinear elliptic equations involving critical Sobolev exponent', *Comm. Pure Appl. Math.* t. **XXXVI** (1983), 437-477.

[3] A. Capozzi, D. Lupo and S. Solimini, 'On the existence of a nontrivial solution to nonlinear problems at resonance', preprint.

[4] A. Capozzi, A. Salvatore and D. Fortunato, 'Periodic solutions of Lagrangian systems with bounded potential', to appear in *Journal of Math. Anal. and Appl.*

[5] A. Capozzi and A. Salvatore, 'On the Hamiltonian systems with bounded potential', in preparation.

[6] V. Coti Zelati, 'Periodic solutions of dynamical systems with bounded potential', to appear in *Journal of Diff. Eq.*

[7] A. Lazer and S. Solimini, 'Nontrivial solutions of operator equations and Morse indices of critical points of min-max type', to appear in *Nonlinear Anal. T.M.A.*

[8] S. Solimini, 'On the solvability of some elliptic partial differential equations with the linear part at resonance', *Journal of Math. Anal. and Appl.* **117**, 1 (1986), 138-152.

AN OLD-FASHIONED METHOD IN THE CALCULUS OF VARIATIONS

Marc Chaperon
Centre de mathématiques
Ecole Polytechnique
91128 Palaiseau Cedex, France

ABSTRACT. The aim of this note is to show that two of Arnol'd's celebrated conjectures in symplectic geometry can be proven without any knowledge of functional analysis. For the convenience of the reader, we shall not use the language of differential geometry before the conclusion, where the advantages of our method are discussed.

0. MORSE-THEORETIC PREREQUISITES

We shall admit the following (classical) result - for a proof, see for instance [ChZ 83], p. 82-95:

Proposition 0. Given a compact manifold M and a finite dimensional real vector space E, let $F : M \times E \to \mathbf{R}$ be a smooth function which is "quadratic at infinity" in the following sense: there exists a non-degenerate quadratic form $Q : E \to \mathbf{R}$ such that the mapping

$$M \times E \ni (a,v) \mapsto \frac{\partial F}{\partial v}(a,v) - dQ(v) \in E^*$$

is bounded. Then, the number of critical points of F is greater than the cup-length $cl(M)$ of M in any case, and at least equal to the sum $SB(M)$ of the Betti numbers of M when none of these critical points is degenerate.

Example. $cl(\mathbf{T}^n) = n$ and $SB(\mathbf{T}^n) = 2^n$.

1. THE CONLEY-ZEHNDER THEOREM

Theorem ([CZ 82]). Let $H : \mathbf{R}^n \times \mathbf{R}^n \times \mathbf{R} \to \mathbf{R}$ be smooth and have period 1 with respect to each of its $2n + 1$ variables. Consider the Hamiltonian system

$$\dot{q} = -\nabla_p H(q,p,t) \quad \text{and} \quad \dot{p} = \nabla_q H(q,p,t) \qquad (1)$$

P. H. Rabinowitz et al. (eds.), Periodic Solutions of Hamiltonian Systems and Related Topics, 93–98.

and the boundary condition

$$q(0) = q(1) \quad \text{and} \quad p(0) = p(1) \ . \tag{2}$$

Then, one can find at least $2n + 1$ solutions of (1)-(2) which are distinct up to $\mathbf{Z}^{2n}$-translation, and at least 2^{2n} if (1)-(2) has no degenerate solution.

Proof. We shall construct a function F as in Proposition 0 with $M = \mathbf{T}^{2n}$, the critical set of which is in $1-1$ correspondence with the mod $\mathbf{Z}^{2n}$ solutions of (1)-(2).

Given a positive integer N and arbitrary points $v_i = (q_i, p_i) \in \mathbf{R}^n \times \mathbf{R}^n$, $0 \leq i \leq N$, let $v_{i+1}^H(v_i) = (q_{i+1}^H(v_i), p_{i+1}^H(v_i)) \in \mathbf{R}^n \times \mathbf{R}^n$, $0 \leq i \leq N$, be given by

$$v_{i+1}^H(v_i) = \left(q\left(\frac{i+1}{N+1}\right), p\left(\frac{i+1}{N+1}\right)\right), \text{ where } (q,p) \text{ satisfies (1)}$$
$$\text{and } \left(q\left(\frac{i}{N+1}\right), p\left(\frac{i}{N+1}\right)\right) = v_i .$$

CHOOSE N LARGE ENOUGH FOR EACH $v_i \mapsto (q_{i+1}^H(v_i), p_i)$ TO BE A DIFFEOMORPHISM (this is possible because H is periodic).

Let V denote the set of all points $v = (v_0, \ldots, v_N) \in (\mathbf{R}^n \times \mathbf{R}^n)^{N+1}$, and let f^H be the (smooth) function on V given by

$$f^H(v) = \oint_{\hat{c}_v} p \cdot dq + \int_0^1 H(c_v(t),t)dt \ ,$$

where the path $c_v : [0,1] \to \mathbf{R}^n \times \mathbf{R}^n$ and the loop $\hat{c}_v$ are defined as follows (see Figure 1):

- on $\left[\frac{j}{N+1}, \frac{j+1}{N+1}\right]$, c_v is the solution $t \mapsto (q(t),p(t))$ of (1) such that $\left(q\left(\frac{j}{N+1}\right), p\left(\frac{j}{N+1}\right)\right) = v_j$;
- the loop $\hat{c}_v$ is obtained by adding "corners" to c_v as in Figure 1, so as to close it.

In other words,

$$f^H(v) = \sum_0^N \int_{\frac{j}{N+1}}^{\frac{j+1}{N+1}} (c_v^*(p \cdot dq) + H(c_v(t),t))dt$$
$$+ \sum_1^{N+1} p_j \cdot (q_j - q_j^H(v_{j-1}))$$

where $q_{N+1} = q_0$ and $p_{N+1} = p_0$. From this and the definition of

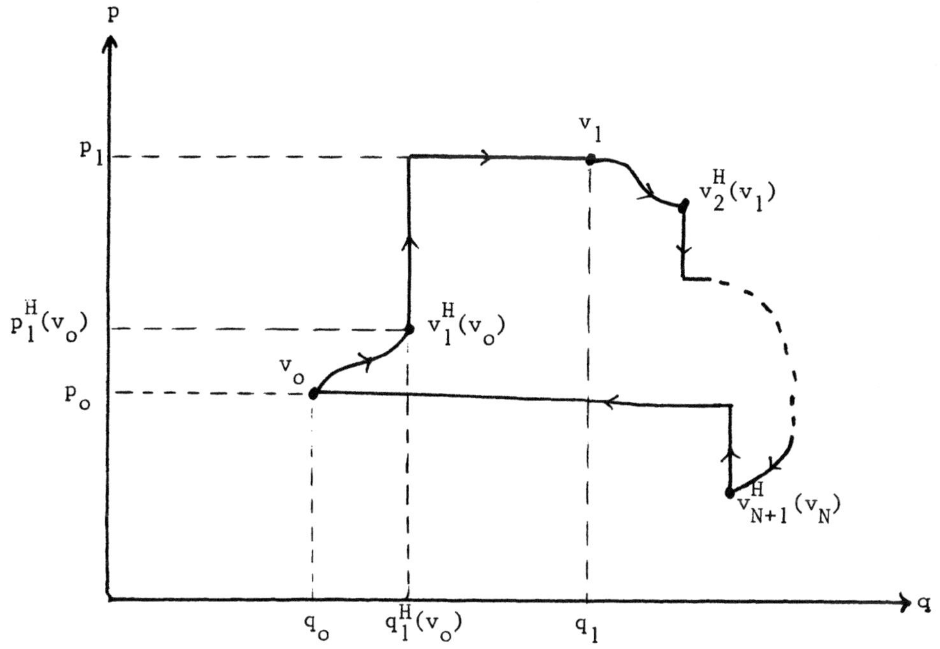

Figure 1

c_v, we deduce that

$$df^H(v) = \sum_0^N \big((p^H_{j+1}(v_j) - p_{j+1}) \cdot dq^H_{j+1}(v_j)$$

$$+ (q_{j+1} - q^H_{j+1}(v_j)) \cdot dp_{j+1}\big) . \tag{3}$$

Therefore, by our choice of N, v is a critical point of f^H if and only if $v_{j+1} = v^H_{j+1}(v_j)$ for $0 \le j < N$ and $v^H_{N+1}(v_N) = v_0$, i.e. if and only if c_v is a solution of (1)-(2).

Let W denote the set of all points $(x,y) \equiv ((x_0,y_0),\dots,(x_N,y_N)) \in (\mathbf{R}^n \times \mathbf{R}^n)^{N+1}$, and let $\tilde{F}^H = f^H \circ h^H : W \to \mathbf{R}$, where $h^H : W \to V$ is defined as follows: for each $(x,y) \in W$, $h^H(x,y) = v$ is given by

$$\left.\begin{aligned} q^H_{j+1}(v_j) &= \sum_0^j x_k, \quad 0 \le j \le N , \\ p_0 = y_0 \quad &\text{and} \quad p_j = y_0 + y_j , \quad 1 \le j \le N . \end{aligned}\right\} \tag{4}$$

By our choice of N, h^H is a well-defined smooth diffeomorphism.

Since $\tilde{F}^H$ is $\mathbf{Z}^{2n}$-periodic with respect to (x_0, y_0), it induces a smooth function $F^H : \mathbf{T}^{2n} \times E \to \mathbf{R}$, where $E = (\mathbf{R}^n \times \mathbf{R}^n)^N$, and we just have to show that F^H is quadratic at infinity in the sense of Proposition 0.

Let $\tilde{F}^0$ and F^0 be defined in the same fashion (with the same N) for $H = 0$. Clearly,

$$\tilde{F}^0(x,y) = \sum_1^N x_j \cdot y_j \ .$$

Therefore, F^0 is of the form $(a,v) \mapsto Q(v)$, where Q is a non-degenerate quadratic form on E, and our theorem will be proven if we can show that $d(\tilde{F}^H - \tilde{F}^0) : W \to W^*$ is bounded. Now, if we set $h^H(x,y) = v$, (3)-(4) yield

$$d(\tilde{F}^H - \tilde{F}^0)(x,y) = \sum_{j=0}^N (p_{j+1}^H(v_j) - p_j) \cdot d\sum_0^j x_k$$
$$+ \sum_{j=1}^N (q_j - q_{j+1}^H(v_j)) \cdot d(y_0 + y_j) + (q_0 - q_1^H(v_0)) \cdot dy_0 \ ,$$

which is bounded because H is periodic.

The fact that non-degenerate critical points of F^H correspond to non-degenerate solutions of (1)-(2) is an easy exercise (this is one of the advantages of the method). □

2. LAGRANGIAN INTERSECTIONS - CONCLUDING REMARKS

Theorem ([Ch 83]). Let $H : \mathbf{R}^n \times \mathbf{R}^n \times [0,1] \to \mathbf{R}$ be smooth and have period 1 with respect to its first n variables. Assume each solution of (1) with $p(0) = 0$ exists for every t, and consider

$$p(0) = p(1) = 0 \ . \tag{2'}$$

Then, one can find at least $n + 1$ solutions of (1)-(2') which are distinct up to $\mathbf{Z}^n \times \{0\}$-translation, and at least 2^n if (1)-(2') has no degenerate solution.

Proof. First, notice that there exists $R > 0$ such that each solution of (1) with $p(0) = 0$ satisfies $|p(t)| < R$ for every $t \in [0,1]$. Therefore, we may assume that $H(q,p,t) = 0$ for $|p| > R$; this allows us to do exactly the same as in 1., except that here v_o lies in $\mathbf{R}^n \times \{0\}$ - hence $p_o = p_{N+1} = y_o = 0$. □

This result, which implies the Conley-Zehnder theorem (see [Ch 83]), is the particular case $M = \mathbf{T}^n$ of a general theorem, which must be formulated in geometric language: given a manifold M, recall that the Liouville form of its cotangent bundle T^*M is the 1-form $p\,dq$ on T^*M such that $\alpha^*(p\,dq) = \alpha$ for every 1-form α on M. An isotopy of T^*M is a smooth family $(g_t)_{0 \leq t \leq 1}$ of smooth diffeomorphisms $g_t : T^*M \circlearrowleft$. Such an isotopy (g_t) is called Hamiltonian if $p\,dq - g_t^*(p\,dq)$ is an exact 1-form for every t - or, equivalently, if (g_t) is obtained by integrating a time-depending Hamiltonian vector field.

Theorem ([H 84]). Let 0_M denote the zero section of the cotangent bundle of a compact manifold M. For every Hamiltonian isotopy (g_t) of T^*M, there are at least $cl(M) + 1$ points in $0_M \cap g_1(0_M)$, and at least $SB(M)$ if all these intersections are transversal.

Hofer's original proof of this result was rather involved. In [LS 85], a much simpler argument was given, based upon the above construction - with an additional idea. Finally, Sikorav expressed this argument in terms of (global) generating phase functions, hence a crystal-clear proof [S 85], in which no real additional idea is needed when the torus is replaced by an arbitrary compact manifold - see [Ch 86] for a presentation when $M = \mathbf{T}^n$.

Our method can be applied to various problems in the calculus of variations in one variable - for example, it provides a proof of Viterbo's celebrated recent theorem, using the same Hamiltonian as in [HZ 86]. When the problem is only slightly non-linear, the choice between this and classical functional analysis as in [HZ 86] is a matter of taste - though our approach may be more suitable to practical computations. In the case of a truly non-linear problem, the comparison between [H 84] and [LS 85]-[S 85] seems to indicate that "solving finite dimensional problems by finite dimensional methods" is not always a bad idea.

REFERENCES

[CZ 82] C. C. Conley and E. Zehnder, 'The Birkhoff-Lewis theorem and a conjecture of V. I. Arnol'd', Inv. Math. **73** (1983), 33-49.

[Ch 83] M. Chaperon, 'Quelques questions de géométrie symplectique', Séminaire Bourbaki, 1982-83, Astérisque 105-106 (1983), 231-249.

[ChZ 83] M. Chaperon and E. Zehnder, 'Quelques résultats globaux en géométrie symplectique', Géométrie symplectique et de contact: autour du théorème de Poincaré-Birkhoff, Travaux en cours, Hermann, Paris (1984), 51-121.

[H 84] H. Hofer, 'Lagrangian embeddings and critical point theory', Ann. Inst. Henri Poincaré, Analyse non linéaire, **2** (1985), 407-462.

[LS 85] F. Laudenbach and J. C. Sikorav, 'Persistance d'intersection avec la section nulle...', Invent. Math. **82** (1985), 349-357.

[S 85] J. C. Sikorav, 'Problèmes d'intersections et de points fixes en géométrie Hamiltonienne', preprint, Orsay (1985).

[Ch 86] M. Chaperon, 'Generating phase functions and Hamiltonian systems', preprint, Ecole Polytechnique (1986).

[HZ 86] H. Hofer and E. Zehnder, 'Periodic solutions on hypersurfaces and a result by C. Viterbo', to appear in Inventiones Math.

OPTIMIZATION AND PERIODIC TRAJECTORIES

Frank H. Clarke
Centre de recherches mathématiques
Université de Montréal
C.P. 6128, Station A
Montréal (Québec) Canada H3C 3J7

ABSTRACT. The application of several, mostly recent techniques of optimization to the study of periodic Hamiltonian trajectories is described. Chief among these are transversality conditions, value functions, and nonsmooth analysis.

1. NECESSARY CONDITIONS

The variational principle has long been a useful tool in the study of many boundary-value problems. The idea in its simplest form is to associate to a given problem an integral functional $I(x)$ in such a way that the Euler equation corresponding to I is identical to (or at least closely related to) the equation being studied. Since the Euler equation is the basic necessary (or stationarity) condition associated to minimizing I, this establishes a link to the theory of optimization. Our purpose here is to illustrate the use of other lesser-known necessary conditions and related optimization techniques in the study of periodic trajectories. Chief among these will be the transversality conditions, the calculus of generalized gradients, and the value-function approach.

The simplest classical situation deals with the integral functional I defined via a <u>Lagrangian</u> L as follows:

$$I(x) := \int_0^T L(x(t),\dot{x}(t))dt \ . \tag{1.1}$$

Here then L is a function mapping $R^n \times R^n$ to R, and x is an <u>arc</u> (an absolutely continuous function from $[0,T]$ to R^n). The <u>Euler</u> equation is the well-known necessary condition that any local minimum x of I must satisfy:

$$\frac{d}{dt}\{L_v(x(t),\dot{x}(t))\} = L_x(x(t),\dot{x}(t)) \quad \text{a.e.} \ . \tag{1.2}$$

P. H. Rabinowitz et al. (eds.), Periodic Solutions of Hamiltonian Systems and Related Topics, 99–110.

(Of course there are hypotheses required for such a result about which we are being consciously vague; the conditions under which the Euler equation holds, even in classical settings, have only recently been clarified [10,11].) If L is not a smooth function, then under different conditions there is available an extended form of the Euler equation, one which asserts the existence of an arc p satisfying

$$(\dot{p}(t),p(t)) \in \partial L(x(t),\dot{x}(t)) \quad \text{a.e.} \; . \tag{1.3}$$

In this "equation" (or inclusion, really) ∂L denotes the <u>generalized gradient</u> of L; the set ∂L reduces to $\{\nabla L\}$ if L is smooth, in which case the reader should confirm that (1.3) yields (1.2).

The explicit introduction of the "adjoint variable" or "generalized momentum" p is useful in making the passage to the Hamiltonian form of the conditions. Classically the <u>Hamiltonian</u> $H(x,p)$ is defined from $L(x,v)$ via the Legendre transform, but a more modern definition stemming from convex analysis, and one offering several important advantages, is the following:

$$H(x,p) := \sup\{\langle p,v\rangle - L(x,v) : v \in R^n\} \; . \tag{1.4}$$

Using H, we can formulate another necessary condition: if x minimizes I, then there is an arc p such that

$$(-\dot{p}(t),\dot{x}(t)) \in \partial H(x(t),p(t)) \quad \text{a.e.} \; . \tag{1.5}$$

Quite often we would expect the functions p in (1.5) and (1.3) to be the same. We shall not dwell on such points here, but instead look upon these various types of necessary conditions as tools available for later use. Details on the preceding and the following appear in the author's book [4].

In considering minima of I we have so far said nothing about possible constraints on the values of $x(0)$ and $x(T)$. If there is at least some freedom in choosing these values, or if we consider functionals I having an explicit dependence on these values, then the necessary conditions have in addition to (1.3) or (1.5) an extra component reflecting the fact. A very useful way to handle in one stroke the myriad possibilities is to consider the <u>generalized problem of Bolza</u>, in which one seeks to minimize the functional I defined by

$$I(x) := \ell(x(0),x(T)) + \int_0^T L(x(t),\dot{x}(t))dt \; . \tag{1.6}$$

It is important to allow ℓ to be an <u>extended-valued</u> function, in order to be able to account for constraints by letting ℓ equal $+\infty$ when they are violated.

For example, if we seek to minimize the original I defined by (1.1) under the condition

$$(x(0),x(T)) \in C \; ,$$

where C is a given subset of $R^n \times R^n$, then we define ℓ to be the <u>indicator</u> of C; i.e., the function

$$\ell(r,s) := \begin{cases} 0 & \text{if } (r,s) \in C \\ +\infty & \text{otherwise .} \end{cases}$$

As another example, suppose that we seek to minimize

$$g(x(T)) + \int_0^T L(x(t),\dot{x}(t))dt$$

subject only to $x(0) = A$. We then define

$$\ell(r,s) := \begin{cases} g(s) & \text{if } r = A \\ +\infty & \text{otherwise .} \end{cases}$$

The necessary condition for the generalized problem of Bolza asserts that corresponding to any solution x is an arc p such that the Hamiltonian inclusion (1.5) holds, and such that

$$(p(0),-p(T)) \in \partial\ell(x(0),x(T)) . \tag{1.7}$$

The relationship (1.7) is called the <u>transversality condition.</u> In the first example above, it reduces to the statement that $(p(0),-p(T))$ is normal (in an appropriate extended sense) to the set C at the point $(x(0),x(T))$ (this condition contains no information if C is a singleton). In the second example, the transversality condition (1.7) is equivalent to

$$-p(T) \in \partial g(x(T)) .$$

We shall see later how such supplementary information, which seems to have been largely unexploited in the use of variational principles, can serve in the study of periodic trajectories. But first we discuss some of the possibilities for the integrand in the variational functional.

2. ACTION FUNCTIONALS

Suppose now that our goal is to produce periodic trajectories $(x(t),p(t))$ of a given Hamiltonian system, which we now write in the form

$$J\dot{z}(t) = \nabla H(z(t)) \quad \text{a.e. ,} \tag{2.1}$$

where $z = (x,p)$ and J is the $2n \times 2n$ matrix $\begin{bmatrix} 0 & -I \\ I & 0 \end{bmatrix}$. The variational method begins by defining an integral functional (an action) whose Euler equation is (2.1). Much of the complexity of the issue is due to the fact there is not a unique possibility, and that various advantages and drawbacks pertain to any given choice.

One approach is to define a Lagrangian $L(x,v)$ which generates the given Hamiltonian via (1.4); we shall not discuss this here, although this approach and some of its variants (e.g., the Jacobi action) have been quite useful in certain contexts.

A second approach defines an action directly in terms of z (i.e., both the variables x and p). The best known choice of this kind is the Hamiltonian action $A_H(z)$, given by

$$A_H(z) := \int_0^T \{\tfrac{1}{2} \langle Jz(t),\dot{z}(t)\rangle + H(z(t))\}dt \ . \tag{2.2}$$

The reader should verify that the Euler equation (1.2) for this functional (which is defined on 2n-dimensional arcs) is precisely (2.1). In producing such extremals (i.e., solutions of Euler's equations) however, optimization and the Hamiltonian action do not seem compatible, due to the fact that (as is easily seen) A_H is highly indefinite: it fails to be either bounded below or above even locally. Rather than optimization therefore, it is critical point theory that must be brought to bear upon A_H; P. Rabinowitz has played a leading role in showing how to do this. But since we wish to apply optimization methods, let us abandon A_H for another action better suited to minimization.

This new action (introduced in [1,2]) incorporates a new function H^*, the dual Hamiltonian, which is defined to be the conjugate of H in the sense of convex analysis. Specifically, we have

$$H^*(\zeta) := \sup\{\langle\zeta,z\rangle - H(z) : z \in R^{2n}\} \ . \tag{2.3}$$

The dual action D is then defined as follows:

$$D(z) := \int_0^T \{\tfrac{1}{2} \langle Jz(t),\dot{z}(t)\rangle + H^*(J\dot{z}(t))\}dt \ . \tag{2.4}$$

Unlike A_H, it turns out to be feasible on occasion to minimize D. And when H is convex, the extremals of D are closely related to the Hamiltonian trajectories we are studying. The particular notion of extremal to be used depends on the context. In general, the integrand defining $D(z)$ is nonsmooth (and nonconvex, because of the bilinear term), and the appropriate notion is provided by (1.3) or (1.5). But to simplify the presentation, let us suppose that H and H^* are both smooth, so that the classical Euler equation (1.2) can be used. (H^* being smooth is equivalent to H being strictly convex.) In the case of D, the equation gives

$$\frac{d}{dt}\left\{\frac{1}{2} Jz - J\nabla H^*(J\dot{z})\right\} = -\frac{1}{2} J\dot{z} \quad \text{a.e.}$$

which implies, for some constant c,

$$\nabla H^*(J\dot{z}(t)) = z(t) + c \quad \text{a.e.}\ , \tag{2.5}$$

which (by the magic of convex analysis — recall that H is convex) is equivalent to

$$\nabla H(z(t) + c) = J\dot{z}(t) \quad \text{a.e.}\ ,$$

which says that the arc $z(\cdot) + c$ is a Hamiltonian trajectory. The basic fact to be retained is this: if z is an extremal of the dual action D (for instance, if z minimizes D), then there is a translate of z by some constant which is a Hamiltonian trajectory.

In the first use of the dual action the problem considered was to find a Hamiltonian trajectory of prescribed energy [2]; in that setting H^* was not only nonsmooth but extended-valued. Let us instead illustrate the use of the dual action in a simple setting in which the period T of the trajectory is prescribed. We suppose that H is convex and differentiable, and that the following (subquadratic) growth conditions hold:

$$c_1|z|^s \leq H(z) \leq c_2|z|^S + B \quad \text{for all} \quad z\ ,$$

where c_1, c_2, s, S, B are given positive constants with s and S lying in $(1,2)$.

Theorem 2.1 [8]. For any $T > 0$ there exists a nonconstant solution $z(\cdot)$ of (2.1) having minimal period T.

The proof begins by observing that the conjugate function H^* satisfies a superquadratic growth condition:

$$\alpha_1|\zeta|^\sigma \leq H^*(\zeta) \leq \alpha_2|\zeta|^\Sigma + \beta \quad \text{for all} \quad \zeta\ ,$$

where σ and Σ are greater than 2. This growth property implies that a solution $\hat{z}$ exists to the problem of minimizing $D(z)$ over the arcs z satisfying $z(0) = z(T) = 0$. As mentioned above, the corresponding necessary condition shows that for some constant c, the arc $\hat{z} + c$ is a Hamiltonian trajectory (and automatically a periodic one, of period T). A simple argument based on the optimality of $\hat{z}$ shows that $\hat{z}$ (and hence $\hat{z} + c$) is nonconstant, and that T is the true or minimal period.

In the next section we shall discuss an alternate approach which uses the last optimization technique we wish to introduce, the value function.

3. A VALUE FUNCTION

Let us imbed the variational problem that lay at the heart of the preceding proof in a family of parametrized problems as follows: the problem $P(\gamma)$ is that of minimizing $D(z)$ over those arcs z satisfying $z(0) = 0$, $z(T) = \gamma$. The corresponding value function V is obtained by defining $V(\gamma)$ to be the value of $P(\gamma)$; i.e. the minimum value in question.

Now let z be any arc satisfying $z(0) = 0$. A moment's thought will confirm that the following important observation is a consequence of the way V is defined:

$$-V(z(T)) + D(z) \geq 0 \ . \tag{3.1}$$

Note that $z(T)$ is completely unconstrained in deriving (3.1). But suppose now that $z = \hat{z}$, where $\hat{z}$ is a solution to $P(0)$ (the problem used to prove Theorem 2.1). Then of course we have

$$\hat{z}(T) = 0, \quad V(0) = D(\hat{z}) \ ,$$

which shows that (3.1) holds with equality. To rephrase then, $\hat{z}$ minimizes the Bolza functional

$$-V(z(T)) + D(z)$$

over those arcs z satisfying $z(0) = 0$.

The next step is to apply the necessary conditions to this problem of Bolza. At this point the nature of V becomes an important point, for a minimal regularity is required to apply known results. When H satisfies the hypotheses put forth for the preceding theorem, as we shall suppose henceforth, it is straightforward to verify that V is Lipschitz near 0, and this is adequate.

The Euler equation gives the same conclusion as it did in the previous section: there is a constant c such that the arc $\hat{z} + c$ is a periodic Hamiltonian trajectory. But now there is a transversality condition as well. To see what (1.7) gives in this case, we note first that (as illustrated in Section 1), we have

$$\ell(r,s) := \begin{cases} -V(s) & \text{if} \quad r = 0 \\ +\infty & \text{otherwise} \ . \end{cases}$$

The adjoint variable is given by the gradient in $\dot{z}$ of the integrand appearing in the dual action, namely

$$\frac{1}{2} Jz - J\nabla H^*(J\dot{z}) \ , \tag{3.2}$$

so that (1.7) yields

$$\frac{1}{2} J\hat{z}(T) - J\nabla H^*(J\dot{\hat{z}}(T)) \in \partial V(0) \ .$$

Since $\hat{z}(T) = 0$, and in view of the Euler equation (2.5), this gives

$$c \in J\partial V(0) . \tag{3.3}$$

The periodic trajectory $\hat{z} + c$ passes through the point c, and now the potential advantage of (3.3) is evident: it provides information about the location of the periodic trajectory, whereas the approach of Section 2 led to no such information.

Before pursuing this further, let us remark that value function methods based on the generalized gradient are seeing increasing application [4,6,9]. The first use in the context of periodic solutions occurred in [3], where a more complex version of the argument outlined above is shown to lead to a controllability result that we now describe.

Let S be a convex energy surface in R^{2n}, and consider the natural Hamiltonian flow on S together with the orbits that it induces. Let M be any symplectic matrix (i.e., $M^*JM = J$). Then there is a point s on S and a nontrivial orbit on S joining s to Ms [3]. (Note that the special case $M = I$ corresponds to periodicity.) We know of no approach other than the value function method that leads to such a conclusion, or to information on the location of periodic trajectories. Other developments and many references can be found in [5,7].

We return now to the value function V and to the import of condition (3.3). As a locally Lipschitz function, V is differentiable a.e. In fact a general theorem of nonsmooth analysis [4, Theorem 2.8.2] yields that $-V$ is regular, a notion which we won't define here but which implies that V has one-sided directional derivatives in the usual sense and that $\partial V(\gamma)$ is a singleton iff $\nabla V(\gamma)$ exists. So if $\nabla V(0)$ exists, then condition (3.3) specifies exactly one point through which passes the periodic trajectory. This may well be viewed as a desirable state of affairs, but we shall now dash these hopes by showing that V is never differentiable at 0, so that $\partial V(0)$ is never a singleton! We shall make amends presently by explaining why it is better thus.

4. A FORMULA FOR $\partial V(0)$

We continue with the hypotheses and notation of the preceding section, so that $\hat{z}$ continues to signify a solution to the problem $P(0)$ and c the corresponding constant so that the arc $\hat{z} + c$ is a (T-periodic) Hamiltonian trajectory. We showed that c belongs to $J\partial V(0)$.

Now consider any $\Delta > 0$ and define

$$\tilde{z}(t) = \hat{z}(t + \Delta) - \hat{z}(\Delta) .$$

Note that $\tilde{z}(0) = \tilde{z}(T) = 0$, so that $\tilde{z}$ is feasible for $P(0)$. We calculate

$$D(\tilde{z}) = \int_0^T \{\frac{1}{2} \langle J\hat{z}(t+\Delta) - J\hat{z}(\Delta), \dot{\hat{z}}(t+\Delta)\rangle + H^*(J\dot{\hat{z}}(t+\Delta))\}dt$$

$$= -\frac{1}{2} J\hat{z}(\Delta) \int_0^T \dot{\hat{z}}(t+\Delta)dt$$

$$+ \int_0^T \{\frac{1}{2} \langle J\hat{z}(t+\Delta), \dot{\hat{z}}(t+\Delta)\rangle + H^*(J\dot{\hat{z}}(t+\Delta))\}dt$$

$$= 0 + D(\hat{z})$$

(the first equality since $\hat{z}(T + \Delta) = \hat{z}(\Delta)$, the second because $\hat{z}$ and $\dot{\hat{z}}$ are T-periodic)

$$= V(0).$$

This calculation shows that $\tilde{z}$ also solves $P(0)$. It follows from the arguments of section 3 therefore that for some constant $\tilde{c}$, belonging to $J\partial V(0)$, the arc $\tilde{z} + \tilde{c}$ is a Hamiltonian trajectory (as well as $\hat{z} + c$). In consequence

$$J\dot{\tilde{z}}(t) = J\dot{\hat{z}}(T + \Delta) = \nabla H(\tilde{z}(t) + \tilde{c}) = \nabla H(\hat{z}(t + \Delta) + c)$$

$$= \nabla H(\hat{z}(t+\Delta) - \hat{z}(\Delta) + \tilde{c}) .$$

It follows that we have

$$\hat{z}(t + \Delta) + c = \hat{z}(t + \Delta) - \hat{z}(\Delta) + \tilde{c} ,$$

which implies that

$$\tilde{c} = \hat{z}(\Delta) + c \in J\partial V(0) .$$

We summarize:

$J\partial V(0)$ <u>contains every point of the form</u> $\hat{z}(t) + c$

<u>for any</u> t <u>in</u> $[0,T]$. (4.1)

Since $\hat{z}$ is nonconstant, this certainly establishes that $J\partial V(0)$ is not a singleton: it contains the entire Hamiltonian trajectory $\hat{z} + c$, in fact the convex hull of all points on the trajectory, since $J\partial V(0)$ is convex. What we now require is an <u>outer</u> estimate for $\partial V(0)$. The key is to examine V at points of differentiability.

<u>Lemma. Let</u> $\nabla V(\gamma)$ <u>exist, and let</u> $\tilde{z}$ <u>solve</u> $P(\gamma)$. <u>Then the arc</u> $\tilde{z} + J\nabla V(\gamma) - \gamma/2$ <u>is a Hamiltonian trajectory on</u> $[0,T]$.

To see this, we parallel the proof in section 3. As before, $\tilde{z}$ minimizes

$$-V(z(T)) + D(z)$$

over the arcs z satisfying $z(0) = 0$. The Euler equation again implies that $\tilde{z} + \tilde{c}$ is a Hamiltonian trajectory for some constant $\tilde{c}$. The transversality condition (1.7) uses as before the adjoint variable (3.2), and leads now to

$$\frac{1}{2} J\gamma - J\nabla H^*(J\dot{\tilde{z}}(T)) \in \partial V(\gamma) = \{\nabla V(\gamma)\}$$

whence

$$\nabla H^*(J\dot{\tilde{z}}(T)) = J\nabla V(\gamma) + \gamma/2 ,$$

which implies

$$J\dot{\tilde{z}}(T) = \nabla H(J\nabla V(\gamma) + \gamma/2) = \nabla H(\tilde{z}(T) + \tilde{c}) = \nabla H(\gamma + \tilde{c}) .$$

It follows that $\tilde{c}$ is equal to $J\nabla V(\gamma) - \gamma/2$, establishing the lemma.

Now suppose that γ_i is a sequence converging to 0 such that $\nabla V(\gamma_i)$ exists and such that

$$\zeta = \lim_{i\to\infty} \nabla V(\gamma_i)$$

exists too. Invoking the lemma, there exist arcs $\tilde{z}_i$ solving $P(\gamma_i)$ such that, for each i, the arc

$$\tilde{z}_i + J\nabla V(\gamma_i) - \gamma_i/2$$

is a Hamiltonian trajectory on $[0,T]$. The arcs $\tilde{z}_i$ form a relatively compact set of arcs (because of an a priori bound on the L^2 norm of their derivatives stemming from the optimality of $\tilde{z}_i$ for $P(\gamma_i)$). It follows that (along a subsequence) there is convergence to an arc $\hat{z}$ solving $P(0)$ and such that $\hat{z} + J\zeta$ is a (T periodic) Hamiltonian trajectory. This shows that $J\zeta$ lies on a T-periodic Hamiltonian trajectory, one which is a translate by a constant of a solution to $P(0)$. Let us denote by Π the set of such orbits, i.e., the set of all points lying on such T-periodic Hamiltonian trajectories. Since $\partial V(0)$ is the convex hull of all points ζ generated as above [2, §2.4], we deduce

$$J\partial V(0) \subset \text{co}\Pi .$$

But (4.1) gives precisely the opposite inclusion, so we have proved:

<u>Theorem 4.1.</u> $\partial V(0) = -J\text{co}\Pi$

Since the extreme points of $\text{co}\Pi$ lie in Π, this formula succeeds in identifying certain points lying on periodic orbits:

Corollary. Through every extreme point of $J\partial V(0)$ there passes a T-periodic trajectory.

Remark 1. For given γ, the value $V(\gamma)$ is that of a standard basic problem in the calculus of variations. Experience indicates that such values can be successfully computed numerically by methods such as that of Rayleigh-Ritz. Thus we would expect to be able to estimate numerically the one-sided directional derivatives $V'(0;d)$ based on the values of V near 0. But since $-V$ is regular, knowing $V'(0,\cdot)$ is equivalent to knowing $\partial V(0)$: $-V'(0;\cdot)$ is the support function of $-\partial V(0)$ [4]. To summarize, $\partial V(0)$ is computable in principle (see the survey article by Mayne and Polak [12], which describes situations in which generalized gradients have been calculated and the algorithms used).

To see the possible application of this to the computation of periodic trajectories, consider the situation in which Π is generated by a single orbit (i.e., there is a single solution $\hat{z}$ to $P(0)$, up to time and space translation). Then $J\partial V(0)$ is the convex hull of the entire orbit. (Note: because H is strictly convex, the extreme points of coΠ consist of those points on the orbit, so no information is lost in knowing the convex hull of Π.) In other words, using local information about V near 0, we can estimate an entire periodic trajectory. This is of particular interest in view of the difficulties associated with lack of stability in the computation of periodic orbits.

Remark 2. We may extend V to take explicit account of the period:

$$V(T,\gamma) := \min \int_0^T \left\{\frac{1}{2} \langle Jz,\dot{z}\rangle + G(J\dot{z})\right\}dt$$

$$\text{s.t. } z(0) = 0,\ z(T) = \gamma .$$

A proof similar to that of Theorem 4.1 can be given to show that $\partial V(T,0)$ is the convex hull of points of the form

$$-[h, J(z(t) + c)] ,$$

where z as before is a solution to $P(T,0)$ and c a constant such that $z(\cdot) + c$ is a Hamiltonian trajectory, and where h is the corresponding energy level:

$$H(z(t) + c) = h \quad \text{on} \quad [0,T] .$$

Another possible route to the calculation of V now suggests itself, based on the fact that V satisfies a form of the Hamilton-Jacobi equation related to the problem whose value function it is. In the present setting, if V were smooth then it would satisfy

$$V_T(t,z) + H(JV_\gamma(t,z) + z/2) = 0 . \qquad (4.2)$$

We do not know whether this equation has been studied before. It contains as a special case the classical Hamilton-Jacobi equation, for if we seek a solution V of the special form

$$V(t,x,p) = -\frac{1}{2}\langle x,p\rangle + \varphi(t,x) ,$$

then φ must satisfy

$$\varphi_t(t,x) + H(x,\varphi_x(t,x)) = 0 .$$

We know from the preceding that V is in fact not differentiable, but it can be shown to satisfy the following generalization [4] of the "dual Hamilton-Jacobi equation" (4.2) via generalized gradients:

$$\min_{(\alpha,\beta)\in\partial V(t,z)} \{\alpha + H(J\beta + z/2)\} = 0 .$$

Another and more recent, yet closely related, type of generalization of the Hamilton-Jacobi equation is that of "viscosity solutions"; the applicability of that theory to the computation of V remains to be explored.

REFERENCES

1. Clarke, F. H., 'Solutions périodiques des équations hamiltoniennes', Comptes Rendus Acad. Sci. Paris **287** (1978), 951-952.
2. Clarke, F. H., 'Periodic solutions of Hamiltonian inclusions', J. Differential Equations **40** (1981), 1-6.
3. Clarke, F. H., 'On Hamiltonian flows and symplectic transformations', SIAM J. Control and Optimization **20** (1982), 355-359.
4. Clarke, F. H., Optimization and Nonsmooth Analysis, Wiley Interscience, New York (1983).
5. Clarke, F. H., 'Hamiltonian trajectories and local minima of the dual action', Trans. Amer. Math. Soc. **287** (1985), 239-251.
6. Clarke, F. H., 'Perturbed optimal control problems', IEEE Trans. on Auto. Cont. **31** (1986), 535-542.
7. Clarke, F. H., 'Action principles and periodic orbits', in Nonlinear Partial Differential Equations: seminar College de France, J.-L. Lions and H. Brézis, eds., Volume 8, to appear.
8. Clarke, F. H. and Ekeland, I., 'Hamiltonian trajectories having prescribed minimal period', Comm. Pure and Appl. Math. **33** (1980), 103-116.
9. Clarke, F. H. and Loewen, P. D., 'The value function in optimal control: sensitivity, controllability and time-optimality', SIAM J. Control and Optimization **24** (1986), 243-263.
10. Clarke, F. H. and Vinter, R. B., 'On the conditions under which the Euler equation or the maximum principle hold', Applied Math. and Optimization **12** (1984), 73-79.

11. Clarke, F. H. and Vinter, R. B., 'Regularity properties of solutions to the basic problem in the calculus of variations', Trans Amer. Math. Soc. **289** (1985), 73-98.
12. Mayne, D. Q. and Polak, E., 'Algorithm models for nondifferentiable optimization', SIAM J. Control and Optimization **23** (1985), 477-491.

PERIODIC SOLUTIONS OF DYNAMICAL SYSTEMS WITH NEWTONIAN TYPE POTENTIALS

Marco Degiovanni
Scuola Normale Superiore
Piazza dei Cavalieri, 7
I 56100 Pisa, Italy

Fabio Giannoni
Dipartimento di Matematica
Università di Roma Tor Vergata
Via Orazio Raimondo
I 00173 Roma, Italy

Antonio Marino
Dipartimento di Matematica
Università
Via Buonarroti, 2
I 56100 Pisa, Italy

ABSTRACT. Periodic solutions of some singular dynamical systems are sought, under assumptions which include the case of Newtonian potential generated by a mass concentrated in a point. A result concerning solutions of minimal period is also given.

INTRODUCTION

Let us consider the problem of finding periodic solutions of a conservative dynamical system

$$\ddot{q} + \text{grad } V(q) = 0 \ . \tag{P}$$

It is well known that the solutions can be found among the critical points of the functional

$$f(q) = (1/2) \int_0^T |\dot{q}|^2 dt - \int_0^T V(q)dt \ ,$$

$$q \in H^1(0,T;\mathbb{R}^n), \quad q(0) = q(T) \ ,$$

where V is the potential energy and T is the period.

Many important studies have been done, when V is regular (see, for instance, the survey paper [11] and references therein).

The case of potentials of Newtonian type (namely $V(x) \simeq -1/|x|$) has brought to the study of (P) under the assumption that V goes to ∞ at the points of a certain "singular" set. Several authors have

P. H. Rabinowitz et al. (eds.), Periodic Solutions of Hamiltonian Systems and Related Topics, 111–115.

faced, under various sets of assumptions, this case (see, for instance, [1,2,3,4,5,8,10]).

In particular, when the potential goes to $-\infty$ on the singular set, several results are available under the so called "strong force" hypothesis (introduced in [8]), which is verified, for example, by $V(x) = -1/|x|^{\alpha}$ when $\alpha \geq 2$, but not when $\alpha < 2$. Indeed, under strong force assumption we have that $f(q) = +\infty$ whenever q is a curve which intersects the singular set. As a consequence, the sublevels $f^c = \{q : f(q) \leq c\}$ $(c \in \mathbb{R})$ are complete (for instance in H^1-metric) and do not contain any curve which intersects the singular set. By means of this fact one can overcome the difficulty involved in the presence of the singularities.

We wish to refer some results of [7] concerning a case in which V goes to $-\infty$ at a certain singular point, under assumptions which include the case $V(x) = -1/|x|^{\alpha}$ with $\alpha \geq 1$.

Therefore, it may happen that $f(q) \in \mathbb{R}$ even if q intersects the singular point. Nevertheless we want to find a periodic solution of (P) which does not meet such a point.

In order to explain the heart of the question, we study both the singular potential (theorem (1.3)) both the symmetric singular potential (theorem (1.4)) under assumptions which include potentials of Newtonian type, but are also rather simple: for instance, the singular set is reduced to a point and the symmetry is the antipodal one.

1. MAIN RESULTS

We wish to consider a class of potentials V like $-b/|x|^{\alpha}$ with $\alpha \geq 1$, $b > 0$.

Let us introduce the following notations:

$$\theta_0(\alpha) = \inf\{(1/2)\int_0^1 \dot{\gamma}^2 dt + \int_0^1 (1/\gamma^{\alpha})dt :$$

$$\gamma \in H_0^1(0,1;\mathbb{R}),\quad \gamma \geq 0\}\ , \tag{1.1}$$

$$\theta_1(\alpha) = \min\{2\pi^2 R^2 + (1/R^{\alpha}) : R > 0\}\ ,$$

$$\varphi(\alpha) = (\theta_0(\alpha)/\theta_1(\alpha))^{(\alpha+2)/2}\ .$$

<u>(1.2) Remark</u>

In this way we have defined a real extended map $\varphi:[1,+\infty[\to [1,+\infty]$ such that $\varphi(1) = 1$, $\varphi(\alpha) > 1$ if $\alpha > 1$, $\varphi(\alpha) = +\infty$ iff $\alpha \geq 2$.

(1.3) Theorem

Let $V \in C^2(\mathbf{R}^n\setminus\{0\};\mathbf{R})$ be such that

i) $\lim_{|x|\to\infty} \operatorname{grad} V(x) = 0, \quad \lim_{|x|\to\infty} V(x) = 0;$

ii) $\exists a > 0,\ \exists \alpha \geq 1$ such that

$$a/|x|^\alpha \leq -V(x) \leq \varphi(\alpha)a/|x|^\alpha, \quad \forall x \in \mathbf{R}^n\setminus\{0\} .$$

Then for every $T > 0$ there exists a T-periodic solution q of (P) such that $q(t) \neq 0$ for any t.

This result can be generalized by substituting $1/|x|^\alpha$ with a convex function of $1/|x|$ (see [7]). However, in such a case the condition corresponding to ii) involves also the period T.

Another result, concerning Newtonian potentials, can be obtained under an evenness assumption on the potential V. Under strong force hypothesis, results for even singular potentials have been obtained in [4].

(1.4) Theorem

Let $V \in C^2(\mathbf{R}^n\setminus\{0\};\mathbf{R})$ be such that

i) $V(x) = V(-x) \quad \forall x \in \mathbf{R}^n\setminus\{0\};$

ii) $\exists a > 0 : a/|x| \leq -V(x) \leq 2a/|x| \quad \forall x \in \mathbf{R}^n\setminus\{0\}.$

Then for every $T > 0$ there exists a T-periodic solution q of (P) which does not cross the origin and has minimal period T.

We remark that we obtain a solution q which is "symmetric" with respect to the origin, that is such that $q(t + T/2) = -q(t)$.

The proof of theorem (1.4) is based on a minimizing argument.

We wish to give an idea of the proof of theorem (1.3).

To this aim, we look for a critical point of the functional $f:H \to \mathbf{R} \cup \{+\infty\}$ defined by

$$f(q) = (1/2) \int_0^1 |\dot{q}|^2 dt - T^2 \int_0^1 V(q)dt$$

where $H = \{q \in H^1(0,1;\mathbf{R}^n):q(0) = q(1)\}$.

Set $X_0 = \{q \in H:\exists t$ such that $q(t) = 0\}$ and denote by $\bar{X}$ the set of the circular trajectories which are parallel to a given one and lie on a suitable ellipsoid centered in the origin.

The main feature is that, by a suitable choice of the ellipsoid, we get the following properties:

i) $\sup\{f(q);q \in \bar{X}\} \leq \inf\{f(q):q \in X_0\};$

ii) $\bar{X} \cap f^\sigma \neq \emptyset,$

where σ is a small positive number such that $f^\sigma \cap X_0 = \emptyset$.

We remark that equality must hold in i), if $V(x) = -b/|x|$. See also the computations which are made in [9] in the case $V(x) = -1/|x|$.

Now, if $\bar{X}$ contains a critical point, the theorem is proved. Otherwise $\bar{X}$ can be deformated by means of the gradient flow associated with f. Since $\inf f > -\infty$ and i) holds, the deformation is globally defined and takes its values in $H\backslash X_0$. Moreover $\bar{X} \cap f^\sigma$ remains in f^σ during the deformation. If, by contradiction, f has no critical point in $H\backslash X_0$, for every $\varepsilon > 0$ we can deformate $\bar{X}$ into a subset X' of f^ε. On the other hand, if ε is sufficiently small, X' can be deformated in f^σ into a subset of the constant trajectories of $\mathbb{R}^n\backslash\{0\}$. This turns out to be impossible by the definition of $\bar{X}$.

For a complete proof and more details, see [6].

2. SOME OPEN PROBLEMS

In theorems (1.3) and (1.4) we have considered rather simple situations, including, however, the Newtonian case.

Therefore, a first problem may consist in weakening in theorem (1.3) the conditions on V near the singularity.

Similarly, one can look for other symmetries of V which can substitute evenness in theorem (1.4).

Another question may concern the extension of theorems (1.3) and (1.4) to potentials V whose singular set is not reduced to a point.

REFERENCES

1. A. Ambrosetti and V. Coti Zelati, 'Solutions with minimal period for Hamiltonian systems in a potential well', Ann. Inst. H. Poincaré. Anal. Non Linéaire, in press.
2. A. Ambrosetti and V. Coti Zelati, 'Critical points with lack of compactness and singular dynamical systems', preprint, Scuola Normale Superiore, Pisa, 1986.
3. V. Benci, 'Normal modes of a Lagrangian system constrained in a potential well', Ann. Inst. H. Poincaré. Anal. Non Linéaire **1** (1984), 379-400.
4. A. Capozzi, C. Greco and A. Salvatore, 'Lagrangian systems in presence of singularities', preprint, Dip. Mat., Bari, 1985.
5. V. Coti Zelati, 'Dynamical systems with effect-like potentials', Nonlinear Anal., in press.
6. M. Degiovanni and F. Giannoni, to appear.
7. M. Degiovanni, F. Giannoni and A. Marino, 'Dynamical systems with Newtonian type potentials', Atti Accad. Naz. Lincei Rend. Cl. Sci. Fis. Mat. Natur., in press.
8. W. B. Gordon, 'Conservative dynamical systems involving strong forces', Trans. Amer. Math. Soc. **204** (1975), 113-135.
9. W. B. Gordon, 'A minimizing property of Keplerian orbits', Amer. J. Math. **99** (1977), 961-971.
10. C. Greco, 'Periodic solutions of some nonlinear ODE with singular nonlinear part', preprint, Dip. Mat., Bari, 1985.

11. P. H. Rabinowitz, 'Periodic solutions of Hamiltonian systems: a survey', SIAM J. Math. Anal. **13** (1982), 343-352.

FAMILIES OF PERIODIC SOLUTIONS NEAR EQUILIBRIUM

G. F. Dell'Antonio*
Department of Mathematics
University of Rome
La Sapienza, Italy

ABSTRACT. We discuss the behavior of a large class of Hamiltonian systems near an equilibrium. We study in particular families of periodic solutions and their limit periods. We prove that in general the minimal periods converge to a minimal period of the linearized system. We describe also the localization of the regular families and illustrate the general theory discussing briefly a case which shows a somewhat unexpected behavior.

INTRODUCTION

We will discuss the behavior of a large class of Hamiltonian systems near an equilibrium. We study in particular families of periodic solutions and their limit periods. In Section 1 we prove that, in general, the minimal periods of a regular family converge to a minimal period of the linearized system. Basic for the proof is a lemma (Lemma 1.1) which holds when the full Hamiltonian is in involution with its quadratic part. In Section 2 we describe the localization of the regular families and in Section 3 we discuss briefly a case which illustrates the general procedure and exhibits a somewhat unexpected behavior.

1. REGULAR FAMILIES AND THEIR LIMIT PERIODS

We consider Hamiltonian systems for which the origin is an equilibrium point and we assume that the Hamiltonian can be written as

$$H = H_0(q,p) + K(q,p), \quad q,p \in R^n \tag{1.1}$$

*CNR, GNFM.

P. H. Rabinowitz et al. (eds.), Periodic Solutions of Hamiltonian Systems and Related Topics, 117–130.

where

$$H_0(q,p) \equiv \frac{1}{2}\sum_1^N \nu_i(q_i^2 + p_i^2), \qquad \nu_i \in R\backslash\{0\} \tag{1.2}$$

and $K(q,p)$ is of class $\mathbf{C}^N$ for some $N > 2$. Moreover K is superquadratic in the following sense: one can find a positive function $h(\varepsilon)$, infinitesimal when $\varepsilon \to 0$, such that

$$K(\varepsilon q,\varepsilon p) = \varepsilon^2 h(\varepsilon)K_\varepsilon(q,p)$$

with $K_0(q,p) \not\equiv 0$.

Since we are interested in the behavior near the origin, we change scale through the (canonical) transformation

$$p = \varepsilon p', \; q = \varepsilon q', \; H(q,p) = \varepsilon^2 H'(q',p') \; . \tag{1.3}$$

From now on we omit the primes, and study for small values of ε the one-parameter family of Hamiltonian systems described by the Hamiltonians

$$H_\varepsilon(q,p) = \frac{1}{2}\sum_1^n \nu_i(q_i^2 + p_i^2) + h(\varepsilon)K_\varepsilon(q,p) \; . \tag{1.4}$$

We shall denote by J the standard symplectic map and by ω the standard symplectic form. If F,G are of class $\mathbf{C}^1$, we denote by $\{F,G\}$ their Poisson bracket, i.e.

$$\{F,G\} = \omega(JdG,JdF) \; .$$

The equations are written in the compact form

$$\dot z = JdH_\varepsilon(z), \qquad z \equiv (q,p) \in R^{2n} \; . \tag{1.5}$$

We study one-parameter families of periodic solutions of (1.5) which are regular in the following sense:

I: The orbits $\gamma_\varepsilon \subset R^{2n}$ satisfy

$$\sup_{z\in\gamma_\varepsilon} |z| = 1 \; . \tag{1.6}$$

II: The minimal periods T_ε are uniformly bounded.

III: One can select $z(\varepsilon) \in \gamma_\varepsilon$ so that $z(\varepsilon)$ depends continuously on ε.

The following lemma will play a basic role in what follows.

Lemma 1.1. Let H_ε be given as in (1.4) and let K_ε be of class $C^{1,1}$ in (q,p), uniformly in $B_1 \times (0,\varepsilon_0)$ where $B_1 \equiv \{q,p|q^2 + p^2 = 1\}$. Assume moreover that, for $0 < \varepsilon < \varepsilon_0$,

$$\{K_\varepsilon, H_0\} = 0 \,, \tag{1.7}$$

and that $\nu_i \nu_K^{-1} \in Z \;\forall\; i,K$. Let $\{\varphi_\varepsilon\}$ be a regular family. One can then find $\varepsilon_1 < \varepsilon_0$ such that, for all $\varepsilon \in (0,\varepsilon_1)$, φ_ε differs only in time-scale from a periodic solution of the linearized system

$$\dot{z} = JdH_0(z) \,. \tag{1.8}$$

Moreover $\lim_{\varepsilon \to 0} T_\varepsilon$ exists and is one of the minimal periods of (1.8). |||

Proof. For $z \in R^{2n}$, let $\xi_\varepsilon(z)$ be the projection of JdK_ε onto the hyperplane perpendicular to JdH_0 at z. We want to prove that

$$z \in \gamma_\varepsilon \implies \xi_\varepsilon(z) = 0, \qquad \varepsilon \in (0,\varepsilon_1) \,. \tag{1.9}$$

Notice that, by (1.7), ξ_ε vanishes on γ_ε if it vanishes at any one point of γ_ε.

We shall argue by contradiction. Suppose that no ε_1 exists for which (1.9) holds.

Let

$$p(\varepsilon) \equiv \min_{z \in \gamma_\varepsilon} |\xi_\varepsilon(z)| \,. \tag{1.10}$$

Consider a Poincaré section Γ_ε transversal to γ_ε at z_ε. For every $z \in \gamma_\varepsilon$, let $\underset{\sim}{Z}(z)$ be the first intersection with Γ_ε of the backward solution of (1.8) which starts at z.

By the tubular neighborhood theorem one can find $\varepsilon_2 < \varepsilon_0$ such that $\underset{\sim}{Z}(z)$ is defined for all $z \in \gamma_\varepsilon$, $\varepsilon \in (0,\varepsilon_2)$, and is a C^1-function of z.

Consider on Γ_ε the curve

$$\chi_\varepsilon \equiv \bigcup_{z \in \gamma_\varepsilon} \underset{\sim}{Z}(z) \,.$$

It follows from (1.7) that $\xi_\varepsilon(\underset{\sim}{Z}(z))$ is the tangent vector to χ_ε at $\underset{\sim}{Z}(z)$ and moreover

$$|\xi_\varepsilon(\underset{\sim}{Z}(z))| = |\xi_\varepsilon(z)| \qquad \forall z \in \gamma_\varepsilon \,. \tag{1.11}$$

Together with II this implies that one can find a constant $c_1 > 1$ such that

$$\max_{z\in\gamma_\varepsilon} |\xi_\varepsilon(z)| < c_1 p(\varepsilon)$$

and therefore

$$p(\varepsilon) < |\xi_\varepsilon(y)| < c_1 p(\varepsilon) \qquad \forall y \in \chi_\varepsilon \ . \tag{1.12}$$

The curve χ_ε cannot then have transversal self-intersections and is in fact the connected union of smooth closed loops χ_ε^i, $i = 1,\ldots,i_0$.

It follows from (1.11), (1.12) that one can find $c_2 > 0$ such that the length $\ell(\chi_\varepsilon)$ of χ_ε satisfies

$$\ell(\chi_\varepsilon) < c_2 h(\varepsilon) p(\varepsilon) \ . \tag{1.13}$$

Let Σ_ε^i be any smooth two-dimensional surface in R^{2n} which has χ_ε^i as boundary and is diffeomorphic to a disk. From (1.13) it follows that one can choose Σ_ε^i in such a way that its diameter $\delta(\Sigma_\varepsilon^i)$ satisfies

$$\delta(\Sigma_\varepsilon^i) < c_3 h(\varepsilon) p(\varepsilon) \ . \tag{1.14}$$

For $z \in \Sigma_\varepsilon^i$, let $\pi_0(\xi_\varepsilon)(z)$ be the orthogonal projection of $\xi_\varepsilon(z)$ onto the plane tangent to Σ_ε^i at z. It is a Lipschitz-continuous vector field on Σ_ε^i and coincides with $\xi_\varepsilon(z)$ at $\partial\Sigma_\varepsilon^i$.

Since ξ_ε never vanishes on $\partial\Sigma_\varepsilon^i$, $\pi_0(\xi_\varepsilon)$ must have at least one zero in Σ_ε^i.

It follows then from (1.14) that one can find $c_4 > 0$ such that

$$\sup_{z\in\gamma_\varepsilon} |\xi_\varepsilon(z)| < c_4 h(\varepsilon) p(\varepsilon) \ .$$

This contradicts (1.10), and proves (1.9).

For $\varepsilon \in (0,\varepsilon_1)$ the curve γ_ε is then a closed orbit of (1.8), and therefore the orbit of a periodic solution. Let $\bar{T}$ be its minimal period. From regular perturbation theory one derives

$$|T_\varepsilon - \bar{T}| = O(h(\varepsilon)) \ . \tag{1.15}$$

This concludes the proof of the lemma. |||

We shall now prove that, under rather general assumptions on K_ε, the statement about the family of periods still holds when assumption (1.7) is dropped.

Let

$$P_\varepsilon(z,T) = 0 \tag{1.16}$$

be the periodicity condition for (1.2). By this we mean that (z_0,T_0) solves (1.16) for $\varepsilon = \varepsilon_0$ precisely if z_0 is the initial datum of a periodic solution of (1.5) (with $\varepsilon = \varepsilon_0$) with minimal period T_0.

Use of the "variation of constants" formula leads to

$$P_\varepsilon(z,T) = (e^{AT} - I)z - h(\varepsilon) \int_0^T e^{A(T-t)} JdK_\varepsilon(\varphi(s,z))ds \qquad (1.17)$$

where $Az \doteq JdH_0(z)$ and $\varphi(s,z)$ is the solution of (1.5) starting at z.

From (1.17) one verifies that all limit points of $\{T_\varepsilon\}$ for a regular family must be multiples of one of the minimal periods of (1.8).

A straightforward application of the Implicit Function Theorem allows to restrict one's attention to the case

$$\nu_i \nu_j^{-1} \in Q \quad \forall\ i,j = 1,\dots,n . \qquad (1.18)$$

Condition (1.18) is of course equivalent to

$$\exists \nu_0 \in R^+ \text{ such that } \nu_i = n_i \nu_0, \quad n_i \in Z . \qquad (1.19)$$

We shall assume (1.19) from now on.

One can then prove the following lemma on normal forms.

<u>Lemma 1.2.</u> Let H_ε be given by (1.4) with $K_\varepsilon \in \mathbf{C}^{m_0+1}$, $m_0 \geq 1$, uniformly in $\varepsilon \in (0,\varepsilon_1)$ and on the ball of radius two at the origin in R^{2n}. Assume (1.19). For each $\varepsilon \in (0,\varepsilon_1)$ and $1 \leq m \leq m_0$ one can find a symplectic transformation $\varphi_{m,\varepsilon}$, asymptotic to the identity when $\varepsilon \downarrow 0$, such that

$$H_\varepsilon \circ \varphi_{m,\varepsilon} = H_0 + h(\varepsilon)N_{m,\varepsilon}(z) + h^m(\varepsilon)R_{m,\varepsilon}(z) \qquad (1.20)$$

where $\{N_{m,\varepsilon},H_0\} = 0$ and $R_{m,\varepsilon}$, $N_{m,\varepsilon}$ are of class $\mathbf{C}^{m_0-m+2}$. $R_{m,\varepsilon}$ is infinitesimal when $\varepsilon \downarrow 0$ uniformly in the unit ball.

The function $N_{m,\varepsilon}$ will be called normal form of H_ε to order m. |||

The construction is done in m steps; at each step the order of the normal form is increased by one. At each step one can, e.g., determine the canonical transformation φ as the time-one map for a suitable Hamiltonian G. To find G, one must solve an equation of the form

$$\{G,H_0\} = B$$

where B is known, and of course has no component in the kernel of $\{\cdot,H_0\}$.

No small divisor problem arises since, by (1.19), if $\sum_K m_K\nu_K \neq 0$, $m \in Z$, then $|\sum_K m_K\nu_K| > \nu_0$.

On the other hand, at each step one loses in general one order of differentiability, since the map φ has the same smoothness as the vector field JdG.

From now on, we shall assume that the canonical transformation $\varphi_{m,\varepsilon}$ has been performed, and therefore the Hamiltonians H_ε have the form given by the right-hand side of (1.20).

The periodicity condition for the Hamiltonian system

$$\dot{z} = JdH_0(z) + h(\varepsilon)JdN_{m,\varepsilon}(z) \tag{1.21}$$

is

$$0 = P_{m,\varepsilon}(z,T) \equiv (e^{AT} - I)z$$
$$- h(\varepsilon) \int_0^T e^{A(T-t)}(JdN_{m,\varepsilon})(\varphi_{m,\varepsilon}(s,z))ds \tag{1.22}$$

where $\varphi_{m,\varepsilon}$ is the solution of (1.21) starting at z.

From (1.20) and regular perturbation theory it follows that, for any fixed $\tau > 0$

$$\sup_{0 \leqslant t \leqslant \tau} |\varphi_\varepsilon(t,z) - \varphi_{m,\varepsilon}(t,z)| = \theta(h^m(\varepsilon)) .$$

Therefore

$$\sup_{|z| \leqslant 1} \sup_{0 \leqslant t \leqslant \tau} |P_\varepsilon(t,z) - P_{m,\varepsilon}(t,z)| = \theta(h^m(\varepsilon)) . \tag{1.23}$$

Notice that P_ε and $P_{m,\varepsilon}$ are maps from $R^{2n} \times R^+$ to R^{2n}. Since (1.5) and (1.21) are autonomous, P_ε and $P_{m,\varepsilon}$ are equivariant under the respective flows. If Γ_{z_0} is a Poincaré section for the flows of (1.5), (1.21) at z_0, one can then regard P_ε and $P_{m,\varepsilon}$ as maps (still denoted by the same symbol) from $\Gamma_{z_0} \times R^+$ to R^{2n}.

Using conservation of energy one can further reduce the study of P_ε (and similarly for $P_{m,\varepsilon}$) to that of a map $\mathcal{P}_\varepsilon$ from $\Sigma_{z_0,\varepsilon}$ to itself, where $\Sigma_{z_0,\varepsilon}$ is the plane tangent to

$$\Gamma_{z_0} \cap \{z | H_\varepsilon(z) = H_\varepsilon(z_0)\} .$$

This is often done (see e.g. [2]) by setting

$$p_{k_0} = \rho\cos\theta, \qquad q_{k_0} = \rho\sin\theta$$

for a suitable choice of the index K_0, and writing then (1.16) as a

2π-periodicity condition in θ. One can neglect the periodicity condition on ρ, since it is then automatically satisfied by conservation of energy.

By the Inverse Function Theorem, every solution of $P_\varepsilon(z) = 0$ can be lifted uniquely to a solution of (1.16).

From Lemma 1.1 one concludes that, if $\varepsilon \in (0,\varepsilon_1)$, the solutions of (1.22) in a ball of radius one of R^{2n} are solutions of

$$TJdN_{m,\varepsilon}(z) - (e^{AT} - I)z = 0 .$$

Denote by $\Sigma^m_{z(\varepsilon)}$ the plane through $z(\varepsilon)$ perpendicular both to the orbit of (1.5) and to $\{z|H_{\varepsilon,m}(z) - H_{\varepsilon,m}(z_0)\}$ where

$$H_{\varepsilon,m} \equiv H_0 + h(\varepsilon)N_{m,\varepsilon} .$$

One can then prove (see e.g. [2]) that the Jacobian matrix $J(\underset{\sim}{P}_{m,\varepsilon})$ at the point $z(\varepsilon) \in \gamma_m(\varepsilon)$ satisfies

$$J(\underset{\sim}{P}_{m,\varepsilon})(z(\varepsilon)) = (\mathrm{Hess}_0 N_{m,\varepsilon})(z(\varepsilon)) \cdot J_0^t, \quad \varepsilon \in (0,\varepsilon_1) \tag{1.24}$$

where J_0 is the restriction J to $\Sigma^m_{z(\varepsilon)}$ and $\mathrm{Hess}_0 N_{m,\varepsilon}$ is the restriction of the Hessian matrix of $N_{m,\varepsilon}$ to $\Sigma^m_{z(\varepsilon)}$. Here $\gamma_m(\varepsilon)$ is the orbit of (1.21) through $z(\varepsilon)$.

We are now in a position to analyze the regular families of periodic solutions of (1.2).

Definition 1.3. A Hamiltonian H is regular of order m if one can choose canonical coordinates such that H has the form described on the right-hand side of (1.20) and moreover, whenever $\{z(\varepsilon),T(\varepsilon)\}$ is a solution of (1.16) one can find $c > 0$ such that

$$d(\{0\},\sigma(\mathrm{Hess}_0 N_{m,\varepsilon}(z(\varepsilon)))) > ch^m(\varepsilon), \quad \varepsilon \in (0,\varepsilon_1) \tag{1.25}$$

where d is the distance between two subsets of R and $\sigma(A)$ is the spectrum of A. |||

We shall verify in Section 2 that if H is sufficiently many times differentiable (depending on $\nu_i\nu_j^{-1}$, $i,j = 1,\dots,n$) the property of being regular is generic in a natural sense [2].

Let H be regular. From (1.23) it follows, using the Inverse Function Theorem, that one can find $\varepsilon_2 < \varepsilon_1$ and, for $\varepsilon \in (0,\varepsilon_2)$, positive numbers T'_ε and a family of points $z'(\varepsilon) \in R^{2n}$ such that $(z'(\varepsilon),T'_\varepsilon)$ solve (1.22) and moreover

$$|z'(\varepsilon) - z(\varepsilon)| = O(h(\varepsilon)), \qquad |T'_\varepsilon - T_\varepsilon| = O(h(\varepsilon)) .$$

From Lemma 1.1 we know that the T'_ε converge, when $\varepsilon \downarrow 0$, to a minimal period of (1.8). The same conclusion holds therefore for the family T_ε. We summarize this analysis in a proposition.

Proposition 1.4. Suppose that the Hamiltonian H is regular of order m in the sense of Definition 1.3. Let γ_ε, T_ε be the orbits and minimal periods of a regular family of periodic solutions of (1.5). Then the sequence T_ε converges to a minimal period of the linearized system and γ_ε is $O(h(\varepsilon))$ near a periodic solution of a Hamiltonian system with Hamiltonian H_0. |||

2. DETERMINATION OF REGULAR FAMILIES

For a large class of Hamiltonian systems the regular families of periodic solutions can be localized by finding the critical points of a function on a $2n-1$ dimensional manifold in R^{2n}. In favorable cases this procedure leads to a rather sharp localization.

One follows the same steps as in the proof of Proposition 1.4, now starting from solutions of (1.21) to construct uniquely solutions of (1.16).

We shall always consider Hamiltonians of the type (1.1), (1.2). As remarked in Section 1, there is no loss of generality in assuming $\nu_i \nu_j^{-1} \in Q\forall(i,j)$. One has then

Proposition 2.1. Assume that, after scaling, one can find canonical coordinates $z \equiv (q,p)$ such that

$$\varepsilon^{-2}H(\varepsilon z) \equiv H_\varepsilon(z) = H_0(z) + h(\varepsilon)N_\varepsilon(z) + h^m(\varepsilon)R_\varepsilon(z) \tag{2.1}$$

where $\{N_\varepsilon,H_0\} = 0$, $h(\varepsilon)$ and $R_\varepsilon(z)$ are infinitesimal when $\varepsilon \to 0$, N_ε is of class C^2 and R_ε is of class C^1, uniformly in $\{z\,|\,|z| < 2\}$.

Assume moreover that one can find $\varepsilon_2 > 0$, $c > 0$ such that

$$dN_\varepsilon(z_1(\varepsilon)) = \lambda_1(\varepsilon)dH_0(z_1(\varepsilon l)), \quad \lambda_1 \in R \tag{2.2}$$

implies

$$d(\{0\},\sigma(\mathrm{Hess}_0 N_\varepsilon(z_1(\varepsilon l)))) > ch^m(\varepsilon) \ . \tag{2.3}$$

Then to each continuous family γ_ε^0 of closed orbits of (1.8) at which (2.2) is satisfied corresponds a regular family γ_ε of periodic solutions of (1.5), with minimal period T_ε. If $R_\varepsilon(z)$ is of class C^2 (i.e. if H is regular), then correspondence is one-to-one if one considers only regular families for which T_ε is uniformly bounded for $\varepsilon \in (0,\varepsilon_2)$. Moreover, one has

$$d(\gamma_\varepsilon,\gamma_\varepsilon^0) = \theta(h^m(\varepsilon)) \tag{2.4}$$

and

$$\lim_{\varepsilon \to 0} T_\varepsilon = T^0 \tag{2.5}$$

where T^0 is one of the minimal periods of (1.8). |||

Proof. From (2.2) it follows that (1.22) is satisfied by $(z_1(\varepsilon), T_1(\varepsilon))$, where $T_1(\varepsilon) = \tau_\varepsilon(1 + O(h(\varepsilon)))$, τ_ε being the period of the solution of (1.8) starting at $z_1(\varepsilon)$.

The first part of Proposition 2.1 follows then from (2.3), since one can use the Inverse Function Theorem to find a unique solution $(z(\varepsilon), T(\varepsilon))$ of the periodicity condition (1.16).

If $R_\varepsilon(z)$ is of class $\mathbf{C}^2$, and ε is sufficiently small, it follows from (2.3) that the Jacobian matrix of the map $\mathcal{P}_\varepsilon$ at $z(\varepsilon) \in \gamma_\varepsilon$ satisfies

$$J(\mathcal{P}_\varepsilon)^{-1}(z(\varepsilon)) > ch^m(\varepsilon), \qquad \varepsilon \in (0, \varepsilon_2)$$

for some constant $c > 0$.

The Inverse Function Theorem can be used to construct uniquely solutions of (1.22), and the second part of Proposition 2.1 follows then from Lemma 1.1. |||

Every Hamiltonian of the form (1.1), (1.2), with $\nu_i \nu_j^{-1} \in Q$ $\forall i,j = 1,\dots,n$, and $K(z)$ of class C^N, $N > 3$, can be written in the form (2.1) for $m < N - 2$. More precisely one has

$$H_\varepsilon(z) = H_0(z) + \sum_{m=1}^{m_0} P_m(z)\varepsilon^m + \varepsilon^{m_0} R(z,\varepsilon), \quad m_0 < N - 2 \tag{2.6}$$

where $R(\cdot,\varepsilon)$ is of class C^{N-m_0} and P_m are polynomials of order $> m + 2$ which satisfy $\{P_m, H_0\} = 0$.

Consider the $U(1)^n$ action on R^{2n} given by independent rotations in the planes (q_i, p_i), $i = 1,\dots,n$. Let G_m be the subgroup which leaves P_m invariant, and denote by $\mathcal{G}_m$ the corresponding Lie algebra. For every m one has $JdH_0 \in \mathcal{G}_m$.

A necessary condition for H to be regular of order m_0 is that

$$\bigcap_{m=1}^{m_0} \mathcal{G}_m = \{\lambda J dH_0, \ \lambda \in R\} . \tag{2.7}$$

If $\nu_i \nu_j^{-1} \in Q$, one can find $N_0 \in Z$ such that (2.7) is satisfied if the P_m are a basis in the linear space of all polynomials of order not greater than N_0 which are in involution with H_0.

The condition of regularity of H takes then the form $\varphi(\beta_1,\dots,\beta_A) \neq 0$, where the β's are the coefficients of the terms of order $< N_0$ in the polynomials P_m, and φ is a function of class $\mathbf{C}^1$. In this sense, the property of being regular is generic for Hamiltonians of the form (1.1), (1.2) with $K(z) \in \mathbf{C}^{N_0}$.

We close this section with some remarks on condition (2.2) and on Lemma 1.1.

Remark 2.2. Equation (2.2) can be regarded alternatively as the equation for the critical points of the function N_ε on the manifolds $E_0(c) \equiv \{z|H_0(z) = c\}$ or as the equation for the critical points of the function H_0 on the manifold $E'_{\varepsilon,c} \equiv \{z|N_\varepsilon(z) = c\}$. More generally, it is the equation for the critical points of a function $F(H_0,N_\varepsilon)$ on the level sets of a function $G(H_0,N_\varepsilon)$, provided the Jacobian of the map $H_0,N_\varepsilon \to F,G$ is nonsingular.

All these functions and their level sets are invariant under the flow of JdH_0. If $\nu_i\nu_j^{-1} \in Q$ all orbits are closed, and one has a fibration of the invariant manifolds. If the Hamiltonian H is regular the critical orbits are nondegenerate. One can then use the equivariant Morse theory or the category theory of Lyusternik and Schnirelman to give a lower bound on the number of critical orbits. In particular, n is a lower bound if $\nu_i > 0\ \forall i$, so that $E_0(c)$ is convex for all c [3], and also if the signs of the ν_i's are arbitrary but $E'_{\varepsilon,c}$ is convex for $\varepsilon \in (0,\varepsilon_1)$ and c sufficiently small [4].

Remark 2.3. Inspection of the proof of Lemma 1.1 shows that the condition $T_\varepsilon < T^0$, $\varepsilon \in (0,\varepsilon_1)$ can be weakened to become

$$T_\varepsilon < \alpha^{-1}(\varepsilon), \qquad \varepsilon \in (0,\varepsilon_1)$$

where $\alpha(\varepsilon)$ is infinitesimal at the origin of order smaller than $h(\varepsilon)$.

This remark applies then also to Proposition 1.4 and Proposition 2.1.

One has also a partial converse. If the Hamiltonian is regular and under some further technical assumptions (satisfied generically if H is of class C^N for N large enough) one can prove the following. One can find a positive function $\beta(\varepsilon)$, infinitesimal for $\varepsilon \to 0$, and a domain

$$\Omega_\varepsilon \subset B_\varepsilon \equiv \{z||z| < \varepsilon\}, \quad \lim_{\varepsilon\to 0} \varepsilon^{-2n}(\mathrm{Vol}(\Omega_\varepsilon)) = 0$$

such that no solution of (1.2) with period smaller than $\beta^{-1}(\varepsilon)$ can enter the domain $B_\varepsilon - \bar{\Omega}_\varepsilon$.

The domain Ω_ε is a neighborhood of the set of regular families of periodic solutions. This result can be obtained along the following lines: write H after scaling in the form (2.1) and assume that $|JdN_\varepsilon(z)|$ is bounded below in B_1. Let $\Omega_1 \equiv \{z|\varepsilon z \in \Omega_\varepsilon\}$. For every $z_0 \in B_1 - \Omega_1$ one can find a function S_ε, with $\{S_\varepsilon,H_0\} = 0$, $\{S_\varepsilon,N_\varepsilon\}(z_0) > 0$, such that $\ddot{S}_\varepsilon$ is infinitesimal with respect to $\dot{S}_\varepsilon$ for $t \in (0,\delta_\varepsilon^{-1})$, where δ_ε is infinitesimal. We have denoted by $\dot{A}$ the derivative of the function A along the solution $\varphi_\varepsilon(t,z_0)$ of (1.5) starting at z_0.

If $\varphi_\varepsilon(t,z_0)$ is periodic of period T_ε, then $\dot{S}_\varepsilon(w_\varepsilon(t,z_0))$ is periodic of period T_ε and has zero mean. Therefore, there exists $\tau_\varepsilon \in (0,T_\varepsilon)$ such that $\dot{S}_\varepsilon(\varphi_\varepsilon(\tau_\varepsilon,z_0)) < 0$. Since $S_\varepsilon(z_0) > 0$ for ε small enough, this implies that τ_ε^{-1} is infinitesimal, and so is T_ε^{-1}.

Details will be given elsewhere; a rather complete analysis for the case $n = 2$, $\nu_2 = -2\nu$, is contained in [5].

3. A CASE STUDY

Equation (2.2) can be solved explicitly in some special cases. In [2] we discussed some cases in which

$$h(\varepsilon)N_\varepsilon(z) = \varepsilon^{m_0}P_{m_0}(z) \tag{3.1}$$

where P_{m_0} is a homogeneous polynomial of order m_0. In this section, we outline the study of a case when (3.1) does not hold. The regular families can be found explicitly and turn out to be localized in a somewhat unexpected way.

We consider the case $n = 3$, $\nu_2 = 3\nu_1$, $\nu_3 = 2\nu_2$. If $K(z)$ is of class C^4, after scaling and performing a canonical transformation one obtains

$$H_\varepsilon(z) = H_0(z) + \varepsilon P_\varepsilon(z) + \varepsilon^2 R(z,\varepsilon) \tag{3.2}$$

where $R(z,\varepsilon)$ is infinitesimal in ε and of class C^2 in z. We have used complex notations

$$z = \frac{1}{\sqrt{2}}(q + ip)$$

and one has

$$P_\varepsilon(z) = c_1(z_2^2\bar{z}_3 + \bar{z}_2^2 z_3) + \varepsilon c_2(z_1^3\bar{z}_2 + \bar{z}_1^3 z_2) + \varepsilon S(z,\bar{z}) \tag{3.3}$$

where $c_i \in R$ and S is a homogeneous polynomial of order two in the variables $|z_i|^2$.

We discuss first the case $c_1 \neq 0$. Equations (2.2) become

$$\begin{aligned}
&3\varepsilon^2 c_2 \bar{z}_1^2 z_2 + \varepsilon^2 \frac{\partial S}{\partial \bar{z}_1} = \lambda z_1 , \\
&2\varepsilon c_1 \bar{z}_2 z_3 + \varepsilon^2 c_2 z_1^3 + \varepsilon^2 \frac{\partial S}{\partial \bar{z}_2} = \lambda z_2 \\
&\varepsilon c_1 \bar{z}_2^2 + \varepsilon^2 \frac{\partial S}{\partial \bar{z}_3} = \lambda z_3
\end{aligned} \tag{3.4}$$

and we look for solutions of the form

$$z_i(\varepsilon) = w_i + \varepsilon\tau_i + \theta(\varepsilon)$$
$$\lambda(\varepsilon) = \varepsilon\lambda_0 + \varepsilon^2\lambda_1 + \theta(\varepsilon^2) \ . \tag{3.5}$$

Substituting in (3.4) and equating terms of the same order in ε one derives the following:

A) If $w_1 \neq 0$, then $\lambda_0 = 0$, $w_2 = 0$.
Moreover, writing explicitly

$$S(z)|_{z_2=0} = \frac{1}{2}\, a|z_1|^4 + b|z_1|^2|z_2|^2 + \frac{1}{2}\, c|z_3|^2$$

one must have

$$(a - b)|w_1|^2 = (c - b)|w_3|^2 \ , \tag{3.6$_a$}$$

$$\lambda_1 = a|w_1|^2 + b|w_3|^2 \ , \tag{3.6$_b$}$$

$$\bar{\tau}_2 = - \frac{2c_2}{c_1} \frac{w_1^2}{w_3} \ , \tag{3.6$_c$}$$

$$|w_1|^2 + 6|w_3|^2 = 1 \tag{3.6$_d$}$$

where $(3.6)_d$ is obtained by fixing the scale of the energy as a function of ε^2.

Equation $(3.6)_a$ can be satisfied only if $(a - b)(c - b) \geqslant 0$. Moreover, if

$$(a - b)(c - b) > 0 \tag{3.7}$$

the system (3.6) can be solved for $\lambda_1, w_1, w_3, \tau_2$.

It can be verified that the kernel of the Jacobian is precisely JdH_0. Therefore, if (3.7) holds, to each real solution of (3.6) there corresponds a one-parameter family of periodic solutions of (1.5). Since $w_1 \neq 0$, the limit period is $2\pi\nu_1^{-1}$. There are two distinct families, since (3.6) is invariant under

$$\tau_2 \to \tau_2,\ w_1 \to w_1,\ w_3 \to -w_3 \tag{3.8}$$

and this map cannot be obtained through the flow of JdH_0.

It should be remarked that these families lie close to the hyperplane $q_2 = p_2 = 0$, at odds with a natural first guess from perturbation theory, since the variables q_2, p_2 enter the term of order one in H_ε while q_1, p_1 enter only the term of order two.

B) If $w_1 = 0$, the only solutions of the form (3.5) are such that $z_1 = 0$.

Setting $z_1 = 0$ in (3.4), one can solve for w_2 and w_3. One finds solutions for which $w_2 = 0$ and other solutions for which $w_2 \neq 0$ and $w_3 \neq 0$. For a generic choice of coefficients in the polynomial S, the kernel of the Jacobian at each solution is precisely JdH_0. Therefore to each real solution of (3.4) with $z_1 = 0$ there corresponds a family of periodic solutions of (1.5). The solutions for which $w_2 = 0$ correspond to a one-parameter family with limit period $\frac{2\pi}{\nu_3}$. The solutions for which $w_2 \neq 0$ correspond to two distinct families, with limit period $\frac{2\pi}{\nu_2}$. The doubling is due to the fact that, when $z_1 = 0$, (3.4) is invariant under the map

$$z_2 \to z_2, \quad z_3 \to -z_3$$

and this map cannot be obtained through the flow of JdH_0.

These results for $c_1 \neq 0$ should be compared with the ones one obtains when $c_1 = 0$.

In this case, one can study separately the Hamiltonians $H(z)|_{z_3=0}$ and $H(z)|_{z_1=z_2=0}$.

For a generic choice of c_2 and S one finds a family of solutions which has limit period $2\pi\nu_3^{-1}$ and lies near the hyperplane $q_1 = q_2 = p_1 = p_2 = 0$. Also, one family which has limit period $2\pi\nu_2^{-1}$ and lies near the hyperplane $q_1 = p_1 = q_3 = p_3 = 0$, and one or three families (depending on c_2 and S) which have limit period $2\pi\nu_1^{-1}$ and lie near a cone in the hyperplane $q_3 = p_3 = 0$.

It would be interesting to study in detail, for fixed $c_2 \neq 0$, the behavior of the regular families using c_1 as a "bifurcation" parameter.

We conclude this section with a remark and a conjecture.

Remark. The families of periodic solutions which have the largest limit period (i.e. $\frac{2\pi}{\nu_1}$) are absent for an open set in parameter space. The lower bound implied by the theorems of Moser and Weinstein is saturated by the families with smaller limit period (i.e. $\frac{2\pi}{\nu_2}$ or $\frac{2\pi}{\nu_3}$).

In [2], a similar result was obtained for the case $d = 3$, $\nu_3 = 2\nu_2$, $\nu_2 = 2\nu_1$.

One is then led to the following

<u>Conjecture</u>. If the Hamiltonian H is sufficiently many times differentiable and one has a full positive resonance ($\frac{\nu_i}{\nu_j} \in Z^+$ if $i > j$), one can find a positive integer $N_0 > 1$ such that the lower

bound of Moser and Weinstein is saturated by families with limit period $< \frac{2\pi}{\nu_{N_0}}$, while the families with limit period $> \frac{2\pi}{\nu_{N_0}}$ are absent in an open set Σ_1 in the space Σ which parameterizes the normal forms of H.

REFERENCES

[1] C. Churcill, M. Kummer and D. Rod, 'On averaging, reduction and symmetry in Hamiltonian systems', Journ. Diff. Equations **49** (1983), 359-414.

[2] G. F. Dell'Antonio and B. D'Onofrio, 'Periodic solutions of Hamiltonian systems near equilibrium II', submitted to Journ. Diff. Equations.

[3] J. Moser, 'Periodic solutions near equilibrium and a theorem by A. Weinstein', Comm. Pure Appl. Math. **29** (1976), 727-747.

[4] G. F. Dell'Antonio, 'Topological methods in the study of periodic solutions of Hamiltonian systems', Geometrodynamics Proceedings, Prastaro, ed., World Scientific Publishing Co., Singapore (1985).

[5] M. L. Bertotti, University of Trento preprint (1986).

VITERBO'S PROOF OF WEINSTEIN'S CONJECTURE IN $\mathbb{R}^{2n}$

Ivar Ekeland
Université Paris Dauphine
Ceremade
75775 Paris, Cedex 16
France

ABSTRACT. This is a sketch of Viterbo's recent proof of the Weinstein conjecture: a hypersurface of contact type in $\mathbb{R}^{2n}$ carries at least one closed trajectory.

I. THE STATEMENT

Endow the linear space $\mathbb{R}^{2n}$ with the 2-form ω defined by:

$$\omega(x,y) := \sum_{i=1}^{n} (x_i y_{i+n} - x_{i+n} y_i)$$

$$:= (Jx,y)$$

where J, as usual, denotes the matrix:

$$J = \begin{pmatrix} 0 & -I_n \\ I_n & 0 \end{pmatrix}.$$

This is a standard form for all linear symplectic spaces.

Now consider in $\mathbb{R}^{2n}$ a compact C^2 hypersurface Σ. With each point x on Σ, we associate the kernel K_x of the pullback ω_Σ of ω to Σ, that is, the set of tangent vectors ξ in $T_x\Sigma$ such that:

$$(J\xi,\eta) = 0 \quad \text{for all} \quad \eta \in T_x\Sigma \ .$$

K_x is spanned by $Jn(x)$, where $n(x)$ is any non-zero normal vector to Σ at x. In this way, we have defined a one-dimensional distribution of Σ. Any integral curve of the distribution, that is, any C^1 curve $c(t)$ such that $\dot{c}(t) \in K_{c(t)}$ for every t is called a <u>characteristic</u> of the surface Σ.

P. H. Rabinowitz et al. (eds.), Periodic Solutions of Hamiltonian Systems and Related Topics, 131–137.

As always, when one studies a foliation, the first basic question is: does there exist a compact leaf? In other words, does there exist a closed characteristic?

This problem has an immediate interpretation in terms of Hamiltonian mechanics. Choose any C^2 function H on $\mathbb{R}^{2n}$ which admits Σ as a level set:

$$\Sigma = \{x \mid H(x) = 1\} \tag{1}$$

and assume that $H'(x)$ does not vanish on Σ. Then closed characteristics are exactly the trajectories of periodic solutions of the Hamiltonian system $\dot{x} = JH'(x)$ lying on Σ. So the problem of finding closed characteristics of Σ can be written in the following way: find (x,T) such that

$$\begin{aligned} &\dot{x} = JH'(x) \\ &x(0) = x(T) \\ &H(x) = 1 \ . \end{aligned} \tag{2}$$

This last constraint is meaningful, since the Hamiltonian H is a first integral of the equation. It should be noted that the choice of H is free, provided only that it satisfies (1) and $H'(x)$ does not vanish on Σ.

Of course, when the system happens to be completely integrable, periodic solutions can be found explicitly. The first existence result going beyond this is due to Seifert, who proved that if H was of the classical form $H(p,q) = p^2/2 + V(q)$, with $V : \mathbb{R}^{2n} \to \mathbb{R}$ convex, then problem (2) has at least one periodic solution [S].

In a famous paper [R], Rabinowitz proved that if Σ is strongly star-shaped, that is, if there exists a point $x_0 \in \mathbb{R}^{2n}$, for instance the origin, such that $(x - x_0, n(x)) > 0$ for all $x \in \Sigma$ (here $n(x)$ denotes any continuous field of non-vanishing normal vectors), then there is at least one closed characteristic. Simultaneously, Weinstein extended Seifert's method to prove the same result under the stronger assumption that Σ be strictly convex and C^2 [W1]. Clarke later showed that, in the convex case, the existence result holds without any regularity at all: Σ can be the boundary of any convex compact set with non-empty interior, and the Hamiltonian equation $\dot{x} = JH'(x)$ then becomes a differential inclusion, $\dot{x} \in J\partial H(x)$, where ∂ denotes the subdifferential in the sense of convex analysis [C].

Setting aside the question of regularity, Rabinowitz's result was the best available. However, a strange fact became clear almost immediately to all of us who were working on this question, namely that the assumptions were not canonical invariants, as they should have been. Indeed, consider a canonical diffeomorphism $\varphi : \mathbb{R}^{2n} \to \mathbb{R}^{2n}$, that is a diffeomorphism with $D\rho(x)^* J D\rho(x) = J]$ for every $x \in \mathbb{R}^{2n}$. Let Σ be a hypersurface satisfying Rabinowitz's star-shapedness condition, so that it carries at least one closed characteristic. Then $\varphi(\Sigma)$ will no longer satisfy Rabinowitz's condition, but will still carry a closed characteristic, since φ carries the foliation of Σ into the foliation of $\varphi(\Sigma)$.

Motivated by this, Weinstein [W2] sought and found the simplest condition that would be a canonical invariant and would be a natural extension of star-shapedness. He defines a hypersurface Σ in $\mathbb{R}^{2n}$ to be of contact type if there is a one-form Ω on Σ such that:

$$d\Omega = \omega_\Sigma \tag{3}$$

$$\Omega_x \text{ does not vanish on } K_x . \tag{4}$$

The point is that condition (3) does not define Ω: if Ω satisfies (3), then so does $\Omega + df$, for any function f on Σ. The gauge df can be chosen to accommodate the geometry of the surface. If for instance we choose $\Omega_x(\xi) := (Jx,\xi)$, then condition (4) becomes simply $(Jx,Jn(x)) \neq 0$ for all $x \in \Sigma$. This is precisely Rabinowitz's star-shapedness condition. On the other hand, in [W2], Weinstein gives an example of a hypersurface Σ in $\mathbb{R}^{2n}$ which is diffeomorphic to S^{2n-1} but is not of contact type.

More generally, Weinstein defines a hypersurface Σ in a symplectic manifold (M,ω) to be of contact type if and only if there is a 1-form Ω on Σ such that $d\Omega = j^*\omega$ (that is, $d\Omega$ is the pullback of ω by the inclusion map $j : \Sigma \to M$) and $\Omega \wedge (d\Omega)^{n-1}$ is a non-vanishing $(2n-1)$-form on Σ (that is, a volume). In the particular case when (M,ω) is $\mathbb{R}^{2n}$ endowed with the canonical 2-form, this reduces to (3) and (4). He then formulates the following conjecture: if Σ is a compact C^2 hypersurface of contact type in a symplectic manifold and $H^1(\Sigma;\mathbb{R}) = 0$, then Σ carries a closed characteristic. Viterbo has been able to prove this conjecture in the case when (M,ω) is $\mathbb{R}^{2n}$ with the standard linear symplectic structure:

Theorem [V]. If Σ is a compact hypersurface of contact type in $\mathbb{R}^{2n}$, then Σ carries at least one closed characteristic. □

II. VITERBO'S TRICK

It is easy to thicken Σ to a strip $\Sigma_\varepsilon := \Sigma \times]1-\varepsilon, 1+\varepsilon[$ and to use $(\sigma,s) \in \Sigma_\varepsilon$ as a coordinate system in some tubular neighborhood of Σ. There is a natural 1-form on Σ_ε, namely $t\Omega$, and its differential $d(s\Omega)$ defines a symplectic structure on Σ_ε, provided Σ is of contact type. We now ask that the change of variables turns $d(s\Omega)$ into ω. By the results of Weinstein [W2], this is possible. More precisely, we have the following lemma:

Lemma 1. There is an $\varepsilon > 0$, a tubular neighborhood U of Σ, and a diffeomorphism φ of $\Sigma \times]1-\varepsilon, 1+\varepsilon[$ onto U such that:

$$\varphi(\sigma,1) = \sigma \quad \text{for all} \quad \sigma \in \Sigma$$
$$\varphi^*(sd\Omega + ds \wedge \Omega) = \omega \,. \qquad \square$$

Viterbo's trick consists in writing Σ as the level set of some convenient Hamiltonian H. Since the work of Rabinowitz [R], this has become a standard trick in the business, but up to now the idea had been to use Hamiltonians which were positively homogeneous of some degree $\alpha \neq 2$, and to take advantage of the star-shapedness to check the Palais-Smale condition. Viterbo's idea is completely different: use a "staircase" Hamiltonian, which is locally constant except on thin strips near Σ and copies of Σ, and use its constancy on the "steps" to prove Palais-Smale.

Viterbo's Hamiltonian is constructed as follows. First choose $k > 1$ sufficiently large, so that U and kU are disjoint. Then the k^pU are pairwise disjoint, for all $k > 1$. Using the coordinate system Σ_ε in U, we then define H on the union of all the strips k^pU by:

$$H(k^p(\sigma,s)) = k^{2p}h(s) \,.$$

The function h in the preceding formula is chosen so that:

(i) $h : [1-\varepsilon, 1+\varepsilon] \to \mathbf{R}$ is increasing
(ii) $\operatorname{Inf} h(s) > 0$
(iii) $h(1+\varepsilon) = k^2h(1-\varepsilon)$
(iv) h is C^2 and $h'(1-\varepsilon) = 0 = h'(1+\varepsilon)$.

By condition (iii) we can extend H by a constant in the domain between the pth and the $(p+1)^{th}$ strip, namely $k^{2p}h(1+\varepsilon)$. This would be the $(p+1)^{th}$ "step". The overall growth of the function H is quadratic.

<u>Lemma 2</u>. If the Hamiltonian system $\dot{x} = JH'(x)$ has a non-constant periodic solution, then Σ carries a closed characteristic. $\square$

Indeed, the domains between the strips k^pU consist only of fixed points for the Hamiltonian system. If $x(t)$ is a non-constant solution, it cannot cross any of these domains, so it must lie entirely within one of these strips.

Let us say it lies in the strip $k^{2p}U$. The Hamiltonian system can be written $\omega(\xi,\dot{x}) = dH(\xi)$. In the local chart φ of lemma 1, the 2-form ω becomes $d(s\Omega)$, and the 1-form dH becomes $k^{2p}h(s)$. In the new coordinates, $x(t) = (\sigma(t),s(t))$, the system then splits in two:

$$\dot{s} = 0$$
$$\Omega(\dot{\sigma}) = h'(s) \quad \text{and} \quad d\Omega(\dot{\sigma},\tau) = 0 \quad \forall \tau \in T_\sigma\Sigma \,.$$

In other words, $\dot{\sigma}(t) \in K_{\sigma(t)}$, that is, $\sigma(t)$ is a characteristic of Σ.

III. THE DUAL ACTION FUNCTIONAL

We now have to find periodic solutions of $\dot{x} = JH'(x)$. Let us prescribe some period T. It is well-known that T-periodic solutions are critical points of the action functional:

$$\Phi(x) := \int_0^T [(J\dot{x},x)/2 + H(x)]dt .$$

Instead of handling this functional directly, Viterbo chooses a procedure which was initiated by Clarke [C] and adapted by Ekeland and Lasry [EL] (see also [BLRM]) to this kind of situation. Choose some number $K > 0$ and introduce the function:

$$H_K(x) := Kx^2/2 + H(x) .$$

Assume K is so large that the function H_K is strictly convex. Its Legendre-Fenchel transform H_{K*} is defined by:

$$H_{K*}(y) := \mathrm{Max}\{(x,y) - H_K(x) \mid x \in \mathbb{R}^{2n}\} .$$

The action functional can now be written:

$$\Phi(x) := \int_0^T [(J\dot{x} - Kx,x)/2 + H_K(x)]dt .$$

If T and K have been chosen in such a way that $KT/2\pi$ is not an integer, the critical points of Φ over $H^1(\mathbb{R}/T\mathbb{Z};\mathbb{R}^{2n})$ are precisely the critical points of Ψ over the same space, where Ψ is the dual action functional, defined by:

$$\Psi(x) := \int_0^T [(J\dot{x} - Kx,x)/2 + H_{K*}(-J\dot{x} + Kx)]dt .$$

It should be noted right now that there is a natural S^1-action over $H^1(\mathbb{R}/T\mathbb{Z};\mathbb{R}^{2n})$:

$$\forall\theta \in S^1,\ x_\theta(t) := x(t + \theta T)$$

and that the function Ψ is invariant for that action:

$$\forall\theta \in S^1,\ \Psi(x_\theta) = \Psi(x) .$$

The problem is now to find non-zero critical points of Ψ over $H^1(\mathbb{R}/T\mathbb{Z};\mathbb{R}^{2n})$. It splits in two parts: first show that Ψ satisfies the Palais-Smale condition; then show that there is enough change in the topology of the level sets $\Psi^c := \{x \mid \Psi(x) < c\}$ to force the existence of a critical point. In carrying out this program, various technical difficulties appear, for instance the fact that Ψ is twice differentiable but not C^2. These difficulties are now more or less standard in the business, and they are usually overcome by reducing the problem to finite dimension, as in [E]. The procedure used in this paper was akin to the broken-geodesic procedure which is classical in Riemannian geometry, and had the disadvantage that the S^1-action got lost in the process. Viterbo uses another kind of reduction, namely a Liapounov-Schmidt procedure, which does not suffer from this defect: there is a finite-dimensional subspace E of $H^1(\mathbf{R}/T\mathbf{Z};\mathbf{R}^{2n})$, which is invariant for the S^1-action, and an equivariant map $\varphi : E \to E^\perp$ such that, setting $F(x) := \Psi(x + \varphi(x))$, the critical points of F on E are in one-to-one correspondence with the critical points of Ψ on $H^1(\mathbf{R}/T\mathbf{Z};\mathbf{R}^{2n})$: if $x \in E$ is a critical point of F, then $x + \varphi(x)$ is a critical point of Ψ, and conversely. In addition, $\varphi(0) = 0$, so that non-trivial critical points correspond to non-trivial critical points.

The program that could not be carried out on Ψ can now be carried out on F. Viterbo first shows that if the Palais-Smale condition does not hold for F, then one can construct a T-periodic solution for the equation $\dot{x} = JH'(x)$ (drawing heavily on the particular properties of the Hamiltonian H), and the problem is solved. So one may assume that F satisfies Palais-Smale on E. Its topology is relatively easy to investigate: there are two invariant subspaces V and W, constants $\alpha > 0$, γ and C such that:

$$V \supset W \supset \{x \in E \mid x_\theta = x\} \quad \text{and} \quad V \neq W$$
$$x \in W^\perp \quad \text{and} \quad \|x\| = \alpha \Rightarrow F(x) > \gamma$$
$$x \in V \Rightarrow F(x) < C .$$

This, together with the S^1-invariance, shows that there exists a critical value between γ and C, and hence a critical circle.

REFERENCES

[C] F. Clarke, 'Periodic solutions of Hamiltonian inclusions', J. Diff. Eq. **40** (1981), 1-6.

[BLRM] H. Berestycki, J. M. Lasry, B. Ruf, and G. Mancini, 'Existence of multiple periodic solutions on star-shaped Hamiltonian surfaces', Comm. Pure App. Math.

[E] I. Ekeland, 'Une théorie de Morse pour les systémes hamiltoniens convexees', Ann. IHP Analyse non linéaire (1984), 19-78.

[EL] I. Ekeland and J. M. Lasry, 'Problèmes variationnels non convexes en dualité', CRAS Paris **291** (1980), 493-495.

[R] P. Rabinowitz, 'Periodic solutions of a Hamiltonian system on a prescribed energy surface', J. Diff. Eq. **33** (1979), 336-352.

[V] C. Viterbo, 'A proof of the Weinstein conjecture in $\mathbf{R}^{2n}$, to appear, Annales IHP "Analyse non linéaire", 1987.

[W1] A. Weinstein, 'Periodic orbits for convex Hamiltonian systems', Annals of Math. **108** (1978), 507-508.
[W2] A. Weinstein, 'On the hypothesis of Rabinowitz' periodic orbit theorem', J. Diff. Eq. **33** (1978), 353-358.

GLOBAL AND LOCAL INVARIANTS FOR CONVEX ENERGY SURFACES AND THEIR PERIODIC TRAJECTORIES: A SURVEY

I. Ekeland
Universite Paris Dauphine
Ceremade
75775 Paris, Cedex 16
France

H. Hofer*
Department of Mathematics
Rutgers University
New Brunswick, NJ 08903

1. INTRODUCTION

Denote by $\langle\cdot,\cdot\rangle$ the usual inner product in $\mathbb{R}^{2n}$ and let J be the standard complex structure on $\mathbb{R}^{2n}$ given by the matrix

$$J = \begin{bmatrix} 0_n & -1_n \\ 1_n & 0_n \end{bmatrix} .$$

Associated to J and $\langle\cdot,\cdot\rangle$ is the canonical 2-form Ω defined by

$$\Omega = \langle J\cdot,\cdot\rangle .$$

Given a compact smooth hypersurface S in $\mathbb{R}^{2n}$ we see by Alexander-Duality that $\mathbb{R}^{2n}\backslash S$ has a bounded component B and an unbounded component A. Consequently we can speak about an outward pointing normal vectorfield $x \to N(x)$ on S. Since $\Omega(N(x), N(x)) = 0$ we infer that $\xi(x) := JN(x)$, $x \in S$, defines a vectorfield on S, which spans the kernel of $\omega = \Omega|S$. More precisely

$$\begin{aligned} \mathrm{kern}(\Omega|S) &:= \{(x,v) \in S\times\mathbb{R}^{2n} \mid v \in T_xS \\ &\qquad \text{and } \Omega(v,w) = 0 \text{ for all } w \in T_xS\} \\ &= \{(x,t\xi(x)) \mid x \in S,\ t \in \mathbb{R}\} . \end{aligned}$$

Hence we obtain a canonical line bundle $L_S \to S$ with $L_S = \mathrm{kern}(\Omega|S)$,

*Research partially supported by National Science Foundation Grant No. DMS-8603149.

P. H. Rabinowitz et al. (eds.), Periodic Solutions of Hamiltonian Systems and Related Topics, 139–146.

which carries a canonical orientation induced by ξ. Since $L_S \subset TS$ we have a one-dimensional distribution on S which of course is integrable.

Definition 1. Let S be given as above. A periodic Hamiltonian trajectory on S is a one-dimensional submanifold P of S satisfying

(i) P is diffeomorphic to S^1

(ii) $TP = L_S|P$.

We denote the collection of all periodic Hamiltonian trajectories on S by $\mathcal{P}(S)$. A well-known open problem is the following:

Problem 1. Assume $S \subset \mathbb{R}^{2n}$ is a compact smooth hypersurface in $\mathbb{R}^{2n}$, is then $\mathcal{P}(S) \neq \emptyset$?

Quite general results are known, see the article by Benci-Hofer-Rabinowitz, [1] in this volume or see a recent paper by C. Viterbo [9].

Here we are not particularly interested in the existence question for periodic Hamiltonian trajectories but in the problem of their multiplicity at least for a class of hypersurfaces S for which the fact $\mathcal{P}(S) \neq \emptyset$ can be easily established.

We denote by $\mathcal{H}$ the collection of all compact smooth hypersurfaces S in $\mathbb{R}^{2n}$ bounding a strongly convex domain (i.e., Gausscurv > 0). We also assume as some kind of normalization that S encloses $0 \in \mathbb{R}^{2n}$. By results due to P. Rabinowitz [8], and A. Weinstein [10], it is well-known that $\mathcal{P}(S) \neq \emptyset$ for $S \in \mathcal{H}$.

We define a metric d on $\mathcal{H}$ by

$$d(S,T) = \max_{x \in S, y \in T} (\operatorname{dist}(x,T) + \operatorname{dist}(y,S)) .$$

This is of course the well-known Hausdorff metric. If we write $S < T$ if T encloses S, then a basis for the topology on $\mathcal{H}$ is given by sets of the form

$$W_{S,T} = \{R \in \mathcal{H} \mid S < R < T\} .$$

The fact that some $T \in \mathcal{H}$ is close to some $S \in \mathcal{H}$ is just a pinching condition.

We want to find lower bounds for $\#\mathcal{P}(S)$ for $S \in \mathcal{H}$. A question whose investigation would clearly give some insight is the following. It is concerned with some kind of surgery on energy surfaces.

Definition 2. A local convex energy surface in $\mathbb{R}^{2n}$ with a single periodic trajectory is an open subset U of some $S \in \mathcal{H}$ containing a single periodic trajectory $P \in \mathcal{P}(S)$. We shall write $(U,P) \subset S$.

We call a collection $\{(U_i,P_i)\}_i = 1\ldots k$ complete if there is $S \in \mathcal{H}$ such that the (U_i,P_i) can be "symplectically" implanted into S such that

$$\mathcal{P}(S) = \{P_1,\ldots,P_k\} .$$

One can ask for a mechanism to decide the question if a collection $\{(U_i,P_i)\}$ is not complete. Here is an example for $n = 2$. Let $\{\alpha_1,\alpha_2\}$ and $\{\beta_1,\beta_2\}$ be independent over $\mathbf{Z}$ and $\alpha_i > 0$, $\beta_i > 0$. Define

$$S := S(\alpha_1,\alpha_2) = \{x \in \mathbf{R}^4 \mid \frac{1}{2}\sum_{i=1}^{2} \alpha_i(x_i^2 + x_{i+2}^2) = 1\}$$

and similarly $S' := S(\beta_1,\beta_2)$. It is easy to show that

$$P(S) = \{P_1,P_2\} \text{ and } P(S') = \{P_1',P_2'\} .$$

Let $(U,P) \subset S$ and $(U',P') \subset S'$. As we shall see later the following surprising fact holds:
If $\alpha_1 + \alpha_2 \neq \beta_1 + \beta_2$ then $\{(U,P),(U',P')\}$ is not complete.

In the following we shall associate to $S \in H$ a global index $\sigma(S)$ and to all $P \in P(S)$ local indices $\gamma(P)$ and $\bar{\gamma}(P)$ and we shall show that the quantities $\sigma(S)$, $\gamma(P)$, $\bar{\gamma}(P)$ can not be independent.

2. GLOBAL INDICES

Define

$$\tilde{H} = \{H \in C(\mathbf{R}^{2n},\mathbf{R}) \cap C^\infty(\mathbf{R}^{2n}\setminus\{0\},R) \mid H$$
$$\text{is positively 2-homogeneous and strongly convex}\} .$$

There is a natural bijection $\tilde{H} \to \tilde{H}$ given by

$$S \to H_S, H_S^{-1}(1) = S .$$

For $H \in \tilde{H}$ we can define the Fenchel-conjugate $H^* \in \tilde{H}$ by

$$H^*(y) = \max_{x \in \mathbf{R}^{2n}} (\langle x,y\rangle - H(x)) .$$

We introduce now a Hilbert space E by

$$E = \{x : \mathbf{R}/\mathbf{Z} \to \mathbf{R}^{2n} \text{ of class } H^1 \text{ with } \int_0^1 x(t)dt = 0\} .$$

We define a smooth map $a : E \to \mathbf{R}$ which we call the action by

$$a(x) = \frac{1}{2}\int_0^1 \langle Jx(t),\dot{x}(t)\rangle dt .$$

Assume now $S \in \mathcal{H}$ and $H = H_S \in \tilde{\mathcal{H}}$. We associate to S a $C^{1,1}$-Hilbert-submanifold of E by

$$M_S = \{x \in E \mid \int_0^1 H^*(-J\dot{x}(t))dt = 1 \text{ and } a(x) < 0\} .$$

$S^1 = \mathbb{R}/\mathbb{Z}$ acts in the obvious way by "phase-shift" on M_S. Clearly the S^1-action is not free, in fact we have isotopy groups of arbitrary order. For $d < 0$ we define

$$M_S^d = \{x \in M_S | a(x) < d\} .$$

Then M_S^d is S^1-invariant and

$$M_S = \bigcup_{d<0} M_S^d .$$

Hence

$$d \to M_S^d$$

defines a continuous filtration of the S^1-space M_S. We want to make use of the classifying-spaces-theory for group actions. Recall that the universal S^1-bundle is $(S^\infty, p = \text{projection}, CP^\infty)$, see Husemoller [6]. Consider the product $M_S^d \times S^\infty$ with the obvious S^1-action. We obtain a principal S^1-bundle

$$M_S \times S^\infty \to (M_S \times S^\infty)/S^1$$

which is "continuously" filtered by

$$d \to [M_S^d \times S^\infty \to (M_S^d \times S^\infty)/S^1] .$$

Principal bundles determine up to homotopy uniquely classifying maps. In our case this procedure yields classifying maps

$$(M_S^d \times S^\infty)/S^1 \xrightarrow{\pi_d} CP^\infty .$$

Passing to Alexander-Spanier-Cohomology with rational coefficients Q we obtain

$$\pi_d^* : \bar{H}(CP^\infty) \simeq Q[\eta] \to \bar{H}_{S^1}(M_S^d) := \bar{H}((M_S^d \times S^\infty)/S^1)$$

where $\deg(\eta) = 2$ and $H_{S^1}(M_S^d)$ is just Borel's construction of an S^1-cohomology theory. One can show, [4], that for $d < 0$ the image of π_d^* is a truncated polynomial algebra. So we can define the truncation coefficient $\alpha_S(d)$ by

$$\alpha_S(d) = \inf\{k \in \mathbf{N} = \{0,1,\ldots\} \mid \pi_d^*(\eta^k) = 0\} .$$

Note that $\alpha_S(d)$ is just the Fadell-Rabinowitz index of M_S^d, [5].

Definition 3. The index of S, denoted by $\sigma(S)$ is the subinterval of $[0,+\infty)$ consisting of all t such that

$$\liminf_{d\downarrow 0} \alpha_S(d)|d| \leq t \leq \limsup_{d\downarrow 0} \alpha_S(d)|d| .$$

In the following denote by ℓ the set of all compact intervals in $(0,+\infty)$ and equip ℓ with the Hausdorff metric. The first result is the following, see [4].

Theorem 1. (i) For $S \in H$ we have $\sigma(S) \in \ell$; (ii) $\sigma : H \to \ell$ is continuous; (iii) If $\sigma(S) \neq$ point then $\#P(S) = \infty$.

Problem 2. Does there exist $S \in H$ with $\sigma(S) \neq \{\text{point}\}$?

If such a surface exists, then by Theorem 1 we find $\delta \in (0,1)$ such that for every $T \in H$ with

$$\delta S < T < \delta^{-1} S$$

we have $\#P(T) = \infty$. Another problem is

Problem 3. Is there a finite-dimensional definition for $\sigma(S)$ avoiding "loop spaces" and equivariant cohomology?

Next we introduce local indices for the P in $P(S)$, $S \in H$.

3. LOCAL INDICES

Fix $S \in H$ and put $H = H_S$. Given $P \in P(S)$ let $x : \mathbf{R} \to \mathbf{R}^{2n}$ be a solution of the problem

$$\text{(HS)} \quad -J\dot{x} = H'(x), \quad x(0) \in P .$$

Then x is periodic and has minimal period $T > 0$, say. Moreover $x(\mathbf{R}) = P$. We compute easily

$$\frac{1}{2}\int_0^T \langle Jx(t),\dot{x}(t)\rangle dt = \frac{1}{2}\int_0^T \langle H'(x(t)),x(t)\rangle dt$$

$$= \frac{1}{2}\int_0^T 2H(x(t))dt = \frac{1}{2}\int_0^T 2dt = T .$$

Definition 4. The action of P is the number

$$\tilde{a}(P) := \int \lambda|P$$

where $\lambda(x)h := \frac{1}{2}\langle Jx,h\rangle$.

One easily verifies that

$$a(x) = T = \tilde{a}(P) .$$

Here we use the orientation of $L_S|P$ as the orientation of P to define $\int \lambda|P$. Now we linearize the Hamiltonian system (HS) along x to obtain the following time-dependent linear Hamiltonian system with T-periodic coefficients

$$\text{(LHS)} \quad -J\dot{R}(t) = H''(x(t))R(t)$$

$$R(0) = Id \,.$$

The fundamental solution R of (LHS) defines a smooth map

$$(\mathbb{R},0) \to (Sp(n,\mathbb{R}),Id) \,.$$

We carry out a Polar-decomposition

$$R(t) = U(t) \cdot B(t)$$

where the unitary part $U(t)$ is symplectic and commutes with J. $U(t)$ has a complex determinant if we consider $\mathbb{R}^{2n}$ as a complex vectorspace with complex multiplication defined by

$$ix := Jx \,.$$

We define a continuous map $\Delta : (\mathbb{R},0) \to (\mathbb{R},0)$ by

$$\det{}_J U(t) = e^{2\pi i \Delta(t)}$$

$$\Delta(0) = 0 \,.$$

Definition 5. The rotation at P is the number

$$\gamma(P) := \Delta(T) = \Delta(\tilde{a}(P)) \,.$$

The rotation per action unit is the number defined by

$$\bar{\gamma}(P) := \gamma(P)/\tilde{a}(P) \,.$$

Problem 4. Is there a more intrinsic way of defining $\gamma(P)$ and $\bar{\gamma}(P)$ without reference to the embedding of S in $(\mathbb{R}^{2n},\omega)$, without choice of a Hamiltonian, etc.

Now we come to the main result, see [4]. We also list again some results from Theorem 1.

Theorem 2. Let $n \geq 2$ and $S \in H$. Then

(i) The global index $\sigma : H \to \ell$ is continuous.

(ii) For $t \in \sigma(S)$ there exists a sequence $(P_k) \subset P(S)$ with $\bar{\gamma}(P_k) \to t$.

(iii) For $P \in P(S)$ we have $\gamma(P) > 1$.

(iv) Given $\varepsilon > 0$ the following inequality holds

$$\sum_{P \in P(S), \bar{\gamma}(P) \in \sigma_\varepsilon(S)} \gamma(P)^{-1} \geq 1 \,.$$

Here $\sigma_\varepsilon(S)$ denotes the open ε-ball around $\sigma(S)$. An immediate Corollary is the following. Denote by H_f the subset of S consisting of all $S \in H$ with $\#P(S) < \infty$.

Corollary. (i) There exists a continuous function $\sigma : H_f \to (0,+\infty)$ such that

$$\sum_{\bar{\gamma}(P)=\sigma(S)} \gamma(P)^{-1} \geq 1 .$$

Moreover we have

(ii) If $S \in H_f$ there exist $P,Q \in P(S)$ with $\bar{\gamma}(P) = \bar{\gamma}(Q)$.

In particular we have a result due to Ekeland-Lassoued, [11]

(iii) $\#P(S) \geq 2$ for all $S \in H$.

Remark 1. Theorem 2 (iv) is sharp. There exists S with $\sum_{P \in P(S)} \gamma(P)^{-1} = 1$. In fact let $\alpha_1,\dots,\alpha_n > 0$ be independent over **Z**. Put $S = S(\alpha_1,\dots,\alpha_n)$. Then

$$P(S) = \{P_1,\dots,P_n\}$$

$$\gamma_i = \sum_{j=1}^{n} \frac{\alpha_j}{\alpha_i}$$

$$\bar{\gamma}_i = \frac{1}{2\pi} \sum_{j=1}^{n} \alpha_j$$

$$\sigma(S) = \{\frac{1}{2\pi} \sum_{j=1}^{n} \alpha_j\} = \{I\}$$

and therefore

$$\sum_{\gamma_i=I} \gamma_i^{-1} = \Sigma_{\text{all } P_i} \gamma_i^{-1} = 1 .$$

Remark 2. Coming back to the question in the introduction where we asked for completeness of $\{(U,P),(U',P')\}$. If $\alpha_1 + \alpha_2 \neq \beta_1 + \beta_2$ then $\bar{\gamma}(P) \neq \bar{\gamma}(P)$. Consequently the noncompleteness follows from Theorem 2.

Problem 5. Is it true if $P(S) = \{P_1,\dots,P_n\}$ then $\bar{\gamma}(P_i) = \bar{\gamma}(P_j)$ for $i \neq j$?

Problem 6. How does $\bar{\gamma}$ behave near to a twist type elliptic P_0?

We end with some conjectures.

Conjecture 1. For $S \in H$ we have $\#P(S) \geq n$. This conjecture is correct for $n = 1,2$.

As a subset $C^\infty(\mathbb{R}^{2n}\setminus\{0\},\mathbb{R})$ we can equip $\tilde{H}$ with the weak Whitney C^∞-topology. This new topology on $\tilde{H}$ induces a topology on H. We write H_+ for H with this topology.

Conjecture 2. For a residual set H_1 of H_+ the following holds: If $S \in H_1$ then

$\bar{\gamma} : \mathcal{P}(S) \to \mathbf{R}$ is injective.

(Since $\bar{\gamma}$ is not injective if $\#\mathcal{P}(S) < \infty$ we infer that $\#\mathcal{P}(S) = \infty$ for $S \in \mathcal{H}_1$.)

Conjecture 2 should be a consequence of known generic results for Hamiltonian systems.

4. REFERENCES

[1] V. Benci, H. Hofer and P. Rabinowitz, 'A remark on a priori bounds and existence for periodic solutions of Hamiltonian systems', (this volume).

[2] A. Borel, 'Seminar on transformation groups', Ann. Math. Studies 46, Princeton University Press, 1961.

[3] C. Conley and E. Zehnder, 'Morse-type index theory for flows and periodic solutions for Hamiltonian equations', Comm. Pure Appl. Math. **27** (1984), 211-253.

[4] I. Ekeland and H. Hofer, 'Convex Hamiltonian energy surfaces and their periodic trajectories', (to appear).

[5] E. Fadell and P. Rabinowitz, 'Generalized cohomological index theories for Lie group actions with an application to bifurcation questions for Hamiltonian systems', Inv. Math. **45** (1978), 139-174.

[6] D. Husemoller, 'Fibre-Bundles', Springer Graduate Texts in Math. **20**, 2nd Ed.

[7] H. Hofer and E. Zehnder, 'Periodic solutions on hypersurfaces and a result by C. Viterbo', (to appear).

[8] P. Rabinowitz, 'Periodic solutions of Hamiltonian systems', Comm. Pure Appl. Math. **31** (1978), 157-184.

[9] C. Viterbo, 'A proof of the Weinstein conjecture in $\mathbf{R}^{2n}$', (to appear).

[10] A. Weinstein, 'Periodic orbits for convex Hamiltonian systems', Ann. Math. **108** (1978), 507-518.

[11] I. Ekeland and L. Lassoued, 'Multiplicite des trajectories fermes de systemes Hamiltoniens convexes', (to appear).

VITERBO'S INDEX AND THE MORSE INDEX FOR THE SYMPLECTIC ACTION

Andreas Floer
Courant Institute of Mathematical Sciences
New York University
251 Mercer Street
New York, NY 10012

ABSTRACT. We define a relative Morse index of nondegenerate critical points of the symplectic action functional, in the case of paths connecting two Lagrangian submanifolds.

Let (P,ω) be a symplectic manifold, i.e. ω is a closed and nondegenerate 2-form on P. Let L and L' be two Lagrangian submanifolds of P, i.e. L and L' have half the dimension of P and ω restricted to L and L' vanishes. Then in [1], we consider the variational theory of the action functional on the space

$$\Omega = \Omega(L,L') = \{z \in C^\infty([0,1],P) \mid z(0) \in L \text{ and } z(1) \in L'\} . \tag{1}$$

The symplectic action is the locally defined real valued function A on Ω satisfying

$$dA(z)\xi = \int_0^1 \omega(\xi,\dot{z}) . \tag{2}$$

This means that A is defined on the universal covering of Ω. In fact, it is easy to see that A is uniquely determined by (2) up to additive constants. Then the critical points of a on Ω are exactly the constant paths $\hat{x}(t) \equiv x$, where x is an intersection of L and L'. Moreover, $\hat{x}$ is a nondegenerate critical point if and only if x is a transversal intersection.

Let us define a smooth path in Ω between critical points x and y as a smooth map

$$u : [0,1]^2 \to P \tag{3}$$

so that $u(\tau)(t) = u(\tau,t)$ defines for each $\tau \in [0,1]$ an element of Ω and so that $u(0) = x$ and $u(1) = y$. In particular, u maps

P. H. Rabinowitz et al. (eds.), Periodic Solutions of Hamiltonian Systems and Related Topics, 147–152.

the boundary of $[0,1]^2$ into $L \cup L'$, defining arcs in L and L' connecting x and y. In this situation, Viterbo [4] defines an integer $m(u)$ by means of the Maslov class

$$\mu \in H^1(\Lambda_n,\mathbf{Z}) \ , \tag{4}$$

where $\Lambda_n \subset G(\mathbf{R}^n,\mathbf{R}^{2n})$ is the set of Lagrangian planes in $\mathbf{R}^{2n} = \mathbb{C}^n$ with respect to the standard symplectic structure $\omega_0 = \sum_{i=1}^{n} dx_i \wedge dx_{i+n}$. To be more precise, consider the pullback bundle u^*TP. Since the base space $[0,1]^2$ is contractible we can find a trivialization

$$\Phi : u^*TP \to [0,1]^2 \times \mathbb{C}^n \tag{5}$$

mapping the symplectic form on each fibre into the standard form. Moreover, we can choose this trivialization so that it is constant on $\{0\} \times [0,1]$ and on $\{1\} \times [0,1]$, and so that $\Phi(T_xL') = iT_xL'$ and $\Phi(T_yL') = i(T_yL)$. Then define $m(u)$ by evaluating the Maslov class over the closed loop

$$\begin{aligned}
&\partial[0,1]^2 \to \Lambda_n \\
&(\tau,0) \to \Phi(T_{u(\tau,0)}L) \\
&(1,t) \to e^{it}\Phi(T_xL) \\
&(\tau,1) \to \Phi(T_{u(\tau,1)}L') \\
&(0,t) \to e^{-it}\Phi(T_xL') \ .
\end{aligned} \tag{6}$$

In order to relate this geometric construction to Morse theory, choose an almost complex structure J on P satisfying the additional condition that the bilinear form

$$g = \omega(J\cdot,\cdot) \tag{7}$$

is a metric, i.e. is symmetric and positive. Since $d\omega = 0$, (7) defines an almost Kähler structure on P (see [2]) i.e. g,ω and J satisfy all relations of a Kähler structure except that J is not integrable. Such J always exist for any symplectic structures ω. The L^2-gradient of the function A of (2) with respect to the metric g is then given by

$$\nabla_g A(z) = J\dot{z} \ . \tag{8}$$

Strictly speaking, $J\dot{z}$ does not define a vector field on Ω, since it is not necessarily tangential to L and L' at the for $t = 0$ and $t = 1$. However, we can define the set of trajectories between

x and y by

$$\mathcal{M}(x,y) = \{v : \mathbf{R} \times [0,1] \to P \mid$$

$$(1)\quad v(\tau,0) \in L,\ v(\tau,1) \in L' \quad \text{for all} \quad \tau \in \mathbf{R}$$

$$(2)\quad \lim_{\tau\to\infty} v(\tau,t) = y, \qquad \lim_{\tau\to-\infty} v(\tau,t) = x$$

$$(3)\quad \frac{\partial v}{\partial \tau} + J(v(\tau,t))\,\frac{\partial v}{\partial t} = 0\} . \tag{9}$$

Of course, this space depends on the choice of the Kähler structure J. It turns out (see [1], Theorem 5) that one can always find such a J so that $\mathcal{M}(x,y)$ is a smooth finite dimensional manifold. By Theorem 4 of [1], its dimension can be identified with the Fredholm index of the following linear operator. Let u be as in (3), and define $\Theta = \mathbf{R} \times [0,1]$

$$\tilde{u} : \Theta \to P : \tilde{u}(\tau,t) = u(\beta(\tau),t) . \tag{10}$$

Here, $\beta : \mathbf{R} \to [0,1]$ is smooth, increasing, and is equal to 0 on $(-\infty,0]$ and equal to $1.$ on $[1,\infty)$. Now for each compactly supported section $\xi \in C_0^\infty(\tilde{u}^*TP)$ of the pullpack of the tangent bundle of P, we can define the norms

$$\|\xi\|_L = \left(\int_H |\xi(\tau,t)|^2 d\tau dt\right)^{1/2} = \|\xi\|_2 , \tag{11}$$

$$\|\xi\|_W = \|\nabla_\tau \xi\|_2 + \|\nabla_t \xi\|_2 + \|\xi\|_2 . \tag{12}$$

We denote by $L(u)$ the closure of $C_0^\infty(\tilde{u}^*TP)$ under the norm $\|\ \|_L$, and by $W(u)$ the closure under $\|\ \|_W$ of the subspace

$$\{\xi \in C_0^\infty(\tilde{u}^*TP) \mid \xi(\tau,0) \in T_{\tilde{u}(\tau,0)}L$$
$$\xi(\tau,1) \in T_{\tilde{u}(\tau,1)}L'\} . \tag{13}$$

<u>Theorem</u>: $m(u)$ <u>is the Fredholm index of the operator</u>

$$E_u = \nabla_\tau + J\nabla_t : W(u) \to L(u) . \tag{14}$$

That (14) is Fredholm follows from Theorem 4 of [1], see also Theorem 1.3 of [3] for a related case. In fact, any operator of the form (14) is Fredholm with respect to Lagrangian boundary conditions if outside a compact set, the boundary conditions are given by a constant pair λ_0, λ_1 of Lagrangian subspaces with respect to which the self-adjoint operator $J\nabla_t$ has no kernel.

As opposed to the finite dimensional case, the relative Morse index $m(u)$ may depend not only on the end points x and y, but also on the path u. However, Viterbo's geometrical construction yields the following

Corollary: Let $\varphi : P \to P$ be a smooth diffeotopy so that $\varphi_0 = \mathrm{id}$, $\varphi_1(L) = L'$, and each φ_t is symplectic. Assume moreover that $\pi_2(P,L) = 0$. Then $m(u)$ depends only on $x,y \in L \cap L'$.

Proof: Let $u,v : [0,1]^2 \to P$ be two paths in Ω connecting x and y. Then the map

$$w : S^1 \times [0,1] = (\mathbb{R}/2\mathbb{Z}) \times [0,1] \to P , \tag{15}$$

$$w(\tau,t) = \begin{cases} u(-\tau,t) & \text{for } \tau < 0 \\ v(\tau,t) & \text{for } \tau > 0 \end{cases} \tag{16}$$

defines a closed loop in Ω. The bundle w^*TP is trivial as a symplectic bundle, since the group of linear symplectic isomorphisms of $\mathbb{R}^{2n}$ is connected. Every symplectic trivialization defines two closed loops $w_0,w_1 : S^1 \to \Lambda(\mathbb{R}^{2n})$. Let us define the index of a loop in Ω as the difference $\mu[w_1] - \mu[w_0]$. It coincides with $m(u) - m(v)$, since the contributions of the endpoints in (6) cancel. To show that it vanishes, we first deform w by means of the symplectic diffeotopy φ_t to the loop

$$\tilde{w}(\tau,t) = \varphi_t(w(\tau,t)) \tag{17}$$

in $\Omega(L,L)$. Its index is the same as the index of w. Now since $\pi_2(P,L) = 0$, $\tilde{w}$ is homotopic within the loops in $\Omega(L,L)$ to a map $\hat{w}(\tau,t) = x(\tau)$ for some smooth loop $x : S^1 \to L$. Again, the index does not change under such a homotopy. But now $\hat{w}_0$ and $\hat{w}_1$ are identical for any trivialization of $\hat{w}^*TP$ which is constant in t, so that the index of $\hat{w}$ vanishes. This completes the proof of the corollary.

Proof of the Theorem: In order to relate the Maslov index construction according to Viterbo to the analytical index, we consider the trivialization Φ of u^*TP defined in (5). Via (10), it yields a trivialization $\tilde{\Phi}$ of $\tilde{u}^*TP$, which is constant for $\tau \notin (0,1)$. Note that for $\tau \in (0,1)$, Φ is not complex linear. However, since both i and J are sections of the bundle of Hermitian structures of $\mathbb{C}^n$ with respect to the standard symplectic form ω, and since this bundle has contractible fibres, the operator $\Phi E_u \Phi^{-1}$ can be deformed within the Cauchy Riemann operators into the standard Cauchy Riemann operator ∂_0. During this deformation, $\Phi E_u \Phi^{-1}$ is not changed for $\tau \notin (0,1)$. Moreover, the boundary conditions remain elliptic, so that by Theorem 4 of [1], the deformation remains within the set of Fredholm operators from $W(u)$ to $L(u)$. Now note that the inclusions of $\Lambda(\mathbb{C}^n)$ in $\Lambda(\mathbb{C}^n)$ defined by the direct sum with $\mathbb{R}^{n-1}$ or $i\mathbb{R}^{n-1}$ induces an isomorphism on the fundamental group $\pi_1(\Lambda(\mathbb{C})) =$

$\pi_1(\Lambda(C^n)) = Z$. In fact, the Maslov class (4) is characterized by

$$\mu[\lambda] = 1 \quad \text{for} \quad \lambda : S^1 = \mathbf{R}/\pi\mathbf{Z} \to \Lambda(C^n) :$$
$$\lambda(s) = e^{i\pi s}\mathbf{R} \oplus \mathbf{R}^{n-1} \ . \tag{18}$$

We can therefore deform the boundary conditions for $\tau \in (0,1)$ through the Lagrangian boundary conditions into the form $\alpha_0(\tau)R^n \oplus R^{n-1}$ for $t = 0$ and $\alpha_1(\tau)R \oplus (iR^{n-1})$ for $t = 1$. Here, $\alpha_i = R \times S^1 \to C$ are phase factors satisfying $\alpha_0(\tau) = 1$ and $\alpha_1(\tau) = i$ for $\tau \notin (0,1)$. Since all Lagrangian boundary conditions are elliptic for $\bar\partial_0$, this defines a deformation of $\Phi E_u \Phi^{-1}$ within the Fredholm operators into $E_u' = E_u^1 + E_u^2$, where

$$E_u^1 = \bar\partial_0 : W_1^p(\Theta,\mathbb{C}^{n-1}) \to L^p(\Theta,\mathbb{C}^{n-1}) \tag{19}$$

is translationally invariant in τ. By means of a Fourier transformation in τ, one can construct an inverse to E_u^2, see Lemma 4.2 of [1]. Hence the index of E_u' coincides with the index of

$$E_u^2 = \bar\partial_0 : W_1^p(\Theta,\mathbb{C}) \to L^p(\Theta,\mathbb{C}) \tag{20}$$

with boundary conditions given by α_0 and α_1. In order to calculate this index, we deform the boundary conditions continuously according to the formula

$$\alpha_i^\varepsilon(\tau) = \alpha_i(\varepsilon\beta(\tau)) \ . \tag{21}$$

Clearly, the boundary conditions α_i^1 yield a Fredholm operator ∂_0 with the same index as E_u^1. If $\varepsilon < 1$, however, $\bar\partial_0$ with boundary conditions α_i^ε is Fredholm if and only if $\delta(\varepsilon)$ with

$$e^{i\pi\delta(\varepsilon)} = \alpha_1(\varepsilon)\alpha_0(\varepsilon) \tag{22}$$

is not an integer, since otherwise the operator $i\,\frac{d}{dt}$ on $L^2([0,1],\mathbb{C})$ with boundary conditions $\xi(0) \in \alpha_0(\tau)\mathbf{R}$ and $\xi(1) \in \alpha_1(\tau)\mathbf{R}$ has a nontrivial kernel. However, if we define the function

$$e_\varepsilon(\tau,t) = e^{\beta(\tau)\delta(\varepsilon)} \ , \tag{23}$$

then the operator $E_u^\varepsilon = e_\varepsilon^{-1}\partial_0 e_\varepsilon$ is Fredholm with respect to the boundary conditions α_i^ε for all $\varepsilon \in [0,1]$. This defines a continuous deformation of E_u^1 into E_u^0 within the class of Fredholm operators. Moreover, E_u^0 is equivalent via conjugation by e_0 to the standard Cauchy Riemann operator ∂_0 with constant boundary conditions between the Banach spaces with norms $\|e_0\cdot\|_{1,p}$ and $\|e_0\cdot\|_p$. Since this operator is translational invariant, one can again use Fourier transformation to find that E_u^0 is surjective with a kernel of dimension $\delta(0) - \delta(1)$ if this number is positive, and that it is injective with a cokernel of dimension $\delta(1) - \delta(0)$, otherwise. This completes the proof of the theorem, since it follows from the definition of the Maslov class that $m(u) = \delta(0) - \delta(1)$.

REFERENCES

[1] Floer, A., 'The unregularized gradient flow of the symplectic action', to appear.
[2] Helgason, S., 'Differential geometry, Lie groups, and symmetric spaces', Academic Press, New York, 1978.
[3] Lockhard, R. and Mc Owen, R., 'Elliptic opertors on manifolds with cylindrical ends', to appear.
[4] Viterbo, C., 'Intersection de sous-varietes Lagrangiennes, foncionnelles d'action et indice des systemes Hamiltoniens', to appear.

SOME PROBLEMS ON THE HAMILTON-JACOBI EQUATION

Giovanni Gallavotti*
Dipartimento di Matematica
II^a Universitá di Roma
Via Raimondo, 00173 Roma, Italia

I shall devote attention to some old, but still open, problems on the Hamilton-Jacobi equation reviewing the context in which they arise.

One can formulate the Hamilton-Jacobi equation as a rather general conjugacy problem between dynamical systems.

Suppose given two regular (i.e. real analytic) Hamiltonian systems (H,W) and (H_0,W_0), defined respectively by the Hamilton functions H and H_0, regular on some W and W_0 in phase space (i.e. W and W_0 are open subsets in the cotangent bundles to regular Riemannian manifolds). We consider the problem of the existence of a regular completely canonical map C, mapping W_0 onto W, and of a regular function F, defined on $H_0(W) = \{\text{range of } H_0\}$, such that:

i) $C : W_0 \leftrightarrow W$

ii) $$H(C(p',q')) = F(H_0(p',q')) \qquad (p',q') \in W_0 \tag{1}$$

iii) there is a regular function Φ defined on the graph $G(C) = \{p,q,p',q' | (p,q) \times (p',q') \in W \times W_0, (p,q) = C(p',q')\}$ such that:

$$p \cdot dq = p' \cdot dq' + d\Phi \quad \text{on} \quad G(C) \tag{2}$$

i.e. C is "action preserving".

The properties i), ii), iii) are a global way of giving a meaning to the Hamilton-Jacobi equation, so that it is relevant for applications to mechanical problems. It is easy to check that the function on $G(C)$

$$S(p',q) = p' \cdot q + \Phi(p',q)$$

---

*Partially supported by National Science Foundation Grant 85-3333 and by Ministero della Publica Istruzione.

P. H. Rabinowitz et al. (eds.), Periodic Solutions of Hamiltonian Systems and Related Topics, 153–160.

defined wherever this makes sense (i.e. (p',q) determine a point $(p,q) \times (p',q') \in G(C))$, verifies the classical Hamilton-Jacobi equation

$$H\left(\frac{\partial S}{\partial q}(p',q),q\right) = F\left(H_0\left(p', \frac{\partial S}{\partial q}(p',q)\right)\right) \tag{3}$$

and one can regard i), ii), iii) as a precise way of stating the requirements on S to consider it an interesting solution of (3). Often one considers only the special case in which H_0 depends only on p'; then (3) has a slightly similar form [1].

The meaning of the equation (3), or better of i), ii), iii), is easy to understand. It simply implies that the Hamiltonian flows (H,W) and (H_0,W_0) are pointwise isomorphic, up to a trivial time rescaling (in fact C maps orbits into orbits together with the motion on them, up to a time rescaling equal to $\frac{dF}{dE}(E)$ if E is the orbit energy, ($H_0 \equiv E$ on the orbit)).

The above formulation of (3) is not very natural when (H_0,W_0) has nontrivial regular constants of motion $A_1(p',q') \equiv H_0(p',q)$, $A_2(p',q'),\ldots,A_r(p',q')$ independent and in involution. In this case the equation (3) is more naturally generalized to a form which, interpreted more precisely, amounts at replacing ii) above by:

ii') $$H(C(p',q')) = F(A_1(p',q'),\ldots,A_r(p',q')) \tag{4}$$

with F now regular on the domain $A(W_0)$ image of $(p',q') \to (A_1(p',q'),\ldots,A_r(p',q'))$. The replacement of ii) by ii') in i), ii), iii) means that (H,W) is a flow which is a linear combination with "constant coefficients" of the (commuting) flows generated by $A_1,\ldots,A_r$.

Here $r \leq n$, if n is the number of degrees of freedom: so if (H_0,W_0) is ergodic on the energy surfaces it is necessarily $r = 1$, while if (H_0,W_0) is integrable and $A_1,\ldots,A_n$ are the n action variables, it is $r = n$ and the solubility of i), ii'), iii) means that (H,W) is also integrable and its motions are quasiperiodic with frequencies and invariant tori trivially related to those of (H_0,W_0).

Let me recall here that (H_0,W_0) is said to be integrable if in suitable canonical coordinates W_0 can be given the form $V \times T^n$ with V open in R^n and T^n = n-dimensional torus, so that if $(A,\varphi) \in V \times T^n$ it is $H_0(A,\varphi) = h_0(A)$ for some h_0 [1].

A concrete example of the case when $r = 1$ is the Hamiltonian flow corresponding to the geodesic flow on a compact surface of constant negative curvature: in this case $H_0(p,q) = \frac{1}{2} g(q)p \cdot p$ is the kinetic energy (g being the metric) and for $0 < E_- < E_+ < \infty$

$$W_0 = \{(p,q) | H_0(p,q) \in (E_-,E_+)\} .$$

We consider now the case in which $H \equiv H_\varepsilon = H_0 + \varepsilon f_\varepsilon$ where f_ε is regular in ε too for $|\varepsilon|$ small: i.e. we consider the situation arising in "perturbation theory" in which one is given a one parameter family of Hamiltonians.

One can look for solutions of i), ii'), iii) of the form of a power series in ε:

$$\begin{aligned} C &= \text{identity} + \varepsilon C_1 + \varepsilon^2 C_2 + \cdots \quad \text{on} \quad W_\varepsilon \subset W_0 \\ F &= A_1 + \varepsilon F_1 + \varepsilon^2 F_2 + \cdots \quad \text{on} \quad J_\varepsilon \subset A(W_0) \end{aligned} \tag{5}$$

with domains W_ε, J_ε being also unknown.

One says that perturbation theory is "well defined" on W_0 if one can find a formal power series (5) solving i), ii'), iii) on W_0 and $J_0 = A(W_0)$.

In general perturbation theory is not well defined.

For instance the condition i), ii) and iii) of existence of a globally defined generating function Φ implies that, if $\gamma_i(\varepsilon,E), \gamma_i(\varepsilon,E)$ are two isoenergetic periodic solutions to the system (H,W) with energy $H = E$ and depending analytically on ε for ε small enough, then (if i), ii'), iii) are verified):

$$\frac{\oint_{\gamma_i(\varepsilon,E)} p \cdot dq}{\oint_{\gamma_i(\varepsilon,E)} p \cdot dq} = \text{very special } \varepsilon\text{-dependence} \tag{6}$$

as it is easy to check. Expanding (6) in powers of ε one finds a family of conditions on (H,W) which have to be necessarily verified, if formal perturbation theory exists.

An application arises, for instance, in the above mentioned case of the geodesic flow on a surface of constant negative curvature: in this case (H_0,W_0) is, for every $H_0 = E$, an Anosov flow and admits a dense set of periodic orbits which can be numbered $\gamma_1(E), \gamma_2(E), \ldots$: they have nonzero Lyapunov exponents and, by structural stability, persist under perturbation for $|\varepsilon|$ small enough. Hence they can be continued analytically in ε into families of orbits $\gamma_i(\varepsilon,E)$ periodic for (H,W).

Then, in this case, (6) becomes

$$\delta_{ij}(\varepsilon,E) = \frac{\oint_{\gamma_j(\varepsilon,E)} p \cdot dq}{\oint_{\gamma_i(\varepsilon,E)} p \cdot dq} = \varepsilon\text{-independent} = \frac{\ell_i}{\ell_j} \tag{7}$$

if ℓ_i is the length of the closed geodesic on which the i-th periodic orbit runs with energy E (in fact $\oint_{\gamma_j(0,E)} p \cdot dq = \sqrt{2E}\, \ell_j$, as is easy to compute).

In fact one can prove [2]:

Proposition 1: (Collet, Epstein, G.): If (H_0,W_0) is a geodesic flow on a surface of constant negative curvature then the condition that for every pair i,j of periodic geodesics the nontrivial Taylor coefficients of the expansion in ε of (6) vanish is necessary and sufficient for the existence of perturbation theory for a given family (H,W) of perturbations.

A comparably simple result in the case in which (H_0,W_0) is integrable is not easy to formulate: if we use action-angle coordinates so that

$$W_0 = V \times T^n \quad V \text{ open in } R^n$$
$$H_0(A,\varphi) = h_0(A) \qquad (A,\varphi) \in V \times T^n \tag{8}$$

the only simple criterion of existence of perturbation theory is, ("Birkhoff theorem"):

$$h_0(A) = \omega_0 \cdot A \text{ and } \omega_0 \text{ such that for some } C_0,\delta_0 > 0:$$
$$|\omega_0 \cdot v| > \frac{1}{C_0|v|^{\delta_0}}, \quad \forall v \in z^n,\ v \neq 0\ , \tag{9}$$

which physically corresponds to a "harmonic oscillator" of "nonresonant type" [3,1].

The nonintegrability theorem of Poincaré can be formulated, on the other hand, as a criterion for nonexistence of perturbation theory: if (H_0,W_0) has the form (8) and

$$\det \frac{\partial^2 h_0(A)}{\partial A \partial A} \neq 0 \tag{10}$$

then, "generically" on the perturbation f_ε (e.g. if $f_0(A,\varphi)$ has nonvanishing Fourier coefficients in φ), perturbation theory does not exist, ([4], and for an expository article [5]).

We come now to the question of convergence of perturbation theory. It is well known that, even if existent, perturbation series (5) need not be convergent. For instance in the above nonresonant harmonic oscillators one easily builds a counter example by choosing $n = 2$ and

$$f(A,\varphi) = (A_1 + g(\varphi_1)g(\varphi_2)) \tag{11}$$

which can be shown to be nonintegrable unless g is very special, and yields formal perturbation series which can be formally summed into functions which exhibit a simple but remarkable structure of dense singularities in ε, [6,5]. For a general class of results on nonconvergence, see [3].

There is, however, a (restrictive) convergence criterion working in the above nonresonant oscillators case [7]:

Proposition 2: (Rüssmann): if $F_0, F_1, F_2, \ldots$ in (5) have the form $F_j(A) = \psi_j(\omega_0, A)$, for some ψ_j, then the series in (5) converge in domains $W_\varepsilon, J_\varepsilon$ with boundaries close within $O(\varepsilon)$ to those of W_0, J_0.

As we shall see later it is remarkable that the above criterion is really relevant for some nontrivial applications. Unfortunately no satisfactory necessary and sufficient criterion is known for convergence, of perturbation theory of integrable systems even in the above nonresonant harmonic oscillators case.

The situation has to be contrasted with the following result [2]:

Proposition 3: (Collet, Epstein, G.): in the case of proposition 1 perturbation theory is convergent, whenever it exists, for ε small.

In other words the adiabatic invariants (6) associated with the pairs of isoenergetic periodic orbits form a complete set of invariants for the conjugacy problem posed by the Hamilton-Jacobi equation, i.e. by i), ii'), iii) above.

The above result extends a rigidity theorem of Guillemin-Kazhdan dealing with the geodesic flows on (variable) negative curvature compact surfaces perturbed by a (very) special perturbation, namely by a perturbation quadratic in the p-variables [8]. It has been extended to surfaces of variable negative curvature, with rather different methods, by De la Llave; Marco, Moryon [9].

A natural problem related with the above proposition is the following: suppose that $\varepsilon \equiv 1$, i.e. that (H,W) is not a family of Hamiltonians but, rather, it is a given fixed system. Assume that the periodic orbits of (H,W), (H_0,W_0) can be labeled by $i = 1,2,\ldots$ and by their energy as $\gamma_i(E)$. Suppose that

$$\frac{\oint_{\gamma_i(E)} p \cdot dq}{\oint_{\gamma_j(E)} p \cdot dq} = \frac{\ell_i}{\ell_j}, \qquad i,j = 1,2,\ldots \tag{12}$$

where $\gamma_i(E)$ is the i-th periodic orbit on the surface $H = E$, for (H,W), and ℓ_i is the length of the closed geodesic on which the corresponding periodic orbit for (H_0,W_0) runs. Then: are (H,W), (H_0,W_0) conjugated canonically in the sense i), ii), iii)?

This problem is very hard as it cannot be attacked by perturbation theory even if H is very close to H_0 (but is not, a priori, a member of a one parameter family of perturbations for which perturbation theory exists), and really new ideas seem necessary.

De la Llave has made progress in questions analogous to the latter in the case of conjugacy problems for maps [10].

It is amusing to remark that one can think of a lot of other necessary conditions for the solubility of the conjugacy problem i), ii), iii), for instance it is clear that special ε-dependence has to be fulfilled also by the ratios of the periods or of the Lyapunov exponents of corresponding isoenergetic periodic orbits [2]: however such invariants may not be sufficient to ensure the solubility of i), ii), iii): a counterexample is discussed in [2].

The problem i), ii)', iii) is only one example of a class of problems related to the Hamilton-Jacobi equation.

I wish to mention here a further extension of the equation (4), which is inspired by renormalization theory in the theory of fields [11].

Suppose (H,W) to be a family of perturbations of (H_0,W_0), as above, for which perturbation theory does not exist or, even if existent, is not convergent.

Then given a subset $C \subset \{\text{analytic functions on } W_0\}$ one can ask: can one find a family $N_\varepsilon \in C$, regular in ε too, such that $(H + N_\varepsilon, W)$ admits a convergent perturbation theory with respect to (H_0,W_0) in the sense of i), ii'), iii) and (5). And one can either prescribe F or leave it free; thus defining two related problems.

We say that the above problems are well posed in C if, up to an additive constant, there is one and only one formal power series solution for N_ε.

To show the interest of the above question let me present an interesting case in which it arises naturally (which, unfortunately, is also the only case known to me).

Suppose that (H_0,W_0) is a nonresonant harmonic oscillator and $C = \{\text{set of linearly A-dependent functions on } N_0\}$. Then the above problem becomes whether or not one can find a function $N_\varepsilon(A)$ analytic on W_0 and ε for ε small such that

$$\omega_0 \cdot A + \varepsilon F_\varepsilon(A,\varphi) + N_\varepsilon(A) \quad \text{on} \quad V \times T^n W_0 \tag{13}$$

is conjugate to $\omega_0 \cdot A$ in the sense i), ii), iii) (here F_ε is prescribed).

It can be shown, easily, that the problem is well-posed (in the above sense): however, the question of convergence is rather unclear.

It can be shown, in fact:

Proposition 4: (G., Chierchia): The above problem is well posed and yields convergent power series solutions for $N \in C$, as well as for C and F. Therefore N can be written

$$N_\varepsilon(A) = A \cdot a(\varepsilon) \tag{14}$$

and $a(\varepsilon)$ is analytic in ε near $\varepsilon = 0$.

The proof of the above proposition [6,5,12] is, in essence, a repetition of the proofs of the Rüssmann proposition 2 quoted above, and of the main proofs in the theory of Dinaburg, Sinai, [13], Rüssmann [14] for the (apparently unrelated) quasiperiodic one-dimensional Schrödinger equation.

Basically one determines $N_\varepsilon(A)$ by imposing that the criterion of proposition 2 for convergence of Birkhoff series is verified order by order in ε and, simultaneously, one proves the convergence of the algorithm [6,5,12].

The interest of the result is its mentioned relation with the Schrödinger equation [6].

Consider the problem

$$-q'' + (\varepsilon V(\omega_0 t + \varphi) - E)q = 0, \qquad t \in R \tag{15}$$

where V is periodic, analytic, on T^n, $\varphi \in T^n$ (e.g. $\varphi = 0$) and $\omega_0 \in R^n$ verifies the nonresonance condition (9): $q'' = \dfrac{d^2}{dt^2}$.

It is easy to check that (15) are the Hamilton equations for a Hamiltonian system described by canonically conjugated pairs of variables $(p,q),(B_1,\varphi_1),\dots,(B_n,\varphi_n)$ in R^2 for (P,φ) and in $R^n \times T^n$ for (B,φ). The Hamiltonian is, in fact:

$$\frac{P^2}{2} + \frac{1}{2}(E + \varepsilon V(\varphi))q^2 + \omega_0 \cdot B . \tag{16}$$

Replacing (P,q) with the action angle variables (A,ψ) for the oscillator $(P^2 + Eq^2)/2$ one describes (16) via the Hamiltonian

$$\sqrt{E}\,A + \omega_0 \cdot B + \frac{A}{\sqrt{E}}\varepsilon V(\varphi)\cos^2\psi . \tag{17}$$

It is easy to see that an invariant torus for the system (17) corresponds to a quasiperiodic solution of (15), i.e. the value E is in the continuum spectrum of the Schrödinger operator.

However, (17) needs not be integrable: nevertheless the above proposition 4, applied to the present case [6,5], yields the existence of a function $a(\varepsilon,E,\lambda)$, analytic in ε such that

$$\lambda A + \omega_0 \cdot B + \frac{A\varepsilon}{\sqrt{E}} V(\varphi)\cos^2\psi + Aa(\varepsilon,E,\lambda) \tag{18}$$

is integrable and conjugated to $\lambda A + \omega_0 \cdot B$ if $(\lambda,\omega_0) = \omega \in R^{n+1}$ also verifies a nonresonance condition like (9).

It follows that for such λ's the value E_ε such that

$$\sqrt{E_\varepsilon} = \lambda + a(\varepsilon,E_\varepsilon,\lambda) \tag{19}$$

is in the continuous spectrum of the Schrödinger operator. On this remark one can build the theory of the continuous spectrum of the quasiperiodic Schrödinger equation developed by Dinaburg, Sinai, Rüssmann (see Chierchia [12]).

It seems plausible that the type of questions like "is it possible to add to H a term δH of specified form so that $H + \delta H$ becomes conjugate to a given H_0" may arise in applications other than the above and therefore it would be nice to know more results in this direction.

REFERENCES

[1] G. Gallavotti, The Elements of Mechanics, Springer, N. Y., 1983.

[2] P. Collet, H. Epstein and G. Gallavotti, 'Perturbations of geodesic flows on surfaces of constant negative curvature and their mixing properties', Comm. Math. Phys. **95** (1984), 61-112.

[3] J. Moser, 'Stable and Random Motions', Princeton University Press, Annals of Math. Studies, vol. **77** (1973).

[4] H. Poincaré, Methodes Nonvelles de la Mécanique Celeste, Geuthier-Villar, 1897.

[5] G. Gallavotti, Les Houches Notes, 1984; 'Quasi integrable mechanical systems', In Phénomenes Critiques, Systèmes aleatoires, théorie de Jange, K. Ostervalder, R. Stora, and Les Houches, eds., session XLIII, part II, pagine 539-623, Reidel, 1986.

[6] G. Gallavotti, 'Classical Mechanics and Renormalization Group', In Regular and Chaotic Motions in Dynamic Systems, G. Velo and A. Wightman, eds., Erice, 1983, pp. 185-232, Plenum, 1985.

[7] H. Russmann, 'Uber normalform analytischer hamiltonscher differentialgleitchungen in der nähe einer gleichgewichtlösung', Math. Annalen **169** (1980), 55.

[8] V. Guillemin, A. 'Kazhdan', Topology **19** (1980), 291-299 and 301-312.

[9] R. De la Llave, J. Marco and R. Moryon, Ann. Math. (1985).

[10] R. De la Llave, private comm.

[11] G. Gallavotti, 'A criterion of integrability', Comm. Math. Phys. **87** (1982), 365-383.

[12] L. Chierchia, Thesis, published in Quaderni del C.N.R.-G.N.F.M. (1986).

[13] E. Dinaburg and Y. Sinai, 'The one dimensional Schrödinger equation with a quasi periodic potential', Funct. Analysis Appl. **9** (1975), 279.

[14] H. Rüssmann, 'On the one dimensional Schrödinger equation with a quasi periodic potential', Ann. N.Y. Acad. Sci. **90** (1979), 197.

SOME RESULTS ON PERIODIC SOLUTIONS OF MOUNTAIN PASS TYPE FOR HAMILTONIAN SYSTEMS

Mario Girardi and Michele Matzeu
Dipartimento Matematico
dell'Universita di Roma I
I - 00185 Roma, Italy

ABSTRACT. Some results on the existence of periodic solutions of Hamiltonian systems, having prescribed minimal period, are presented. They are found as critical points of Mountain Pass type of a suitable functional and some estimates on the energy behaviour are shown. The main techniques used are the dual action principle and the Morse index theory.

INTRODUCTION

In a recent paper [6] Ekeland and Hofer proved the existence of periodic solutions, having arbitrarily fixed minimal periods, for Hamiltonian systems related to a convex Hamiltonian function H of superquadratic growth. The arguments given in [6] are based on the use of the dual action principle by Clarke and Ekeland (see [2],[3]) and the Morse index theory (see also [5] in the framework of Hamiltonian systems).

By using a suitable modified version of the duality principle (see also [8]) and developing some ideas contained in [6], the authors of the present paper gave some generalizations of this result to the case that H is given by the sum of a superquadratic term plus a quadratic one (see [9],[10]).

More precisely let us consider the Hamiltonian system

$$J\dot{z} = H'(z) \tag{H}$$

where $J(x,y) = (y,-x)$ for any $(x,y) \in \mathbf{R}^N \times \mathbf{R}^N$, $H(z) = \frac{1}{2} \langle Qz,z\rangle + \hat{H}(z)$, Q is the $2N \times 2N$ matrix $\begin{pmatrix} Q_0 & 0 \\ 0 & Q_0 \end{pmatrix}$, with $Q_0 = \begin{pmatrix} \omega_1 & & 0 \\ & \ddots & \\ 0 & & \omega_N \end{pmatrix}$, $0 < \omega_1 \leq \dots \leq \omega_N$, and $\hat{H} \in C^2(\mathbf{R}^{2N};\mathbf{R})$ is strictly convex and has a superquadratic behaviour.

In [9], [10] the existence of a T-periodic solution of (H) for any fixed T in the interval $(0,2\pi/\omega_N)$ and, under the strong non-

P. H. Rabinowitz et al. (eds.), Periodic Solutions of Hamiltonian Systems and Related Topics, 161–168.

resonance assumption $\omega_i/\omega_j \notin \mathbf{Q}$ for $i \neq j$, the existence of a T-periodic solution for any T in a left neighbourhood of $2\pi/\omega_j$ are shown (here we mean T as minimal period).

In the present paper another result in this last direction is shown, in the sense that we get the same thesis under the weaker non-resonance assumption $\omega_i/\omega_j \notin \mathbf{N}$, in case that a suitable condition on $\hat{H}''$ is given.

We recall that this result can be obtained by bifurcation techniques too, but we point out that here we follow the variational approach, which enables us to get some estimates on the energy behaviour too and it seems to be useful for further developments.

THE RESULTS

Let the Hamiltonian system (H) be given. One can prove the following theorems.

Theorem 1 (see [8],[11]). *Let $\hat{H}$ satisfy the following assumptions:*

$$\hat{H}(z)|z|^{-2} \to 0 \quad \text{as} \quad |z| \to 0 \tag{1}$$

$$\exists r > 0,\ \beta > 2 : \langle \hat{H}'(z), z\rangle \geq \beta \hat{H}(z) \quad \text{if} \quad |z| > r\,. \tag{2}$$

Then for any $T > 0$ there exists a T-periodic solution z_T of (H). Moreover, if $\hat{H}$ satisfies the further assumptions:

$$\hat{H}(z) \geqq a_1|z|^\beta \quad \text{for all} \quad z \in \mathbf{R}^{2N},\ a_1 > 0 \tag{3}$$

$$\hat{H}(z) \leqq a_2|z|^\beta \quad \text{for all} \quad z \in \mathbf{R}^{2N},\ a_2 > 0 \tag{4}$$

$$\langle \hat{H}'(z), z\rangle \geq \beta \hat{H}(z) \quad \text{for all} \quad z \in \mathbf{R}^{2N} \tag{5}$$

then one has:

$$\lim_{T\to 0^+} \left(\inf_{t\in[0,T]} |z_T(t)|\right) = +\infty \tag{6}$$

$$\lim_{T\to (2k\pi/\omega_j)^-} \left(\sup_{t\in[0,T]} |z_T(t)|\right) = 0, \quad \text{for all} \quad k \in \mathbf{N},$$

$$j \in \{1,\dots,N\}\,. \tag{7}$$

Theorem 2 (see [9]). *Let $\hat{H}$ satisfy assumptions* (1), (2). *Then, for any $T < 2\pi/\omega_N$, the T-periodic solution z_T of* (H) *given by Theorem 1 has T as its minimal period.*

Theorem 3 (see [10]). *Let $\hat{H}$ satisfy assumptions* (1),...,(5). *Let, for a fixed* $j \in \{1,\dots,N\}$,

$$\omega_i/\omega_j \notin \mathbf{Q} \quad \text{for} \quad i \neq j \ . \tag{8}$$

Then there exists some $\varepsilon > 0$ *such that, for any*

$T \in (2\pi/\omega_j - \varepsilon, 2\pi/\omega_j)$, *the T-periodic solution* z_T *of* (H) *given by Theorem* 1 *has* T *as its minimal period.*

Theorem 4. *Let* $\hat{H}$ *satisfy assumptions* (1),...,(5) *and let*

$$\langle \hat{H}''(z)\zeta,\zeta\rangle \leq a_3|z|^{\beta-2} \quad \text{for all} \quad z \in \mathbf{R}^{2N}, \ \zeta \in \mathbf{R}^{2N},$$
$$|\zeta| = 1, \ a_3 > 0 \ . \tag{9}$$

Let, for a fixed $j \in \{1,\ldots,N\}$,

$$\omega_i/\omega_j \notin \mathbf{N} \quad \text{for} \quad i \neq j \ . \tag{10}$$

Then the same thesis of Theorem 3 *holds.*

PROOF OF THEOREM 4

Let us recall some of the main ideas used in the proofs of Theorems 1 and 2, which are necessary for the proof of Theorem 4.

First of all, for a fixed $T \neq 2k\pi/\omega_i$ ($k \in \mathbf{N}$, $i \in \{1,\ldots,N\}$), we define the functional F_T on the space $L^\alpha(0,T;\mathbf{R}^{2N})$(0) as

$$F_T(v) = \int_0^T \hat{G}(v) - \frac{1}{2}\int_0^T \langle \mathcal{L}_T^{-1}v,v\rangle$$

where $\hat{G}$ is the Legendre transform of $\hat{H}$, given by

$$\hat{G}(v) = \sup\{\hat{H}(z) - \langle z,v\rangle : z \in \mathbf{R}^{2N}\} \quad \text{for all} \quad v \in \mathbf{R}^{2N}$$

and $\mathcal{L}_T : H^{1,\alpha}(\mathbf{R}/T\mathbf{Z};\mathbf{R}^{2N}) \to L^\alpha(0,T;\mathbf{R}^{2N})$ is defined as

$$\mathcal{L}_T(v) = J\dot{v} - Qv \ .$$

Then it is easy to show that $F_T'(u_T) = 0$ if and only if $z_T = \mathcal{L}_T^{-1}u_T$ is a T-periodic solution of (H). At this point one can prove that F_T satisfies all assumptions of the well-known Mountain-Pass theorem by Ambrosetti and Rabinowitz (see [1]). More precisely one proves the existence of a critical point u_T of F_T such that

$$F_T(u_T) = \inf_{\gamma\in\Gamma} \{\sup F_T(v) : v \in \gamma\} \ ,$$

---

(0) $\alpha = \beta/(\beta - 1)$.

where Γ is the set of all continuous path γ, with $\gamma(0) = 0$ and $\gamma(1) = \bar{u}$, a suitable point such that $F_T(u_T) < 0$, belonging to the unit sphere of the eigenspace associated with the eigenvalue of $\mathcal{L}_T^{-1}$ given by

$$T/(2k\pi - T\omega_j) \ ,$$

where, denoting $\Omega = \{2h\pi/\omega_i : h \in \mathbf{N},\ i \in \{1,\dots,N\}\}$, k and j are taken, in $\mathbf{N}$ and $\{1,\dots,N\}$ respectively, in such a way that:

$$2k\pi/\omega_j = \min\{s \in \Omega : s > T\} \ .$$

Also, one can prove that, putting $z_T = \mathcal{L}_T^{-1}u_T$, the family $\{u_T\}_{T\notin\Omega}$ verifies conditions (6), (7), when assumptions (3), (4), (5) are satisfied.

Then we consider, for any fixed $T \notin \Omega$, the family of quadratic forms $\{Q_T^s\}_{s\in(0,T)\setminus\Omega}$ given by

$$Q_T^s(v) = \int_0^s \langle \hat{G}''(u_T)v,v\rangle - \int_0^s \langle \mathcal{L}_s^{-1}v,v\rangle \quad \text{for all}$$

$$v \in L^2(0,s;\mathbf{R}^{2N}) \ .$$

Here we want to point out only two facts about the connection between $\{Q_T^s\}_{s\in(0,T)\setminus\Omega}$ and the minimality of the period T.

A - Given $\bar{s} \in (0,T)\setminus\Omega$, $\bar{s}$ is another period of $z_T = \mathcal{L}_T^{-1}u_T$ only if $\ker Q_T^{\bar{s}}$ is different from $\{0\}$.

B - If $\bar{s} \in (0,2\pi/\omega_N)$ is such that $Q_T^{\bar{s}}$ is positive definite in $L^2(0,\bar{s};\mathbf{R}^{2N})$, then Q_T^s is positive definite in $L^2(0,s;\mathbf{R}^{2N})$ for any $s \in (0,\bar{s})$.

Now the proof of Theorem 4 can be obtained by using the following:

Lemma. *Let assumptions* (1),...,(5), (9) *be satisfied and let* $k \in \mathbf{N}$, $j \in \{1,\dots,N\}$. *Then there exists an evaluable constant number* $C > 0$, *with* $C = C(a_1,a_2,a_3,\beta,\omega_1,\omega_N)$ *such that, for any* T *in a small left neighbourhood of* $2k\pi/\omega_j$, *one has*

$$\min\{\|\hat{G}''(u_T(t))\| : t \in [0,T]\} \geq CT/(2k\pi - T\omega_j) \qquad (11)$$ [(0)]

where u_T *is the Mountain Pass critical point given by Theorem* 1.

[(0)] Here and in the following $\|\cdot\|$ denotes the Euclidean norm in $\mathbf{R}^{2N}$.

Proof of Lemma. By following some standard arguments of convex analysis (see [4],[7]), it is possible to show that (3), (4), (5), (9) imply the existence of some constants a_4,a_5,a_6,a_7,a_8, only depending on a_1,a_2,a_3,β, such that

$$|\hat{H}'(v)| \leq a_4|v|^{\beta-1} \quad \text{for all} \quad v \in \mathbf{R}^{2N},\ a_4 > 0 \tag{12}$$

$$|\hat{G}'(v)| \leq a_5|v|^{\alpha-1} \quad \text{for all} \quad v \in \mathbf{R}^{2N},\ a_5 > 0 \tag{13}$$

$$\hat{G}(v) \leq a_6|v|^{\alpha} \quad \text{for all} \quad v \in \mathbf{R}^{2N},\ a_6 > 0 \tag{14}$$

$$\hat{G}(v) \geq a_7|v|^{\alpha} \quad \text{for all} \quad v \in \mathbf{R}^{2N},\ a_7 > 0 \tag{15}$$

$$\langle\hat{G}'(v),v\rangle \leq \alpha\hat{G}(v) \quad \text{for all} \quad v \in \mathbf{R}^{2N} \tag{16}$$

$$\langle\hat{G}''(v)w,w\rangle \geq a_8|v|^{\alpha-2} \quad \text{for all} \quad v \in \mathbf{R}^{2N},$$
$$\text{for all} \quad w \in \mathbf{R}^{2N},\ |w| = 1,\ a_8 > 0\ . \tag{17}$$

Let now $z_T = \mathcal{L}_T^{-1}u_T$. By (7), there exists a left neighbourhood I of $2k\pi/\omega_j$ such that, for any $T \in I$, one has

$$|z_T(t)| \leq 1 \quad \text{for all} \quad t \in [0,T]\ , \tag{18}$$

so, by (12), (13), (18), one gets, for any $t \in [0,T]$,

$$\begin{aligned}
|u_T(t)| &= |\hat{H}'(z_T)| \leq a_4(a_1|z_T(t)|^{\beta} \\
&\quad + (\omega_1/2)|z_T(t)|^2)^{(\beta-1)/2}(2/\omega_1)^{(\beta-1)/2} \leq \\
&\leq a_4H(z_T(t))^{(\beta-1)/2}(2/\omega_1)^{(\beta-1)/2} = \\
&= a_4\,\frac{(2/\omega_1)^{(\beta-1)/2}}{T}\int_0^T (H(z_T(t))^{(\beta-1)/2} \leq \\
&\leq a_4(2/\omega_1)^{(\beta-1)/2}\,\frac{1}{T}\int_0^T (a_2|z_T(t)|^{\beta} + \\
&\quad + (\omega_N/2)|z_T(t)|^2)^{(\beta-1)/2} \leq \\
&\leq a_4(2/\omega_1)^{(\beta-1)/2}\,\frac{1}{T}\int_0^T (a_2+(\omega_N/2))^{(\beta-1)/2}|z_T(t)|^{\beta-1} = \\
&= a_4(2/\omega_1)^{(\beta-1)/2}(a_2+(\omega_N/2))^{(\beta-1/2)}\frac{1}{T}\int_0^T |\hat{G}'(u_T)|^{\beta-1} \leq \\
&\leq a_4(2/\omega_1)^{(\beta-1)/2}(a_2 + (\omega_N/2))^{(\beta-1)/2}\,\frac{(a_5)^{\beta-1}}{T}\int_0^T |u_T|
\end{aligned}$$

so, by Hölder's inequality,

$$\sup_{t\in[0,T]} |u_T(t)| \leq c_1 \frac{1}{T^{(\beta-1)/\beta}} \left(\int_0^T |u_T|^\alpha\right)^{1/\alpha} \tag{19}$$

where c_1 only depends on superquadraticity coefficients and on ω_1, ω_N.

Now let us give an estimate of the L^α-norm of u_T. By construction of u_T, by (14) and the obvious fact that, for T sufficiently near to $2k\pi/\omega_j$, the maximum eigenvalue of $\mathcal{L}_T^{-1}$ is given by $T/(2k\pi - T\omega_j)$, one can prove the estimate

$$F(u_T) \leqq \max\{a_6\mu^\alpha - \frac{T}{2}\frac{\mu^2}{2k\pi - T\omega_j} : \mu > 0\}$$

so one gets

$$F(u_T) \leqq \frac{2-\alpha}{\alpha}\left(\frac{\alpha a_6}{2}\right)^{2/2-\alpha}\left(\frac{(2k\pi - T\omega_j)^{\alpha/2-\alpha}}{T^{2(\alpha-1)/(2-\alpha)}}\right). \tag{20}$$

On the other side, by (15), (16),

$$F(u_T) = \int_0^T \hat{G}(u_T) - \frac{1}{2}\int_0^T \langle \hat{G}'(u_T), u_T\rangle \geq (1-\alpha/2)a_7 \int_0^T |u_T|^\alpha , \tag{21}$$

hence (20), (21) yield

$$\int_0^T |u_T|^\alpha \leqq c_2 \frac{(2k\pi - T\omega_j)^{\alpha/(2-\alpha)}}{T^{2(\alpha-1)/(2-\alpha)}} \tag{22}$$

where c_2 only depends on superquadraticity coefficients of $\hat{H}$.

Taking into account that $\alpha < 2$, (19) and (22) give

$$\sup_{t\in[0,T]} |u_T(t)|^{\alpha-2} \geq c_3 \frac{T^{2(\alpha-1)/\alpha}}{2k\pi - T\omega_j}\frac{1}{T^{(\beta-1)/\beta}}$$

$$= c_3 \frac{T}{2k\pi - T\omega_j} \tag{23}$$

where c_3 only depends on superquadraticity coefficients and ω_1, ω_N. By virtue of (17), (23) gives the thesis of Lemma. □

Now we are able to conclude the proof of Theorem 4. Let us proceed by steps.

Step 1. Choose $\varepsilon_0 > 0$ such that, for any $T \in (2\pi/\omega_j - \varepsilon_0, 2\pi/\omega_j)$, T/k doesn't belong to Ω. It is possible thanks to assumption (10).

Step 2. Choose $\varepsilon_1 \in (0, \varepsilon_0]$ such that, for any $T \in (2\pi/\omega_j - \varepsilon_1, 2\pi/\omega_j)$, the maximum eigenvalue of $\mathcal{L}_T^{-1}$ is given by $T/(2\pi - T\omega_j)$ and that (11) holds with $k = 1$.

Step 3. For any $k = 2,3,\ldots,[\omega_N/\omega_1],[\omega_N/\omega_1] + 1$(0), choose ε_k in $(0,\varepsilon_1]$ such that, if $T \in (2\pi/\omega_j - \varepsilon_k, 2\pi/\omega_j)$, then T/k is <u>not</u> a period of u_T. It is possible to find such a number ε_k as a consequence of the fact that, if T lies in a sufficiently small left neighbourhood of $2\pi/\omega_j$, then the maximum eigenvalue of $\mathcal{L}_{T/k}^{-1}$ is bounded by a constant number λ_k independent on T, so, by Steps 1 and 2,

$$Q_T^{T/k}(v) \geq \left(\frac{CT}{2\pi - T\omega_j} - \lambda_k\right) \int_0^{T/k} |v|^2 \quad \text{for all}$$

$$v \in L^2(0,T/k;\mathbf{R}^{2N}) \tag{24}$$

and the right member of (24) is positive for all $v \in L^2(0,T/k;\mathbf{R}^{2N})$, if T lies in a small left neighbourhood of $2\pi/\omega_j$. Using statement A, this guarantees that T/k is <u>not</u> a period of u_T.

Step 4. <u>Conclusion</u> - Let $\varepsilon = \min\{\varepsilon_k : k \in \{1,\ldots,[\omega_N/\omega_1] + 1\}\}$. Already we know, by Step 3, that T/k is not a period of u_T, for $k = 1,\ldots,1 + [\omega_N/\omega_1]$. On the other side, for any $k > [\omega_N/\omega_1] + 1$, T/k is less than $2\pi/\omega_N$ so, by statement B, Q_T^s is positive definite for all s in the interval $(0, 2\pi/1 + [\omega_N/\omega_1])$. Therefore, still by statement A, T/k is not a period of u_T also for $k > 1 + [\omega_N/\omega_1]$.

So u_T and z_T have minimal period T and the proof is concluded. □

REFERENCES

[1] A. Ambrosetti and P. Rabinowitz, 'Dual variational methods in critical point theory and applications', <u>J. Funct. Anal</u>. **14** (1979), 349-381.

[2] F. Clarke, 'Periodic solutions to Hamiltonian inclusions', <u>J. Diff. Eq</u>. **40** (1981), 1-6.

---

(0)Here $[x]$ denotes the integer part of x.

[3] F. Clarke and I. Ekeland, 'Hamiltonian trajectories having prescribed minimal period', Comm. Pure and Appl. Math. **33** (1980), 103-116.

[4] I. Ekeland, 'Periodic solutions to Hamiltonian equations and a theorem of P. Rabinowitz', J. Diff. Eq. **34** (1979), 523-534.

[5] I. Ekeland, 'Une théorie de Morse pour les systemes Hamiltoniens convexes', Ann. IHP "Analyse non linéaire" **1** (1984), 19-78.

[6] I. Ekeland and H. Hofer, 'Periodic solutions with prescribed period for convex autonomous Hamiltonian systems', preprint Ceremade n. 8421, Paris (1984).

[7] I. Ekeland and R. Temam, Analyse convexe et problémes variationnelles, Dunod-Gauthier Villars (1974).

[8] M. Girardi and M. Matzeu, 'Some results on solutions of minimal period to Hamiltonian systems', in Nonlinear Oscillations for Conservative Systems, Proceedings, Venice 9-12/1/1985.

[9] M. Girardi and M. Matzeu, 'Periodic solutions of convex autonomous systems with a quadratic growth at the origin and superquadratic at infinity', to appear in Ann. Mat. Pura e Applicata.

[10] M. Girardi and M. Matzeu, 'Solutions of minimal period for Hamiltonian systems with a quadratic growth at the origin and superquadratic at infinity', to appear in Rend. Ist. Mat. Univ. Trieste.

[11] P. Rabinowitz, 'On subharmonic solutions of Hamiltonian systems', Comm. Pure and Appl. Math. **33** (1980), 609-633.

REMARKS ON PERIODIC SOLUTIONS FOR SOME DYNAMICAL SYSTEMS WITH SINGULARITIES*

C. Greco
Dipartimento di Matematica
Università di Bari
70125 Bari, Italy

ABSTRACT. This paper contains some results concerning periodic solutions of dynamical systems with a singular potential. Such results are obtained by variational methods. In particular, dynamical systems constrained in a potential well are examined, as well as the case of the singular potentials.

INTRODUCTION

The aim of this paper is to state some results concerning periodic solutions, with prescribed period, for dynamical systems with singularities. More precisely, we search to T-periodic solutions (for some fixed $T > 0$) of the system:

$$\ddot{u} = -V'(u) , \tag{1}$$

where $V \in C^1(\Omega, \mathbf{R})$, and Ω is an open subset of $\mathbf{R}^N (N \geq 2)$; we suppose that $V(x) \to +\infty$ (or $V(x) \to -\infty$) as $x \to \partial\Omega$, namely V has singularities at $\partial\Omega$. As well known, the critical points, on the Sobolev space of the T-periodic functions $H = H^{1,2}(S_T, \mathbf{R}^N)$, of the action-functional:

$$J(u) = \frac{1}{2} \int_0^T |\dot{u}(t)|^2 dt - \int_0^T V(u(t))dt , \tag{2}$$

are T-periodic solutions of (1). Notice that, in our case, J is defined on the open subset $\Lambda = \{u \in H | Im(u) \subset \Omega\}$ of H.

*Work supported by G.N.A.F.A. of C.N.R. and by Ministero P.I. (40%-60%).

P. H. Rabinowitz et al. (eds.), Periodic Solutions of Hamiltonian Systems and Related Topics, 169–173.

The main difficulties arising from the study of J are: a) lack of compactness; b) lack of a suitable geometric situation as in the usual "linking" theorems; c) presence of "collision orbits". We shall give some examples in which such phenomena occur.

1°) THE POTENTIAL WELL

Let Ω be bounded, with $\partial\Omega \in C^2$, and suppose that:

$$\lim_{x \to \partial\Omega} V(x) = +\infty ; \tag{3}$$

$$\lim_{x \to \partial\Omega} \frac{-V'(x)d'(x)}{V(x)} = +\infty, \quad \text{where} \quad d(x) = \mathrm{dist}(x, \partial\Omega) . \tag{4}$$

In this case, J does not seem to satisfy the well-known (PS) compactness condition of Palais-Smale. To overcome this difficulty, Benci [4] showed that there exists a suitable positive function $\rho \in C^1(\Lambda, \mathbf{R})$, with $\lim_{u \to \partial\Lambda} \rho(u) = +\infty$, such that, for every sequence $(u_n)_n \subset \Lambda$:

(i) if $(\rho(u_n))_n$, $(J(u_n))_n$ are bounded, and $J'(u_n) \to 0$, then there exists a subsequence of $(u_n)_n$ which converge to some $u \in \Lambda$;

(ii) if $\rho(u_n) \to +\infty$ and $J(u_n) \to c \in \mathbf{R}$, then there exists $\bar{\nu} > 0$ such that $\|J'(u_n)\| \geq \nu \|\rho'(u_n)\|$ for n large enough.

By using such a condition (which we can consider a sort of (PS) condition "with respect to the weight-function ρ"), it is possible to get a deformation lemma for J, and then, to apply standard techniques of Calculus of Variations in order to obtain infinitely many critical points of J. We refer to [11] for the case in which Ω is an external domain (namely $\mathbf{R}^N - \Omega$ is bounded) and V grows super (or sub)-quadratically as $|x| \to +\infty$. Finally, under suitable convexity assumptions, we can replace the study of J with the study of a "dual" functional, which (contrarily of J) is defined throughout a Banach space and satisfies (PS)-condition (see [2]). Then informations about the minimal period of solutions of (1) are obtained.

2°) STRONG FORCES

Replace conditions (3), (4) with the following assumptions:

$$\lim_{x \to \partial\Omega} V(x) = -\infty ; \tag{5}$$

strong force condition: there exists $U \in C^1(\Omega,\mathbf{R})$

such that $U(x) \to -\infty$ as $x \to \partial\Omega$, and $-V(x) \geq |U'(x)|^2$ (6)

on a suitable neighborhood of $\partial\Omega$.

An example of function which satisfy such assumptions is $V(x) = -1/|x|^\alpha$, where $\alpha \geq 2$ (notice that, if $1 \leq \alpha < 2$, (6), is not satisfied). This situation was first studied by [8], [10].

The most important consequence of (6) is that $J(u_n) \to +\infty$ as $u_n \to \partial\Omega$, provided $(u_n)_n$ is bounded; we shall use this fact later. As in [8], we suppose that V is bounded from above, and define the following subset Λ^* of Λ:

$u \in \Lambda^* \iff$ for every $c > 0$, there exists $K_c \subset \mathbf{R}^N$,

K_c compact, such that, if $v \in \Lambda$, v is homotopic to u

in Ω, and arc length$(v) \leq c$, then $\mathrm{Im}(v) \subset K_c$.

Then, if $\Lambda^* \neq \emptyset$, J is coercive on every component of Λ^* (there is not lack of compactness, see [8]); a minimizing sequence for J must converge to some $u_0 \in \bar{\Lambda}$. Since "$J(u) = +\infty$" on $\partial\Lambda$, we have $u_0 \in \Lambda$, and u_0 is a critical point of J.

This result by Gordon is generalized in [5] to Lagrangian systems with a potential V which grows subquadratically at infinity (see also [6]).

Notice that $\Lambda^* \neq \emptyset$ is a rather restrictive condition: if, for instance, $\Omega = \mathbf{R}^N - \{0\}$, it satisfied only for $N = 2$. In [12] we deal with the case $\Omega = \mathbf{R}^N - \{0\}$ and $N \geq 3$. In this case, the main problem is to find geometrical obstructions to deformations.

In [12] we obtain, by a direct study of the pseudogradient vector field of J, infinitely many (not geometrically) distinct solutions of (1). More recently, the same problem is studied in [3] (also in the not autonomous case), and infinitely many distinct solutions of (1) are obtained (see also [1] and the contribution of Coti Zelati in these Proceedings). Other results, in the case of "effective-like" potentials (as $V(x) = -1/|x| + 1/|x|^2$) are given in [7]; in this case, $-V$ is the function which satisfies (5) and (6).

3°) WEAK FORCES

The assumption (6) does not hold in many interesting physical problems. In the Kepler problem, for instance, the equation of motion $\ddot{u} = -u/|u|^3$, corresponds to the Newtonian potential $V(x) = -1/|x|$; such a function does not satisfy (6).

So, in many physical problems, additional arguments are required in order to avoid solutions which cross the singularities ("collision orbits"). More precisely, if we set: $c_0 = \inf_{u \in \partial\Lambda} J(u)$, we must search for the critical points of J on $\{J < c_0\}$. In the particular case of the Kepler problem, for instance, J achieves its minimum value on $\{J < c_0\}$; this minimum corresponds to elliptical T-periodic solutions of the equation of motion, with minimum period T (see [9], [10]). The general case is more complicated: we refer to the contribution of Giannoni in these Proceedings (see also the references of [3]).

Finally, we observe that completely different problems arise from the study of boundary value problems for semilinear elliptic equation: $Lu = g(x,u)$ $(x \in \Omega)$, where the nonlinear part $g : \bar{\Omega} \times I \to \mathbf{R}$ is defined only on some interval I of **R**. Some results on this problem are given in [13], by the upper and lower solutions method.

REFERENCES

[1] A. Ambrosetti, 'Sistemi dinamici con potenziali singolari', Proc. of Recent Advances in Hamiltonian Systems, L'Aquila, June 1986.

[2] A. Ambrosetti and V. Coti Zelati, 'Solutions with minimal period for Hamiltonian systems in a potential well', to appear in Ann. I.H.P. "Analyse non lineare".

[3] A. Ambrosetti and V. Coti Zelati, 'Critical points with lack of compactness and singular dynamical systems', preprint Scuola Normale Superiore, Nov. 1986.

[4] V. Benci, 'Normal modes of a Lagrangian system constrained in a potential well', Ann. I.H.P. "Analyse nonlineare" **1** (1984), 379-400.

[5] A. Capozzi, C. Greco and A. Salvatore, 'Lagrangian systems in presence of singularities', preprint.

[6] A. Capozzi and A. Salvatore, 'Periodic solutions of Hamiltonian systems: the case of the singular potential', Proc. NATO-ASI, Singh, ed. (1986), 415-425.

[7] V. Coti Zelati, 'Dynamical systems with effective-like potentials', to appear in Nonlinear Anal. T.M.A.

[8] W. B. Gordon, 'Conservative dynamical systems involving strong forces', Trans. A.M.S. **204** (1975), 113-135.

[9] W. B. Gordon, 'A minimizing property of Keplerian orbits', Amer. J. Math. **99** (1977), 961-971.

[10] W. B. Gordon, 'Very strong forces', in Selected Studies: Physics-Astrophysics, Math., Hist. of Sci., North-Holland, Amsterdam, New York (1982), 79-91.

[11] C. Greco, 'Periodic solutions to second order Hamiltonian systems in an unbounded potential well', to appear in Proc. Royal Soc. Edinburg.

[12] C. Greco, 'Periodic solutions of a class of singular Hamiltonian systems', to appear in Nonlinear Anal. T.M.A.

[13] R. K. Nagle, 'Equations with nonlinearities defined only on subsets', in Lect. Note in Pure Appl. Math. **90**, Dekker, New York (1984), 389-394.

CAUCHY-RIEMANN EQUATION IN LAGRANGE INTERSECTION THEORY

M. Gromov
IHES
Bures-sur-Yvette 91440
France

Double points of an immersed Lagrange manifold L in a symplectic manifold (V,ω) generalize fixed points of symplectic diffeomorphisms. Recall that "Lagrange" signifies $\dim V = 2 \dim L$ and $\omega|L = 0$, where ω denotes the symplectic 2-form on V. The simplest case of the intersection problem (which does not come from fixed points) concerns Lagrange manifolds in

$$(V,\omega) = (\mathbb{R}^{2n}, \omega = \sum_{i=1}^{n} dx_i \wedge dy_i) .$$

(A) Theorem. Every compact without boundary Lagrange manifold L immersed into $\mathbb{R}^{2n}$ has a double point, provided the first homology group $H_1(L;\mathbb{R})$ vanishes.

This follows from

(B) Analytic theorem. Let $L \subset \mathbb{R}^{2n} = \mathbb{C}^n$ be a compact without boundary immersed Lagrange manifold with transversal selfintersection points (if there are any). Then there exists a non-constant holomorphic map $f : D \to \mathbb{C}^n$ of the unit disk $D \subset \mathbb{C}^1$, such that the boundary circle $S^1 = \partial D$ is sent by f into $L \subset \mathbb{C}^n$.

Let us explain how (B) ==> (A). Suppose, contrary to (A), that $L \subset \mathbb{C}^n = \mathbb{R}^{2n}$ has no double (self-intersection) point and look at the map $f : (D,S^1) \to (\mathbb{R}^{2n},L)$ insured by (B). Since f is holomorphic, the induced 2-form $f^*(\omega)$ on D is non-negative at every point in D; since f is non-constant, this form is not identically zero. Hence,

$$\int_D f^*(\omega) \neq 0 . \qquad (*)$$

P. H. Rabinowitz et al. (eds.), Periodic Solutions of Hamiltonian Systems and Related Topics, 175–176.

Next, we consider the (action) 1-form $\alpha = \sum_{i=1}^{n} x_i dy_i$ whose differential $d\alpha = \omega$. Since L is Lagrange, $\alpha|L = 0$. In other words, α restricts to a closed form on L. Now, we integrate α along the curve $C = f(S^1) \subset L$ and apply Stokes' formula,

$$\int_C \alpha = \int_D f^*(\omega) \ .$$

By (*) this integral does not vanish and so C represents a non-zero homology class in L. Q.E.D.

(C) Idea of the proof of (B). Let us look at the $\bar{\partial}$-equation $\bar{\partial} f = g$ for C^∞-maps $f : D \to \mathbf{C}^n$ satisfying the boundary condition $f(S^1) \subset L$, where $g : D \to \mathbf{C}^n$ is a given C^∞-map. If $g = 0$, the solutions of the equations $\bar{\partial} f = g$ are holomorphic maps and our boundary problem is solved by a constant map $f_0 : D \to \ell_0 \in L \subset \mathbf{C}^n$. Now let g be a constant map, $D \to x_1 \in \mathbf{C}^n$. Then $\bar{\partial} f = g_1$ implies that f is a harmonic map. If x_1 is taken far away from L, then the boundary problem $f(S^1) \subset L$ is unsolvable for the equation $\bar{\partial} f = g_1 \equiv x_1$ by the maximum principle for harmonic maps. It follows that there exists $t_0 \in [0,1]$, such that our boundary value problem is solvable for the equation $\bar{\partial} f = tg_1$ for all $t \in [0,t_0)$ but is not solvable for some $t > t_0$.

If L is Lagrange, one can show (and this is the crux of the matter) that solutions f_t of $\bar{\partial} f = tg_1$ blow up as $t \to t_0$ and give birth to a non-constant solution of the (homogeneous) equation $\bar{\partial} f = 0$.

The pattern of the blow-up is intimately related to the double-points of L. This relation seems most transparent for Lagrange intersections coming from symplectic fixed points and according to an announcement by A. Floer, the blowup pattern carries enough topological information to prove a homological version of Arnold's conjecture on fixed points of symplectic diffeomorphism.

REFERENCES

D. Bennequin, 'Problèmes elliptiques, surfaces de Riemann et structures symplectiques', Sem. Bourbaki, Fevr., Soc. Math., France (1986), 1-23.

M. Gromov, 'Pseudo-holomorphic curves in symplectic manifolds', Inv. Math. **82** (1985), 307-347.

M. Gromov, 'Partial differential relation', Springer-Verlag (1986).

M. Gromov, 'Soft and hard symplectic geometry', Proc. ICM-1986, Berkeley, to appear.

MODULUS OF CONTINUITY FOR PEIERLS'S BARRIER

John N. Mather*
Department of Mathematics
Princeton University
Fine Hall - Washington Road
Princeton, New Jersey 08544

ABSTRACT. For an exact area preserving monotone twist diffeomorphism with a uniform lower bound β for the amount of twisting, we will show that Peierls's barrier satisfies

$$|P_{p/q}(\xi) - P_\omega(\xi)| \le C(q^{-1} + |\omega q - p|) ,$$

where $C = (1200)\cot\beta$.

§1. INTRODUCTION

This paper continues our series of papers on area preserving diffeomorphisms of the cylinder ([6] and references therein). In this paper, we will prove a modulus of continuity for Peierls's barrier $P_\omega(\xi)$. This modulus of continuity is the first of the results announced in [7]. Other results announced there will be proved in later papers in this series.

For the case of a monotone twist map, we announced a modulus of continuity similar to that of this paper in Herman's seminar in Palaiseau in June 1985. However, since then, we have made two improvements. First, the constant in the basic estimate is $(1200)\theta = (1200)\cot\beta$ whereas in our previous version it depended on the partial derivatives of first and second order of the generating function h. Second, we can now replace the generating function with a "variational principal" h satisfying certain regularity conditions. The difficult part of this generalization was to find the right regularity conditions, general enough to apply to the interesting cases (compositions of twist maps, geodesics on the two torus), but restrictive enough for the proof to still work.

*Partially supported by National Science Foundation Grant No. DMS85-04984.

P. H. Rabinowitz et al. (eds.), Periodic Solutions of Hamiltonian Systems and Related Topics, 177–202.

The importance of this sort of generalization has been shown by Bangert [3], and in this paper, we will assume his regularity conditions (H_1)-(H_4) in order to apply his results. In addition, we will impose other regularity conditions (H_5) and $(H_{6\theta})$ in order for our main result (§7) to be true. We will show (§5) that all the regularity conditions are satisfied for conjunctions of generating functions of monotone twist mappings.

I would like to thank Charles Fefferman, Sergio Fenley, Michel Herman, Jürgen Moser, David Rana, and David Stuart for listening to oral presentations of various preliminary versions of this paper. Without their interest, I would not have had enough enthusiasm to write this paper. I would like to thank Mrs. Ross for the care she took typing this paper.

§2. MONOTONE TWIST DIFFEOMORPHISMS

The main part of this paper is the study of certain properties of area preserving, monotone twist diffeomorphisms of an infinite cylinder $(\mathbf{R}/\mathbf{Z}) \times \mathbf{R}$. We will use the definitions of [6]. For the convenience of the reader, we repeat those definitions here.

Let $\bar{f}$ be a mapping of $(\mathbf{R}/\mathbf{Z}) \times \mathbf{R}$ into itself. We will say that $\bar{f} \in J$ if the following conditions are satisfied: First, we require that $\bar{f}$ be a C^1 diffeomorphism which maps points near the top end of the cylinder to points near the top end, and likewise for points near the bottom end. Second, setting $\bar{f}(\theta,y) = (\theta',y')$, we require that the form $y'd\theta' - yd\theta$ on $(\mathbf{R}/\mathbf{Z}) \times \mathbf{R}$ be exact. Third, we require that $\bar{f}$ satisfy a positive monotone twist condition, i.e. $\partial\theta'/\partial y > 0$ everywhere. Fourth, we require that $\bar{f}$ twist the cylinder infinitely at either end. To express this condition, we use a lift f of $\bar{f}$ to the universal cover $\mathbf{R}^2$ of $(\mathbf{R}/\mathbf{Z}) \times \mathbf{R}$. It means that for fixed x, we have $x' \to +\infty$ as $y \to +\infty$ and $x' \to -\infty$ as $y \to -\infty$, where we set $f(x,y) = (x',y')$.

The positive monotone twist condition can be expressed in another way which is sometimes convenient. Consider a point $P \in (\mathbf{R}/\mathbf{Z}) \times \mathbf{R}$ and let $v_P = (0,1)$ denote the vertical vector at P. Let $\beta_{\bar{f}}(P)$ denote the angle which $d\bar{f}_P \cdot v_P$ makes with $v_{\bar{f}P}$, counted in the clockwise direction from $v_{\bar{f}P}$. The positive monotone twist condition amounts to the assertion that $0 < \beta_{\bar{f}}(P) < \pi$ everywhere. Likewise, let $\bar{\beta}_{\bar{f}}(P)$ denote the angle which $d\bar{f}_P^{-1} \cdot v_P$ makes with $v_{\bar{f}P}$. The positive monotone twist condition can also be formulated as the assertion that $0 < -\bar{\beta}_{\bar{f}}(P) < \pi$ everywhere.

For a number $\beta > 0$, we define $J_\beta = \{\bar{f} \in J : \beta_{\bar{f}}(P) \geq \beta$ and $-\bar{\beta}_{\bar{f}}(P) \geq \beta$, for all $P \in (\mathbf{R}/\mathbf{Z}) \times \mathbf{R}\}$. In fact, most of the machinery we develop in this paper will be for elements of J_β. Although $\bigcup_{\beta>0} J_\beta$ is a proper subset of J, most of our results will generalize to J without any difficulty. This is because our results concern what happens in a compact region of $(\mathbf{R}/\mathbf{Z}) \times \mathbf{R}$ and because for $\bar{f} \in J$ and

K a compact region of $(\mathbf{R}/\mathbf{Z}) \times \mathbf{R}$, there exists $\beta > 0$ and $\bar{g} \in J_\beta$ such that $\bar{f}|K = \bar{g}|K$.

Throughout this paper, we denote the translation $(x,y) \to (x + 1,y)$ by T. If $\bar{f} \in J$ and f is a lift of $\bar{f}$ to $\mathbf{R}^2$, we have $fT = Tf$. Moreover, there exists a C^2 function $h : \mathbf{R}^2 \to \mathbf{R}$, called a generating function for f, such that for $(x,y,x',y') \in \mathbf{R}^4$, we have $f(x,y) = (x',y')$ if and only if $y = -\partial_1 h(x,x')$ and $y' = \partial_2 h(x,x')$, where $\partial_1 h$ and $\partial_2 h$ denote the first partial derivatives of h with respect to x and x', resp. (cf. [6,§2]).

§3. BANGERT'S SET-UP FOR AUBRY-LE DAERON THEORY

The methods of this paper rely heavily on the theory of minimal configurations, as expounded by Bangert [3]. This theory was developed by Aubry and Le Daeron [1] and, in part, by the author [8] (independently). Bangert's article generalizes this theory in a way which will be useful to us, and gives a beautiful comprehensive exposition of the theory of minimal configurations.

Here, we explain the results expounded in [3] which will be useful for us. For proofs, we refer to [3].

By a configuration, we will mean an element $x = (x_i)_{i \in \mathbf{Z}}$ of the set $\mathbf{R}^{\mathbf{Z}}$ of bi-infinite sequences of real numbers. Given a function $h : \mathbf{R}^2 \to \mathbf{R}$, we extend h to arbitrary finite segments $(x_j,\ldots,x_k)$, $j < k$ of configurations by setting

$$h(x_j,\ldots,x_k) = \sum_{i=j}^{k-1} h(x_i,x_{i+1}) \ .$$

Following Bangert, we say that the segment $(x_j,\ldots,x_k)$ is minimal with respect to h if

$$h(x_j,\ldots,x_k) \leq h(x_j^*,\ldots,x_k^*)$$

for all $(x_j^*,\ldots,x_k^*)$ with $x_j = x_j^*$ and $x_k = x_k^*$. (Such segments were said to be of minimal energy in Aubry and Le Daeron [1] and Mather [6].) We say that a configuration is minimal (with respect to h) if every finite segment of x is minimal (with respect to h).

Bangert [3] generalized the main results of Aubry-Le Daeron theory, proving them under the hypothesis that h is continuous and satisfies the following properties:

$$h(x,x') = h(x + 1,x' + 1), \quad \text{for all } x,x' \in \mathbf{R} \ . \tag{H_1}$$

$$\lim_{|\xi|\to\infty} h(x,x + \xi) = \infty, \quad \text{uniformly in } x \ . \tag{H_2}$$

$$h(x,x') + h(\xi,\xi') < h(x,\xi') + h(\xi,x'),$$
$$\text{if } x < \xi,\ x' < \xi . \tag{H_3}$$

$$\text{If } (\bar{x},x,x'),(\bar{\xi},x,\xi') \text{ are both minimal segments and are distinct, then } (\bar{x} - \bar{\xi})(x' - \xi') < 0 . \tag{H_4}$$

Example. If $\bar{f} \in J$, f is a lift of $\bar{f}$ to the universal cover $\mathbf{R}^2$, and h is a generating function for f, then h satisfies (H_1)-(H_4).

In the verification of these conditions, we use the notation $f(x,y) = (x',y')$. In view of the monotone twist condition, we may take either (x,x') or (x,y) as the independent variables.

The defining condition for the generating function is $dh = y'dx' - ydx$, where we take x and x' as the independent variables. The assumption in the definition of J that $y'd\theta' - yd\theta$ is exact on $(\mathbf{R}/\mathbf{Z}) \times \mathbf{R}$ implies (H_1). Here, $\theta = x(\text{mod. } 1)$ and $\theta' = x'(\text{mod. } 1)$.

Since $\bar{f}$ maps each end of the infinite cylinder to itself, there exists $C > 0$ such that $y > C$ implies $y' > 1$ and $y < -C$ implies $y' < -1$. The positive monotone twist hypothesis together with the hypothesis that $\bar{f}$ twists the cylinder infinitely at each end implies that for each fixed x the mapping $y \to x'$ is an orientation preserving diffeomorphism of $\mathbf{R}$ onto itself. Consequently, so is its inverse. Therefore, for each fixed x, there exists ξ_x such that $x' - x > \xi_x$ implies $y > C$ and $x' - x < -\xi_x$ implies $y < -C$. The smallest such ξ_x depends continuously and periodically (of period 1) on x. Consequently, $\xi_0 = \max \xi_x$ exists. For $x' - x > \xi_0$, we have $y > C$ and $y' > 1$ and for $x' - x < -\xi_0$, we have $y < -C$ and $y' < -1$. By the definition of generating function, we have $y' = \partial_2 h(x,x')$. We have shown that if $x' - x > \xi_0$ then $\partial_2 h(x,x') = y' > 1$ and if $x' - x < -\xi_0$ then $\partial_2 h(x,x') = y' < -1$. Condition (H_2) follows immediately.

The argument above shows for fixed x, we have that y is an increasing function of x' with positive derivative. Since $y = -\partial_1 h(x,x')$ (by the definition of generating function), we therefore get the fundamental inequality

$$\partial_{12} h < 0 .$$

Condition (H_3) follows from this inequality by taking the double integral over the rectangle with vertices (x,x'), (x,ξ'), (ξ,ξ'), and (ξ,x'). Note that h is C^2 because f is C^1.

In Bangert's set-up, h is not necessarily differentiable. We will give in §5 examples where h is not differentiable. Bangert gives other such examples in his article [3]. However, in the example we are discussing now, h is differentiable. In such a case, we will say that a segment $(x_j,\ldots,x_k)$ of a configuration (or a configuration) is <u>stationary</u> if

$$\partial_2 h(x_{i-1},x_i) + \partial_1 h(x_i,x_{i+1}) = 0$$

for $j < i < k$. Obviously, minimal configurations are stationary.

Condition (H_4) may be verified (in our example) by showing that

$$\partial_2 h(\bar{x},x) + \partial_1 h(x,x') \neq \partial_2 h(\bar{\xi},x) + \partial_1 h(x,\xi')$$

if $(\bar{x},x,x')$ and $(\bar{\xi},x,\xi')$ are distinct and $(\bar{x} - \bar{\xi})(x' - \xi') > 0$. For example, if $\bar{x} < \bar{\xi}$ and $x' < \xi'$, then the left side is greater than the right side by $-\partial_{12}h > 0$. The verification of this inequality in other cases is similar.

In this example, stationary configurations $(\ldots,x_i,\ldots)$ correspond to orbits $(\ldots,(x_i,y_i),\ldots)$ under the correspondence $y_i = -\partial_1 h(x_i,x_{i+1})$. Thus, minimal configurations of h correspond to a class of orbits of f, called minimal orbits.

Aubry and Le Daeron [1] give a fairly complete structure theory of minimal configurations for h or, equivalently, minimal orbits of f. Bangert [3] observed that these results generalize to h satisfying (H_1)-(H_4). In the remainder of this section, we quote some of these results. For proofs, we refer to Bangert [3].

The results which we quote in the remainder of this section are valid under the sole hypothesis that the "variational principal" h is a continuous real valued function on $\mathbf{R}^2$ and satisfies (H_1)-(H_4).

The following pictorial representation, due to Aubry, of a configuration $x = \{x_i\}_{i \in \mathbf{Z}} \in \mathbf{R}^{\mathbf{Z}}$ is useful. One joins (i,x_i) and $(i + 1,x_{i+1})$ by a line segment in $\mathbf{R}^2$. The union of all such line segments is a piecewise linear curve in $\mathbf{R}^2$, which we will call the Aubry graph of x. We will say that two configurations cross if their Aubry graphs cross and count the number of crossings as the number of crossings of their Aubry graphs.

If x and x^* are two configurations, we will say that $x < x^*$ if $x_i < x_i^*$, for all i. We will say that x and x^* are comparable if $x < x^*$, $x = x^*$, or $x > x^*$. From (H_4), it follows that any two minimal configurations x and x^* either cross or are comparable.

It is easy to prove that two minimal configurations (or minimal segments of configurations) cross at most once, and if they meet at some $i \in \mathbf{Z}$, they cross there. For the proof see [3, Lemma 3.1].

If x is a minimal configuration, then $\rho(x) = \lim_{|i|\to\infty} x_i/i$ exists [3, Corollary 3.16]. The number $\rho(x)$ is called the rotation number of x.

We let $M = M_h \subset \mathbf{R}^{\mathbf{Z}}$ denote the set of minimal configurations (with respect to h). We provide $\mathbf{R}^{\mathbf{Z}}$ with the product topology and M with the induced topology. By the continuity of h, we have that M is a closed subset of $\mathbf{R}^{\mathbf{Z}}$.

The rotation number is a function $\rho : M \to \mathbf{R}$. It is continuous [3, Corollary 3.16] and surjective [3, Theorem 3.17]. Moreover,

$(pr_0, \rho) : M \to \mathbf{R}^2$ is proper (i.e. the inverse image of every compact set is compact), where $pr_0(x) = x_0$ [3, 3.18].

If $x \in M$ and $\rho(x) = p/q \in Q$ where $q > 0$ and p,q are relatively prime integers, then one of three possibilities holds:

a) $x_{i+q} > x_i + p$, for all i,

b) $x_{i+q} = x_i + p$, for all i,

c) $x_{i+q} < x_i + p$, for all i.

This follows easily from [3, Theorem 5.3 and Lemma 3.9]. For, in the notation of [3, Theorem 5.3] (with $\alpha = p/q$) if $x \in M_\alpha^{per}$, then b) holds, by the definition of M_α^{per} in [3]. If $x \in M_\alpha^+$ or $x \in M_\alpha^-$ and $x_i^* = x_{i+q} - p$, then x and x^* are asymptotic, again by the definitions of these sets in [3]. But then x and x^* do not cross by [3, Lemma 3.9] and we obtain the desired result.

This leads us to introduce the rotation symbol $\tilde{\rho}(x)$ of a minimal configuration x. If the rotation number $\rho(x)$ is an irrational number, we set $\tilde{\rho}(x) = \rho(x)$. On the other hand, if $\rho(x)$ is a rational number p/q, then we set $\tilde{\rho}(x) = p/q+$ in case $x_{i+q} > x_i + p$, we set $\tilde{\rho}(x) = p/q$ in case $x_{i+q} = x_i + p$, and we set $\tilde{\rho}(x) = p/q-$ in case $x_{i+q} < x_i + p$.

We also define a symbol space $\mathbf{S} = (\mathbf{R}\backslash\mathbf{Q}) \cup (\mathbf{Q}+) \cup \mathbf{Q} \cup (\mathbf{Q}-)$, where $\mathbf{Q}+$ denotes the set of all symbols $p/q+$ and $\mathbf{Q}-$ is defined similarly. The symbol space has an obvious order so that $p/q- < p/q < p/q+$ and the map $\mathbf{S} \to \mathbf{R}$ which forgets + and − is weakly order preserving. We provide $\mathbf{S}$ with the order topology, i.e. the set of intervals $(\alpha,\beta) = \{x : \alpha < x < \beta\}$ is a basis for this topology. Note that every rational number p/q is an isolated point in this topology, since it is the unique point in the interval $(p/q-, p/q+)$.

It will be convenient to define a subset $M_\omega = M_{\omega,h}$ of M for each $\omega \in \mathbf{S}$. We give somewhat complicated definitions in order that the basic results should have a simple form. If ω is an irrational number, we let $M_\omega = M_{\omega,h}$ denote the set of $x \in M$ with $\rho(x) = \omega$. We let $M_{p/q} = M_{p/q,h}$ denote the set of $x \in M$ with $\tilde{\rho}(x) = p/q$. We let $M_{p/q+} = M_{p/q+,h}$ denote the set of $x \in M$ with $\tilde{\rho}(x) = p/q+$ or p/q. We let $M_{p/q-} = M_{p/q-,h}$ denote the set of $x \in M$ with $\tilde{\rho}(x) = p/q-$ or p/q. Note that our $M_{p/q}$ differs from Bangert's [3].

In this notation, we can state the following results. For each $\omega \in \mathbf{S}$, M_ω is closed and totally ordered. The mapping $pr_0 : M_\omega \to \mathbf{R}$ is a homeomorphism of M_ω onto its image A_ω which is a closed subset of $\mathbf{R}$.

These results are a combination of various results in [3]. For ω an irrational number, M_ω is closed because ρ is continuous [3, Corollary 3.16]. In this case, M_ω is totally ordered by [3, Theorem 4.1]: our M_ω agrees with Bangert's when ω is an irrational number, but not when ω is a rational number. In fact, when ω is a rational number, Bangert's M_ω (which is our $M_{\omega-} \cup M_{\omega+}$) is not totally ordered, with the rare exception that our A_ω (Bangert's A_ω^{per}) is all of $\mathbf{R}$. See the discussion at the end of §5 in [3]. This discussion also mentions that (in the case ω is a rational number) our $M_{\omega+}$ and $M_{\omega-}$ (which are Bangert's $M_\omega^{per} \cup M_\omega^+$ and his $M_\omega^{per} \cup M_\omega^-$) are closed and totally ordered, a consequence of [3, Theorems (5.1), (5.3), and (5.8)]. Our M_ω (Bangert's M_ω^{per}) is closed and totally ordered, since it is $M_{\omega-} \cap M_{\omega+}$.

For ω an irrational number, the fact that pr_0 maps M_ω homeomorphically onto A_ω and the fact that A_ω is closed is part of [3, (4.2)]. For other ω, these results follow from the discussion in [3, §5].

We end this section by mentioning one other result in [3]: For each rotation symbol ω, there is a mapping $\zeta_\omega : \mathbf{R} \to \mathbf{R}$ which is continuous, strictly increasing, and satisfies $\zeta_\omega(t + 1) = \zeta_\omega(t) + 1$, such that A_ω is invariant, and $\zeta_\omega(x_i) = x_{i+1}$ for $x \in M_\omega$. In fact, this last condition determines ζ_ω uniquely on A_ω. For an irrational number ω, this statement is part of [3, (4.2)] and for other ω, it follows from [3, §5].

§4. FURTHER CONDITIONS ON h

We will prove the main result of this paper (the estimate on Peierls's barrier) under two further assumptions on h. These assumptions hold when h is the generating function of a lift f of an element $\bar{f}$ of J_β. In addition to Bangert's assumptions, we suppose:

There exists a positive continuous function ρ on $\mathbf{R}^2$ such that

$$h(\xi,x') + h(x,\xi') - h(x,x') - h(\xi,\xi') > \int_x^\xi \int_{x'}^{\xi'} \rho, \text{ if } x < \xi \text{ and } x' < \xi'. \tag{H_5}$$

In the case that $\bar{f} \in J$ and h is the generating function of the lift f, we have that (H_5) holds with $\rho = -\partial_{12}h$. In fact, the inequality is an equation in this case.

The remaining condition depends on a positive number θ:

$$x \to \theta x^2/2 - h(x,x') \text{ is convex, for any } x', \text{ and}$$

$$x' \to \theta x'^2/2 - h(x,x') \text{ is convex, for any } x\ . \tag{$H_{6\theta}$}$$

We recall that a function of one variable is convex if the line segment joining two points of its graph lies on or above the graph. Intuitively, our condition $(H_{6\theta})$ amounts to the assertion that $\partial_{11}h$ and $\partial_{22}h$ are bounded above by θ. However, we do not wish to assume that these derivatives exist, because our results have interesting applications in some cases when they do not exist.

We say that h satisfies (H_6) if it satisfies $(H_{6\theta})$ for some $\theta > 0$.

If $\bar{f} \in J_\beta$ and h is the generating function of its lift f, then h satisfies $(H_{6\theta})$ with $\theta = \cot\beta$. In this case, h is C^2, so it is enough to prove that $\partial_{11}h, \partial_{22}h < \theta$, everywhere. Using the notation $f(x,y) = (x',y')$, we have $\partial y'/\partial x' = \partial_{22}h$, if we take x and x' as the independent variables. Here, y' is the height of the point $(x' \times \mathbf{R}) \cap f(x \times \mathbf{R})$. Our assumption that $\beta_{\bar{f}}(P) > \beta$ is equivalent to the assertion that the slope of the curve $f(x \times \mathbf{R})$ is $< \cot\beta$. Thus, $\partial_{22}h(x,x') = \partial y'/\partial x' < \cot\beta$. The inequality $\partial_{11}h(x,x') = \partial y'/\partial x' < \cot\beta$ follows similarly from $-\bar{\beta}_{\bar{f}}(P) > \beta$.

There exist functions satisfying (H_1)-(H_6) which are not C^1. On the other hand, by (H_6), the one-sided partial derivatives $\partial_1 h(x-,x')$, $\partial_1 h(x+,x')$, $\partial_2 h(x,x'-)$, $\partial_2 h(x,x'+)$ exist. Here,

$$\partial_1 h(x-,x') = \lim_{\Delta x \uparrow 0} (h(x + \Delta x) - h(x))/\Delta x ,$$

etc. In addition,

$$\partial_1 h(x-,x') > \partial_1 h(x+,x'), \quad \partial_2 h(x,x'-) > \partial_2 h(x,x'+) .$$

The one-sided partial derivatives of h exist for the same reason that the one-sided partial derivatives of a convex function exist. The inequalities above hold for the same reason that a convex function satisfies the opposite inequalities.

It follows that if h satisfies (H_6) and $(x_j,\ldots,x_k)$ is a minimal segment of a configuration (with respect to h), then $(x_j,\ldots,x_k)$ is <u>stationary</u> (with respect to h) in the sense that for $j < i < k$, both $\partial_2 h(x_{i-1},x_i)$ and $\partial_1 h(x_i,x_{i+1})$ exist and

$$\partial_2 h(x_{i-1},x_i) + \partial_1 h(x_i x_{i+1}) = 0 .$$

For, if one of these derivatives did not exist, we could decrease $h(x_j,\ldots,x_k)$ slightly, by perturbing x_i slightly, by the inequalities above. Obviously, the equation follows, once we know that the derivatives exist.

The conditions (H_1)-(H_6) are not all independent. Obviously, (H_5) implies (H_3). Moreover, (H_5) and (H_6) imply (H_4). For, if $(\bar{x},x,x')$ and $(\bar{\xi},x,\xi')$ are minimal segments, then as we have just seen they are stationary, i.e.

$$\partial_2 h(\bar{x},x) + \partial_1 h(x,x') = 0 = \partial_2 h(\bar{\xi},x) + \partial_1 h(x,\xi') ,$$

so we may verify (H_4) in exactly the same way as for the generating function of a monotone twist diffeomorphism, using (H_5) in place of $-\partial_{12}h > 0$.

§5. CONJUNCTION

Our reason for not restricting our attention to C^2 functions is that we wish to consider not only generating functions associated to monotone twist diffeomorphisms, but also the class of functions generated by such functions by the following operation, which we call <u>conjunction</u>:

$$h_1 * h_2(x,x') = \min_{\xi} (h_1(x,\xi) + h_2(\xi,x')) .$$

This operation may be applied to the class of continuous real valued functions on $\mathbf{R}^2$ which satisfies Bangert's condition (H_2). If h_1 and h_2 are two continuous real valued functions on $\mathbf{R}^2$ which satisfy (H_2), then $h_1 * h_2$ is defined, continuous, and satisfies (H_2). Conjunction is clearly associative.

In general, the conjunction of two C^∞ functions need not be C^1, even if they satisfy our regularity conditions (H_1)-(H_6). However, if h_1 and h_2 both satisfy these regularity conditions, then so does $h_1 * h_2$. In fact, if h_1 and h_2 both satisfy (H_1)-(H_5) and $(H_{6\theta})$, then so does $h_1 * h_2$, with the same θ (cf. Lemma 5.3, below).

For our discussion of $h_1 * h_2$, we need to generalize Bangert's discussion in the following way. Instead of considering one function h on $\mathbf{R}^2$, we consider a bi-infinite sequence $h = (\dots,h_i,\dots)$ of such functions. We modify our previous definition, setting

$$h(x_j,\dots,x_k) = \sum_{i=j}^{k-1} h_i(x_i,x_{i+1}) .$$

As in §2, we say that the segment $(x_j,\dots,x_k)$ is <u>minimal</u> with respect to h if

$$h(x_j,\dots,x_k) \leq h(x_j^*,\dots,x_k^*)$$

for all $(x_j^*,\dots,x_k^*)$ with $x_j = x_j^*$ and $x_k = x_k^*$.

<u>Lemma 5.1. If each</u> h_i <u>satisfies</u> (H_5) <u>and</u> (H_6), <u>then</u> h <u>satisfies</u> (H_4).

<u>Proof</u>. In the last section, we proved that h_i satisfies (H_4). We may prove that h satisfies (H_4) in exactly the same way. Suppose (x_{i-1},x_i,x_{i+1}) and $(\xi_{i-1},x_i,\xi_{i+1})$ are minimal segments, with respect to h. The argument of the last section shows that they are stationary, i.e.

$$\partial_2 h_{i-1}(x_{i-1},x_i) + \partial_1 h_i(x_i,x_{i+1}) = 0 = \partial_2 h_{i-1}(\xi_{i-1},x_i) + \\ + \partial_1 h_i(x_i,\xi_{i+1}) .$$

We may apply the argument of the last section in this context, because both h_{i-1} and h_i satisfy (H_6). Then we may verify (H_4) in exactly the same way as for the generating function of a monotone twist diffeomorphism, using (H_5) for h_{i-1} and h_i. □

Lemma 5.2. *If each h_i satisfies (H_1)-(H_6), then two minimal configurations (or minimal segments of configurations) with respect to h cross at most once. If they meet at some $i \in \mathbf{Z}$ (other than the endpoint of a segment), then they cross there. If two minimal segments meet at an endpoint, then they do not meet except at endpoints.*

Proof. This generalizes [3, Lemma 3.1] where the single function h is replaced by a bi-infinite sequence. By Lemma 5.1 above, we have (H_4) for the bi-infinite sequence. Consequently, the proof of [3, Lemma 3.1] applies without change. □

Lemma 5.3. *If h_1 and h_2 are two continuous real valued functions on $\mathbf{R}^2$ which satisfy (H_1)-(H_5) and $(H_{6\theta})$, then $h_1 * h_2$ satisfies (H_1)-(H_5) and $(H_{6\theta})$.*

Proof. We have already remarked that $h_1 * h_2$ is defined, continuous, and satisfies (H_2). That it satisfies (H_1) is obvious.

To say that $x \to \theta x^2/2 - h(x,x')$ is convex is to say that

$$(1 - \lambda)h(x,x') + \lambda h(\xi,x') \leq h(x_\lambda,x') + \theta\lambda(1 - \lambda)(x - \xi)^2/2 ,$$

for $0 \leq \lambda \leq 1$, where $x_\lambda = (1 - \lambda)x + \lambda\xi$. To verify this condition for $h = h_1 * h_2$, we choose x'_λ so that

$$h_1 * h_2(x_\lambda,x') = h_1(x_\lambda,x'_\lambda) + h_2(x'_\lambda,x') .$$

Then

$$\begin{aligned}(1 - \lambda)h_1 &* h_2(x,x') + \lambda h_1 * h_2(\xi,x') \\ &\leq (1 - \lambda)[h_1(x,x'_\lambda) + h_2(x'_\lambda,x')] + \lambda[h_1(\xi,x'_\lambda) + h_2(x'_\lambda,x')] \\ &\leq h_1(x_\lambda,x'_\lambda) + \theta\lambda(1 - \lambda)(x - \xi)^2/2 + h_2(x'_\lambda,x') \\ &= h_1 * h_2(x_\lambda,x') + \theta\lambda(1 - \lambda)(x - \xi)^2/2 ,\end{aligned}$$

by the hypothesis that h_1 satisfy $(H_{6\theta})$. This proves that $h_1 * h_2$ satisfies the first condition in $(H_{6\theta})$. Similarly, we may prove that it satisfies the second condition, by using the fact that h_2 satisfies $(H_{6\theta})$.

As a first step towards verifying (H_5), we consider the case when h_1 and h_2 satisfy it with ρ a positive constant. We will show that $h_1 * h_2$ satisfies it with ρ replaced by $\rho^2/2\theta$. Consider real numbers $x < \xi$ and x'. It is enough to show that

$$\partial_2 h_1 * h_2(x,x'+) - \partial_2 h_1 * h_2(\xi,x'+) \geq (\xi - x)\rho^2/2\theta ,$$

since (H_5) follows from this inequality, by integration with respect to the second variable. Note that this one-sided partial derivative exists because $h_1 * h_2$ satisfies (H_6), as we have just verified. Let x'' and ξ'' be the largest real numbers such that

$$h_1 * h_2(x,x') = h_1(x,x'') + h_2(x'',x'), \quad \text{and}$$

$$h_1 * h_2(\xi,x') = h_1(\xi,\xi'') + h_2(\xi'',x') \ .$$

In order to compute the one-sided partial derivatives with respect to the second variable, we consider $\xi' > x'$. We hold x, ξ, and x' fixed in what follows and let ξ' approach x' from above. We let x^* and ξ^* be such that

$$h_1 * h_2(x,\xi') = h_1(x,x^*) + h_2(x^*,\xi'), \quad \text{and}$$

$$h_1 * h_2(\xi,\xi') = h_1(\xi,\xi^*) + h_2(\xi^*,\xi') \ .$$

Both (x,x'',x') and (x,x^*,ξ') are minimal segments for $h = (\ldots,h_1,h_2,\ldots)$, so $x^* > x''$ by Lemma 5.2. Similarly, $\xi^* > \xi''$. Lemma 5.2 also implies that x^* and ξ^* are monotone functions of ξ', which decrease as ξ' does. Consequently, x^* and ξ^* approach limits as ξ' approaches x' from above. Since h_1 and h_2 are continuous, it follows that these limits satisfy the defining conditions for x'' and ξ''. Moreover, it is clear that they are the maximum elements satisfying these conditions and hence must be x'' and ξ''. Thus, we have proved that $x^* \downarrow x''$ and $\xi^* \downarrow \xi''$, as $\xi' \downarrow x'$ (for x and ξ fixed).

By Lemma 5.2, $x'' < \xi''$ because $x < \xi$ and (x,x'',x') and (ξ,ξ'',x') are both minimal for $h = (\ldots,h_1,h_2,\ldots)$. Consequently, $x^* < \xi''$ for ξ' close enough to ξ. We have

$$\begin{aligned}
& h_1 * h_2(x,\xi') + h_1 * h_2(\xi,x') - h_1 * h_2(x,x') - h_1 * h_2(\xi,\xi') \\
& \quad \geq h_1(x,x^{\wedge}) + h_2(x^{\wedge},\xi') + h_1(\xi,\xi'') + h_2(\xi'',x') - \\
& \qquad - h_1(x,x^*) - h_2(x^*,x') - h_1(\xi,\xi'') - h_2(\xi'',\xi') \\
& \quad = h_2(x^*,\xi') + h_2(\xi'',x') - h_2(x^*,x') - h_2(\xi'',\xi') \\
& \quad \geq \rho(\xi'' - x^*)(\xi' - x') \ ,
\end{aligned}$$

where the first inequality follows from the definition of conjunction and the last inequality is a consequence of our assumption that h_2 satisfies (H_5) for the positive constant ρ. Dividing by $\xi' - x'$ and taking the limit as $\xi' \downarrow x'$, we see that

$$\partial_2 h_1 * h_2(x,x'+) - \partial_2 h_1 * h_2(\xi,x'+) \geq \rho(\xi'' - x'') \ . \tag{*}$$

Moreover,

$$\partial_2 h_1(x,x'') - \partial_2 h_1(\xi,x''+) \geq \rho(\xi - x) ,$$

since h_1 satisfies (H_5) for the constant ρ, and

$$\partial_2 h_1(\xi,\xi'') + \partial_1 h_2(\xi'',x') - \partial_2 h_1(\xi,x''+) - \partial_1 h_2(x'',x') \leq 2\theta(\xi'' - x'') ,$$

since h_1 and h_2 satisfy ($H_{6\theta}$). Since

$$\partial_2 h_1(x,x'') + \partial_1 h_2(x'',x') = 0 = \partial_2 h_1(\xi,\xi'') + \partial_1 h_2(\xi'',x') ,$$

subtracting the last inequality above from the previous one gives

$$0 \geq \rho(\xi - x) - 2\theta(\xi'' - x'') ,$$

from which we obtain $\xi'' - x'' \geq (\xi - x)\rho/2\theta$. Applying (*), we then find

$$\partial_2 h_1 * h_2(x,x'+) - \partial_2 h_1 * h_2(\xi,x'+) \geq (\xi - x)\rho^2/2\theta ,$$

as required. This completes the proof that (H_5) holds in the case when h_1 and h_2 satisfy it with ρ a positive constant.

To prove (H_5) in general, it is enough to show that for each $(u,u') \in \mathbf{R}^2$, there is a neighborhood of (u,u') in $\mathbf{R}^2$ such that (H_5) holds for (x,x') and (ξ,ξ') in that neighborhood, since both sides of the inequality in (H_5) define additive set functions. Let u^* and u'' be the least and greatest elements v of $\mathbf{R}$ such that

$$h_1 * h_2(u,u') = h_1(u,v) + h_2(v,u') .$$

The above argument shows that if h_1 satisfies (H_5) for the constant ρ when (x,x') and (ξ,ξ') are in a neighborhood of $u \times [u^*,u'']$ and h_2 satisfies (H_5) for the constant ρ when (x,x') and (ξ,ξ') are in a neighborhood of $[u^*,u''] \times u'$, then $h_1 * h_2$ satisfies (H_5) for the constant $\rho^2/2\theta$ when (x,x') and (ξ,ξ') are in a neighborhood of (u,u'). □

<u>Example</u>. If $f_1,\ldots,f_k \in J_\beta$ and $h_1,\ldots,h_k$ are their generating functions, then minimal configurations of $h = h_1 * \cdots * h_k$ correspond to certain orbits of $f = f_k \circ \cdots \circ f_1$, which we call <u>minimal orbits</u>. If $(\ldots,x_i,\ldots)$ is a minimal configuration, let $y_i = -\partial_1 h(x_i,x_{i+1}) = \partial_2 h(x_{i-1},x_i)$; then $f(x_i,y_i) = (x_{i+1},y_{i+1})$, and $(\ldots,(x_i,y_i),\ldots)$ is the corresponding minimal orbit of f. Note that h satisfies (H_1)-(H_5) and ($H_{6\theta}$), with $\theta = \cot\beta$, by the theory which we have developed in this section and the remark we made previously that the generating function of a monotone twist diffeomorphism satisfies these conditions.

To verify that $f(x_i,y_i) = (x_{i+1},y_{i+1})$ in the above example, the following remark is useful. Consider $x,x' \in \mathbf{R}$. Consider the set of sequences $\xi_0,\ldots,\xi_k$ with $\xi_0 = x$, $\xi_k = x'$, and

$$h(x,x') = h_1(\xi_0,\xi_1) + \cdots + h_k(\xi_{k-1},\xi_k) .$$

By Lemma 5.2, the set of such sequences is totally ordered. Furthermore, by (H_2) and the continuity of the h_i's, there is a least such sequence $(\xi_0^{min},\ldots,\xi_k^{min})$ and a greatest such sequence $(\xi_0^{max},\ldots,\xi_k^{max})$. Then

$$\partial_1 h(x-,x') = \partial_1 h_1(\xi_0^{min},\xi_1^{min}) ,$$

$$\partial_1 h(x+,x') = \partial_1 h_1(\xi_0^{max},\xi_1^{max}) ,$$

$$\partial_2 h(x,x'-) = \partial_2 h_k(\xi_{k-1}^{min},\xi_k^{min}) ,$$

$$\partial_2 h(x,x'+) = \partial_2 h_k(\xi_{k-1}^{max},\xi_k^{max}) .$$

We leave the proof of these assertions as an exercise for the reader.

As A. Katok remarked to me, another way of obtaining the class of minimal orbits of f is simply to consider the orbits associated to minimal configurations of the bi-infinite sequence $(\ldots,h_1,\ldots,h_k,\ldots)$ of period k. This avoids the necessity of considering nondifferentiable functions. However, a clear exposition of [1] was needed anyway, and it is useful to have the extra generality that Bangert's article gives us.

Example. The variational principle h for geodesics on a two torus with a smooth Riemannian structure described in [3, §6] satisfies (H_1)-(H_6). In [3, Lemma 6.4], (H_1)-(H_4) are verified; we leave the other conditions for the reader to verify.

§6. PEIERLS'S BARRIER

Consider a rotation symbol ω, i.e. an irrational number, or one of $p/q+$, p/q, or $p/q-$, where p/q is a rational number. Let ξ be a real number. Let h be a real valued function on $\mathbf{R}^2$ which satisfies (H_1)-(H_6). In this section, we will define a number $P_\omega(\xi) = P_{\omega,h}(\xi)$, associated to ω, ξ, and h. The quantity $P_\omega(\xi)$ was called "Peierls's energy barrier" by Aubry, Le Daeron, and André [1], because it is a version of a known notion in solid state physics. We will call $P_\omega(\xi)$ simply "Peierls's barrier". It is related to the quantity ΔW_ω introduced by A. Katok and the author ([4],[5]). The latter is an upper bound for $P_\omega(\xi)$. For the case ω is irrational, $P_\omega(\xi)$ was defined in [6]. In this section, we extend the definition of $P_\omega(\xi)$ to the case of an arbitrary rotation symbol.

At the end of this section, we show that $P_\omega(\xi)$ is Lipschitz in ξ. In the next section, we prove the result announced in the abstract of this paper.

Throughout this section, we will use the notations of §3. We recall that the various sets and other mathematical objects introduced there depend on h, even though h is often omitted from the notation, and that these objects are defined when h satisfies Bangert's conditions (H_1)-(H_4). In this section, we will assume, in addition, that h satisfies (H_5) and $(H_{6\theta})$.

We recall, from the end of §3, that $A_\omega = \mathrm{pr}_0(M_\omega)$ is a closed subset of $\mathbf{R}$, invariant under $t \to t + 1$. For $\xi \in \mathbf{R}$, we let $\xi_{0-} = \xi_{0+} = \xi$ in the case that $\xi \in A_\omega$ and we let (ξ_{0-}, ξ_{0+}) be the complementary interval of A_ω which contains ξ, otherwise. We set $\xi_{i-} = \zeta_\omega^i(\xi_{0-})$ and $\xi_{i+} = \zeta_\omega^i(\xi_{0+})$. (Recall from the end of §3 that ζ_ω is a homeomorphism of $\mathbf{R}$ such that $\zeta_\omega(t + 1) = \zeta_\omega(t) + 1$ and $\zeta_\omega(x_i) = x_{i+1}$ if $x \in M_\omega$.)

Suppose x is a configuration which satisfies $\xi_{i-} \leq x_i \leq \xi_{i+}$ and if ω is a rational number p/q, also $x_{i+q} = x_i + p$. We set

$$G_\omega(x) = \sum_I (h(x_i, x_{i+1}) - h(\xi_{i-}, \xi_{i+1-})) .$$

Here $I = \mathbf{Z}$ if $\omega \in (\mathbf{R}\setminus\mathbf{Q}) \cup (\mathbf{Q}+) \cup (\mathbf{Q}-) = \mathbf{S}\setminus\mathbf{Q}$ and $I = \{0, \ldots, q - 1\}$ if $\omega = p/q$.

Note that ζ_ω has rotation number ω, and

$$\zeta_\omega^q(t) > t + p \quad \text{if} \quad \omega = p/q+ ,$$

$$\zeta_\omega^q(t) = t + p \quad \text{if} \quad \omega = p/q, \quad \text{and}$$

$$\zeta_\omega^q(t) < t + p \quad \text{if} \quad \omega = p/q- ,$$

when $t \in A_\omega$. From these facts and elementary arguments (cf. Bangert [3, §2]), it follows that the intervals $[\xi_{i-}, \xi_{i+}]$, $i \in I$, are disjoint (mod. 1), so

$$\sum_I x_i - \xi_{i-} \leq \sum_I \xi_{i+} - \xi_{i-} \leq 1 .$$

We have seen that it follows from (H_6) that all one-sided first partial derivatives of h exist everywhere.

<u>Lemma 6.1. Suppose</u> $x < \xi$ <u>and</u> $x' < \xi'$. <u>Then we have</u>

$$-\theta \leq \partial_1 h(\xi\pm, x') - \partial_1 h(x\pm, x') \leq (\xi - x)\theta, \quad \underline{\text{and}}$$

$$-\theta \leq \partial_2 h(x, \xi'\pm) - \partial_2 h(x, x'\pm) \leq (\xi' - x')\theta .$$

Proof. In both cases, the upper bound is an obvious consequence of $(H_{6\theta})$. From (H_1) and (H_3), it follows that $\partial_1 h(x + n\pm, x') = \partial_1 h(x\pm, x' - n) > \partial_1 h(x\pm, x')$ if n is a positive integer. Taking $n = [\xi - x] + 1$, and using the first upper bound, applied to $(\xi, x + [\xi - x] + 1)$ in place of (x,ξ), we obtain the first lower bound. The second lower bound may be obtained similarly. □

It follows from this lemma that every one-sided first partial derivative of h is bounded on each compact subset of the plane. Therefore, the sum defining $G_\omega(x)$ is absolutely convergent and the convergence is uniform over the set of x satisfying $\xi_{i-} < x_i < \xi_{i+}$ (in the case I is infinite). Here, we use (H_1), the fact that such x satisfy $[\omega] < x_{i+1} - x_i < [\omega] + 1$, as well as $\sum_I x_i - \xi_{i-} < 1$.

From this uniform convergence, it follows that G_ω is a continuous function on $\prod_I [\xi_{i-}, \xi_{i+}]$. We let $P_\omega(\xi)$ be the minimum of $G_\omega(x)$ over those x such that $x_0 = \xi$, i.e.

$$P_\omega(\xi) = \min\{G_\omega(x) : \xi_{i-} < x_i < \xi_{i+},\ \xi = x_0, \text{ and }$$
$$x_{i+q} = x_i + p \text{ if } \omega = p/q\} .$$

Lemma 6.2. $P_\omega(\xi) > 0$ and we have $P_\omega(\xi) = 0$ if and only if $\xi \in A_\omega$. If (ξ_{0-}, ξ_{0+}) is a complementary interval of A_ω and $\xi \in [\xi_{0-}, \xi_{0+}]$ then $P_\omega(\xi)$ is given by the formula immediately above.

Proof. Suppose $\xi_{i-} < x_i < \xi_{i+}$ and $x_{i+q} = x_i + p$ if $\omega = p/q$. We assert that $G_\omega(x) > 0$ and we have $G_\omega(x) = 0$ if and only if x is minimal. These assertions clearly imply the lemma. Note that the last statement in the lemma is just the definition, except when $\xi = \xi_{0+}$, in which case there is something to prove.

In the case that $\omega = p/q$, our assertions are a special case of a result proved in [3, §3]: if y is a configuration of period (q,p), i.e. $y_{i+q} = y_i + p$, then y is minimal if and only if $\sum_I h(y_i, y_{i+1}) < \sum_I h(x_i, x_{i+1})$ for all configurations x of period (q,p). Since ξ_- is minimal, we obtain our assertions.

When ω is not a rational number, ξ_- and ξ_+ are asymptotic, i.e. $\lim_{|i|\to\infty} |\xi_{i+} - \xi_{i-}| = 0$. Moreover, $[\omega] < |\xi_{i+1\pm} - \xi_{i\pm}| < [\omega] + 1$, h is continuous and satisfies (H_1), so our assertions follow by an elementary argument, which we leave to the reader. □

Lemma 6.3. $|P_\omega(\xi') - P_\omega(\xi)| < 2\theta|\xi' - \xi|$.

Proof. It is enough to prove this when $\xi, \xi' \in [\xi_{0-}, \xi_{0+}]$, where (ξ_{0-}, ξ_{0+}) is a complementary interval of A_ω, since P_ω vanishes on A_ω. It is enough to prove $P_\omega(\xi') - P_\omega(\xi) < 2\theta|\xi' - \xi|$, since we may then interchange ξ and ξ' to get the opposite inequality. By Lemma 6.2, there is a configuration x such that $\xi_{i-} < x_i < \xi_{i+}$, $x_0 = \xi$, $G_\omega(x) = P_\omega(\xi)$ and $x_{i+q} = x_i + p$, in the case $\omega = p/q$. By (H_2), there exists y such that (x_{-1}, y, x_1) is a minimal segment of a configuration. Since $(\xi_{-1-},\xi_{0-},\xi_{1-})$ and $(\xi_{-1+},\xi_{0+},\xi_{1+})$ are also minimal, we obtain that $\xi_{0-} \le y \le \xi_{0+}$, from the fact that minimal segments can cross at most once. Then

$$\partial_2 h(x_{-1},y) + \partial_1 h(y,x_1) = 0 ,$$

so, by Lemma 6.1,

$$|\partial_2 h(x_{-1},u+) + \partial_1 h(u+,x_1)| \le 2\theta$$

for u between ξ and ξ', since we have $\xi_{0-} \le y, u \le \xi_{0+}$ and therefore $|y - u| \le 1$. Hence, $G_\omega(x') \le G_\omega(x) + 2\theta|\xi' - \xi|$, where $x'_0 = \xi'$, $x'_i = x_i$ for $i \in I$, $i \neq 0$, and $x'_{i+q} = x_i + p$ if $\omega = p/q$. Hence,

$$P_\omega(\xi') \le G_\omega(x') \le G_\omega(x) + 2\theta|\xi'-\xi| = P_\omega(\xi) + 2\theta|\xi'-\xi| .\square$$

§7. THE BASIC ESTIMATE

The purpose of this section is to prove the estimate given in Theorem 7.1, below. In the next section, we will use this estimate to obtain a partial modulus of continuity for $P_\omega(\xi)$ as a function of $\omega \in S$, where S is the symbol space, introduced in §2. If $\omega \in S$, we will let ω^* denote the number obtained by forgetting $+$ or $-$, i.e. $\omega^* = \omega$ if ω is irrational, $\omega^* = p/q$ if $\omega = p/q+$, $p/q-$, or p/q. We will call ω^* the <u>number underlying</u> ω.

<u>Theorem 7.1.</u> *If h is a continuous real valued function on $\mathbf{R}^2$ satisfying (H_1)–(H_5) and $(H_{6\theta})$, and $P_\omega(\xi) = P_{\omega h}(\xi)$ denotes Peierls's barrier with respect to h, then*

$$|P_\omega(\xi) - P_{p/q}(\xi)| \le (1200)\theta(q^{-1} + |\omega^* q - p|) ,$$

for any $\xi \in \mathbf{R}$, $p/q \in \mathbf{Q}$ (expressed in lowest terms), and $\omega \in S$.

<u>Proof.</u> We may suppose that if ω is a rational number p'/q', then $q' \ge q$. For, the estimate without this supplementary condition follows by applying the estimate with p/q and p'/q' interchanged. This is because if $q' < q$ then

$$(q')^{-1} + |(p/q)q' - p'| = q^{-1} + (q')^{-1} + q^{-1}\Delta$$

$$< q^{-1} + (q')^{-1} + (q')^{-1}\Delta = q^{-1} + |(p'/q')q - p| ,$$

where $\Delta = |p'q - pq'| - 1$.

Note that

$$P_{p/q}(\xi) = \min\{G_{p/q}(u) : u_0 = \xi \text{ and } u_{i+q} = u_i + p\} , \qquad (7.2)$$

i.e. we may drop the condition $\xi_{i-} \leq u_i \leq \xi_{i+}$ in the definition of $P_{p/q}(\xi)$; the periodicity condition $u_{i+q} = u_i + p$ is enough. This follows from the proof of [3, Lemma 3.1], which shows that the minimizing u in (7.2) satisfies $\xi_{i-} \leq u_i \leq \xi_{i+}$. The point is that the graphs of u and ξ_- (resp. ξ_+) cannot cross twice between u_0 and u_q, by the proof of [3, Lemma 3.1]; therefore, by periodicity, they cannot cross at all. Since $\xi_{0-} \leq \xi \leq \xi_{0+}$, by definition, it then follows that $\xi_{i-} \leq u_i \leq \xi_{i+}$, for all i.

Moreover,

$$G_{p/q}(u) = h(u_0,\dots,u_q) - h(x_0,\dots,x_q) , \tag{7.3}$$

for any minimal configuration x of rotation symbol p/q. In other words, we may replace ξ_- by x in the definition of $G_{p/q}(u)$. For, ξ_- is also a minimal configuration of rotation symbol p/q, and minimal configurations of rotation symbol p/q minimize $h(x_0,\dots,x_q)$ over all configurations x of period (q,p), by results in [3, §3].

Since p/q is expressed in lowest terms, there exist $s,t \in \mathbf{Z}$ such that $tq - sp = 1$. For a configuration y of period (q,p) (i.e. such that $y_{i+q} = y_i + p$), we set $(Uy)_i = y_{i-s} + t$. Note that Uy is independent of the pair (s,t) of integers such that $tq - sp = 1$, since y is of period (q,p). We may iterate U and clearly $U^q y = y + 1$, if y is of period (q,p). Recall that all configurations in $M_{p/q}$ are of period (q,p) and $M_{p/q}$ is totally ordered. (Cf. Bangert [3, Theorem 5.1] and recall that what we call $M_{p/q}$ is what Bangert calls $M_{p/q}^{per}$.) Obviously U preserves $M_{p/q}$, so we have $Uy < y$, $Uy = y$, or $Uy > y$ when $y \in M_{p/q}$. In the first case, it follows that $U^2y < Uy$, etc., so $U^q y < U^{q-1}y < \cdots < Uy < y$, contradicting $U^q y = y + 1$. Likewise $Uy \neq y$. Hence $Uy > y$, $U^2 y > Uy$, etc., and, in fact, $U^\ell y < U^m y$ if and only if $\ell < m$ when $y \in M_{p/q}$.

In the rest of the proof of Theorem 7.1, we will suppose that $\omega > p/q$, the case $\omega < p/q$ being similar. We set $B = [(2q - 1)|q\omega^* - p|] + 1$, where $[x]$ denotes the greatest integer $\leq x$. We set $\varepsilon = q^{-1}(3B + 1)$. We choose $x \in M_{p/q}$. Since $U^\ell x_0$ is an increasing function of ℓ and $U^{\ell+q}x_0 = U^\ell x_0 + 1$, there exists an integer ℓ such that $U^{\ell+3B+1}x_0 \leq U^\ell x_0 + \varepsilon$, by the pigeon hole principal. By replacing x with its translate $U^\ell x$, we may suppose that

$$U^{3B+1}x_0 \leq x_0 + \varepsilon . \tag{7.4}$$

In the rest of the proof of Theorem 7.1, we will let x be a fixed element of $M_{p/q}$ which satisfies (7.4).

It will be convenient to use the notation which we introduced in the definition of $P_\omega(\xi)$, viz. if $\xi \in A_\omega$, we let $\xi_{0-} = \xi_{0+} = \xi$ and if $\xi \notin A_\omega$, we let (ξ_{0-},ξ_{0+}) be the complementary interval of A_ω which contains ξ. We set $\xi_{i-} = \zeta_\omega^i(\xi_{0-})$ and $\xi_{i+} = \zeta_\omega^i(\xi_{0+})$, where ζ_ω is the homeomorphism defined at the end of §3. Then

$\xi_- = (\ldots,\xi_{i-},\ldots)$ and $\xi_+ = (\ldots,\xi_{i+},\ldots)$ are elements of M_ω and $\xi_- < \xi_+$, since M_ω is totally ordered and $\xi_{0-} < \xi_{0+}$.

For $i \in \mathbf{Z}$, we define m_i to be the unique integer m such that ξ_{i-} lies in the interval $[U^m x_i, U^{m+1}x_i)$. (Note that these intervals partition $\mathbf{R}$.) From the fact that the graphs of $U^m x$ and ξ_- cross at most once, it follows that $i \to m_i$ is monotonic. Moreover, since the rotation symbol ω of ξ_- is greater than the rotation symbol p/q of x, it follows that $i \to m_i$ is non-decreasing.

Next, we prove (7.5) below. As a first step towards proving (7.5), we consider any two integers i and j. We let j' be the smallest integer $\geq j$ such that $(j' - i) + s(m_i + 1 - m_j)$ is divisible by q and let n be the quotient, so $j' = i + s(m_j - m_i - 1) + nq$. We have

$$\xi_{j'-} \geq U^{m(j')}x_{j'} \geq U^{m(j)}x_{j'} = U^{m(j)}x_{i+s(m(j)-m(i)-1)} + np$$
$$= U^{m(i)+1}x_i + np + t(m_j-m_i-1) > \xi_{i-} + np + t(m_j-m_i-1) .$$

Here, the first inequality is a consequence of the definition of $m_{j'}$. The second inequality holds because $j' \geq j$ and m is non-decreasing. The first equation is a consequence of the fact that $U^{m(j)}x$ is periodic of period (q,p). The second equation follows from the definition of U. The last inequality follows from the definition of m_i.

Let $\xi^*_k = \xi_{k-j'+i-} + np + t(m_j - m_i - 1)$. Because ξ^* and ξ_- are translates of one another, they have the same rotation symbol and hence are comparable, since M_ω is totally ordered (cf. end of §3). The inequality above then shows $\xi_- > \xi^*$. Hence

$$\xi_{j'+q-} > \xi^*_{j'+q} = \xi_{i+q-} + np + t(m_j - m_i - 1)$$
$$\geq U^{m(i+q)}x_{i+q} + np + t(m_j - m_i - 1)$$
$$= U^{m(i+q)+m(j)-m(i)-1}x_{j'+q} .$$

Consequently, $m_{j'+q} \geq m_j + m_{i+q} - m_i - 1$, and

$$m_{j'+q} - m_j \geq m_{i+q} - m_i - 1 .$$

Since $j \leq j' \leq j + q - 1$ and $i \to m_i$ is non-decreasing, we obtain

$$m_{i+q} - m_i \leq m_{j+2q-1} - m_j + 1 .$$

Let μ be the minimum (over $j \in \mathbf{Z}$) of the right side. Then $m_{i+q} - m_i \leq \mu$.

Since the rotation number of ξ_- is ω^* and the rotation number of x is p/q, we have $\lim_{|i|\to\infty} m_i/i = q\omega^* - p$. Hence, $\mu \leq (2q - 1)|q\omega^* - p| + 1$. Since μ is an integer, $\mu \leq B = [2(q - 1)|q\omega^* - p|] + 1$. Hence,

$$m_{i+q} - m_i \leq B, \quad \text{for all } i \in \mathbf{Z} . \tag{7.5}$$

Next, using (7.5), we will deduce (7.9) below. We set $m = m_0$. We let a be the unique integer satisfying $-q < a \leq 0$ and $a \equiv (m - B)s \pmod{q}$. We let $n = ((B - m)s + a)/q$. Using the (q,p)-periodicity of x and the definition of U, we see that

$$U^{\ell}x_j = U^{\ell+B-m}x_{j-a} - D , \tag{7.6}$$

where we set $-D = np - (B - m)t \in \mathbf{Z}$. Since $-q < a \leq 0$, we have that $m - B \leq m_a \leq m$, by (7.5) and the fact that m_i is a non-decreasing function of i. It follows that

$$U^{m-B}x_a \leq U^{m(a)}x_a \leq \xi_{a-} < U^{m(a)+1}x_a \leq U^{m+1}x_a .$$

By (7.5), we have $m_{a+iq} \leq m_{a+(i+1)q} \leq m_{a+iq} + B$, so that, by induction, $m - B \leq m_{a+iq} \leq m + iB$ and

$$U^{m-B}x_{a+iq} \leq \xi_{a+iq-} < U^{m+iB+1}x_{a+iq} ,$$

for all non-negative integers i. From (7.6), applied with $j = a + iq$ and $\ell = m - B$ or $m + iB + 1$ (according to the case), we then obtain

$$\begin{aligned} x_{iq} &= U^{m-B}x_{a+iq} + D \leq \xi_{a+iq-} + D \\ &< U^{m+iB+1}x_{a+iq} + D = U^{(i+1)B+1}x_{iq} , \end{aligned} \tag{7.7$_i$}$$

for all non-negative integers i.

Since (ξ_{0-},ξ_{0+}) is a complementary interval of A_ω and $\omega > p/q$, we have $\xi_{i-} < \xi_{i+} < \xi_{i+q-} - p$, for all $i \in \mathbf{Z}$. Consequently, $(7.7)_i$, $(7.7)_{i+1}$ and the periodicity of x imply

$$\begin{aligned} x_{iq} &\leq \xi_{a+iq+} + D < \xi_{a+(i+1)q-} + D - p \\ &< U^{(i+2)B+1}x_{(i+1)q} + D - p = U^{(i+2)B+1}x_{iq} , \end{aligned} \tag{7.8$_i$}$$

for all non-negative integers i. In fact, we will need only the following consequences of the last two inequalities:

$$x_0 < \xi_{a-} + D < \xi_{a+} + D < U^{2B+1}x_0 < x_0 + \varepsilon \ , \qquad (7.9a)$$

$$x_q < \xi_{a+q-} + D < \xi_{a+q+} + D < U^{3B+1}x_q < x_q + \varepsilon \ . \qquad (7.9b)$$

The last inequality in (7.9a) (resp. 7.9b) follows from (7.4) (resp. the (q,p) periodicity of x together with (7.4)).

In what follows, we will suppose not only that $\omega > p/q$, as above, but also that $\omega < (p+1)/q$, and prove the estimate of Theorem 7.1 with 1200 replaced by 600. Theorem 7.1 will then follow by an elementary argument which we leave to the reader. The assumption that $p/q < \omega < (p+1)/q$ implies that $B < 2q$ and $\varepsilon < 7$.

The inequalities (7.9) will permit us to compare $P_\omega(\xi)$ and $P_{p/q}(\xi)$. When $a = 0$, the comparison is easy. It is based on the following observation.

Lemma 7.10. Consider a minimal segment $(\nu_{i-1},\nu_i,\nu_{i+1})$ and let $\nu_i' \in \mathbf{R}$. Then

$$0 < h(\nu_{i-1},\nu_i',\nu_{i+1}) - h(\nu_{i-1},\nu_i,\nu_{i+1}) < \theta(\nu_i' - \nu_i)^2 \ .$$

Proof. The first inequality is the definition of what it means for $(\nu_{i-1},\nu_i,\nu_{i+1})$ to be minimal. The second inequality follows from $(H_{6\theta})$. □

We let $w_0 = \xi$, $w_i = \xi_{i-}$, otherwise. Then

$$P_\omega(\xi) < G_\omega(w) < \theta(\xi - \xi_{0-})^2 \ .$$

The first inequality is a consequence of the definition of $P_\omega(\xi)$ and the second follows from Lemma 7.10. Similarly, $P_\omega(\xi) < \theta(\xi_{0+} - \xi)^2$, so combining these two inequalities and using $\xi_{0-} < \xi < \xi_{0+}$, we obtain

$$P_\omega(\xi) < \theta(\xi_{0+} - \xi_{0-})^2/4 < \theta\varepsilon_1^2/4 \ ,$$

where $\varepsilon_1 = \min(\varepsilon,1)$. By (7.9a) and the assumption that $a = 0$, we have $x_0 < \xi < U^{2B+1}x_0 < x_0 + \varepsilon$. An argument similar to that just given implies

$$P_{p/q}(\xi) < \theta\varepsilon_1^2/4 \ .$$

These two inequalities give the required estimate when $a = 0$.

The comparison of $P_\omega(\xi)$ and $P_{p/q}(\xi)$ is more difficult in the remaining cases, i.e. when $-q < a < 0$. We consider a configuration v such that $\xi_{i-} < v_i < \xi_{i+}$, $G_\omega(v) = P_\omega(\xi)$, and $v_0 = \xi$. In addition, if ω is a rational number p'/q', we choose v so that it is (q',p') periodic. Such a v exists by the definition of $P_\omega(\xi)$.

We next define two other configurations v' and v''. When ω is not a rational number, we let $v'_i = v''_i = v_i$ if $a < i < a + q$, $v'_i = v''_i = \xi_{i-}$ if $i = a$ or $a + q$, and $v'_i = v_i$ and $v''_i = \xi_{i-}$, otherwise. In the case that ω is a rational number p'/q', we will assume that $q' > q$. By our first remark in this proof, there is no loss of generality in assuming this. In this case, we define v'_i, v''_i in the same way as before for $i = a, a + 1, \ldots, a + q' - 1$ and extend to all i so that v' and v'' are both of period (q', p').

By definition, $P_\omega(\xi) = G_\omega(v) \leq G_\omega(v'')$. Since ξ_- is minimal and v is asymptotic to ξ_- in the case ω is not a rational number, $G_\omega(v'') \leq G_\omega(v')$. Moreover,

$$G_\omega(v') - G_\omega(v) = \sum_{i=a,a+q} h(v_{i-1}, \xi_{i-}, v_{i+1}) - h(v_{i-1}, v_i, v_{i+1}) .$$

By definition, v is minimal among all configurations satisfying $v_0 = \xi$ and $\xi_{i-} \leq v_i \leq \xi_{i+}$ and, in the case ω is a rational number p'/q', $v_{i+q'} = v_i + p'$. In the case that $\xi \notin A_\omega$, we have $\xi_{i-} < v_i < \xi_{i+}$ for all i (by an argument similar to [3, Lemma 3.1]). It follows that the segments (v_{i-1}, v_i, v_{i+1}) are minimal for $i \neq 0$, in the case that p'/q' is not a rational number, and for $i \not\equiv 0 \pmod{q'}$ when $\omega = p'/q'$. In particular, this is true for $i = a$, $a + q$, since $-q < a < 0$. Therefore, Lemma 7.10 applies, and we get

$$\begin{aligned} G_\omega(v') - G_\omega(v) &\leq \theta[(\xi_{a-} - v_a)^2 + (\xi_{a+q-} - v_{a+q})^2] \\ &\leq \theta[(\xi_{a+} - \xi_{a-})^2 + (\xi_{a+q+} - \xi_{a+q-})^2] \\ &\leq 2\theta\varepsilon_1^2 . \end{aligned}$$

To summarize the above inequalities, we have

$$P_\omega(\xi) = G_\omega(v) \leq G_\omega(v'') \leq G_\omega(v') \leq P_\omega(\xi) + 2\theta\varepsilon_1^2 . \tag{7.11}$$

Consequently, we may use $G_\omega(v'')$ in place of $P_\omega(\xi)$ in estimating $|P_\omega(\xi) - P_{p/q}(\xi)|$. This will be convenient, because the former has the form:

$$G_\omega(v'') = h(v''_a, \ldots, v''_{a+q}) - h(\xi_{a-}, \ldots, \xi_{a+q-}) , \tag{7.12}$$

by the definition of v''.

To estimate $|P_\omega(\xi) - P_{p/q}(\xi)|$, we will use the value of $P_{p/q}(\xi)$ given by (7.2) and (7.3). We will use (7.11) and (7.12) to estimate $P_\omega(\xi)$. It is enough to approximate $h(x_0, \ldots, x_q) - h(\xi_{a-}, \ldots, \xi_{a+q-})$ and $h(v''_a, \ldots, v''_{a+q}) - h(u_0, \ldots, u_q)$. We do the former first. Our approximation is based on the fact that $(x_0, \ldots, x_q)$ and $(\xi_{a-}, \ldots, \xi_{a+q-})$ are both minimal and $|x_0 - \xi_{a-} - D| < \varepsilon$ and $|x_q - \xi_{a+q-} - D| < \varepsilon$ by (7.9).

Let $x_i' = \xi_{i+a-} + D$ for $i = 0,q$ and $x_i' = x_i$ for $0 < i < q$. Then

$$h(\xi_{a-},\dots,\xi_{a+q-}) \le h(x_0',\dots,x_q')$$

by the fact that ξ_- is minimal and by (H_1). By the definition of x', we have

$$h(x_0',\dots,x_q') - h(x_0,\dots,x_q)$$

$$= h(x_0',x_0) - h(x_0,x_1) + h(x_{q-1},x_q') - h(x_{q-1},x_q) .$$

By $(H_{6\theta})$, we have

$$h(x_0',x_1) - h(x_0,x_1) \le \partial_1 h(x_0,x_1)(x_0'-x_0) + \theta(x_0'-x_0)^2/2$$

$$h(x_{q-1},x_q') - h(x_{q-1},x_q) \le \partial_2 h(x_{q-1},x_q)(x_q'-x_q) + \theta(x_q'-x_q)^2/2 .$$

By the periodicity and minimality of x, we have $\partial_2 h(x_{q-1},x_q) + \partial_1 h(x_0,x_1) = \partial_2 h(x_{-1},x_0) + \partial_1 h(x_0,x_1) = 0$. Moreover, $(x_0' - x_0) - (x_q' - x_q) = \xi_{a-} - \xi_{a+q-} + p$. Setting $Y = \partial_1 h(x_0,x_1)(\xi_{a-} - \xi_{a+q-} + p)$, we therefore have $\partial_1 h(x_0,x_1)(x_0' - x_0) + \partial_2 h(x_{q-1},x_q)(x_q' - x_q) = Y$. By (7.9), $|x_0' - x_0|$, $|x_q' - x_q| < \varepsilon$. Combining the above inequalities, we have

$$h(\xi_{a-},\dots,\xi_{a+q-}) \le h(x_0,\dots,x_q) + Y + \theta\varepsilon^2 . \tag{7.13}$$

To obtain the opposite inequality, we consider $\xi_i' = x_i$, for $i = 0,q$, and $\xi_i' = \xi_{i+a-} + D$ for $0 < i < q$. Then

$$h(x_0,\dots,x_q) \le h(\xi_0',\dots,\xi_q')$$

by the fact that x is minimal. By definition of ξ' and (H_1), we have

$$h(\xi_0',\dots,\xi_q') - h(\xi_{a-},\dots,\xi_{a+q-}) = h(\xi_0',\xi_1') -$$

$$- h(\xi_{a-},\xi_{a+1-}) + h(\xi_{q-1}',\xi_q') - h(\xi_{a+q-1-},\xi_{a+q-}) .$$

By Lemma 6.1, we have

$$h(\xi_0',\xi_1') - h(\xi_{a-},\xi_{a+1-}) = h(x_0,\xi_{a+1-}+D) - h(\xi_{a-}+D,\xi_{a+1-}+D)$$

$$\le \partial_1 h(x_0+,\xi_{a+1-}+D)(x_0-\xi_{a-}-D) + \theta(x_0-\xi_{a-}-D) .$$

Likewise,

$$h(\xi_{q-1}',\xi_q') - h(\xi_{a+q-1-},\xi_{a+q-})$$

$$\le \partial_2 h(\xi_{a+q-1-}+D,x_q+)(x_q-\xi_{a+q-}+D) + \theta(x_q-\xi_{a+q-}-D) .$$

Let $n = [\varepsilon] + 1$. We have $\partial_1 h(x_0, x_1) \le \partial_1 h(x_0 - n+, x_1) + n\theta$ $= \partial_1 h(x_0+, x_1 + n) + n\theta \le \partial_1 h(x_0+, \xi_{a+1-} + D) + n\theta$, by $(H_{6\theta})$, (H_1), (H_3), and the fact that $\xi_{a+1-} + D \le x_1 + n$. The last follows from (7.9), (H_1) and the fact that minimal segments do not cross [3, Lemma 3.1]. By 7.9, we have $-\varepsilon \le x_0 - \xi_{a-} - D$, $x_q - \xi_{a+q-} - D \le 0$, so the above inequalities yield

$$h(\xi_0', \xi_1') - h(\xi_{a-}, \xi_{a+1-}) \le \partial_1 h(x_0, x_1)(x_0 - \xi_{a-} - D) + \theta(n\varepsilon + \varepsilon) ,$$

$$h(\xi_{q-1}', \xi_q') - h(\xi_{a+q-1-}, \xi_{a+q-}) \le \partial_2 h(x_{q-1}, x_q)(x_q - \xi_{a+q-} - D) + $$
$$+ \theta(n\varepsilon + \varepsilon) ,$$

where $n = [\varepsilon] + 1$. Therefore

$$h(\xi_0', \ldots, \xi_q') - h(\xi_{a-}, \ldots, \xi_{a+q-}) \le -Y + \theta(2n\varepsilon + 2\varepsilon) .$$

Combining this with the upper bound we found previously for $h(x_0, \ldots, x_q)$, we obtain

$$h(x_0, \ldots, x_q) + Y \le h(\xi_{a-}, \ldots, \xi_{a+q-}) + \theta(2[\varepsilon] + 4)\varepsilon .$$

Combining (7.13) and the above inequality, we obtain

$$|h(x_0, \ldots, x_q) + Y - h(\xi_{a-}, \ldots, \xi_{a+q-})| \le 23\theta\varepsilon , \qquad (7.14)$$

since $\varepsilon \le 7$.

To finish, it is enough to approximate $h(v_a'', \ldots, v_{a+q}'') - h(u_0, \ldots, u_q)$. We do this by approximating $h(v_a, \ldots, v_{a+q}) - h(u_0, \ldots, u_q)$ and $h(v_a'', \ldots, v_{a+q}'') - h(v_a, \ldots, v_{a+q})$. The procedure to approximate the first is similar to the procedure we just used to obtain (7.14). Note that $(v_a, \ldots, v_{a+q})$ is minimal, subject to the condition $v_0 = \xi$. For, by definition, v is minimal subject to this condition and the further condition $\xi_{i-} \le v_i \le \xi_{i+}$; moreover, it still minimizes over the larger set where we drop this supplementary condition, by an argument similar to [3, Lemma 3.1], since ξ_- and ξ_+ are minimal. Note that by (7.9),

$$x_0 \le v_a + D \le x_0 + \varepsilon , \qquad (7.15a)$$

$$x_q \le v_{a+q} + D \le x_q + \varepsilon , \qquad (7.15b)$$

and, using [3, Lemma 3.1], in addition, we have

$$x_{-a} \le \xi_{0-} + D \le \xi + D \le \xi_{0+} + D \le U^{3B+1} x_{-a} .$$

It follows, by yet another application of the proof of [3, Lemma 3.1] that

$$x_i \le u_{i+a} + D \le U^{3B+1} x_i ,$$

for all integers i. This is because x and $U^{3B+1}x$ are minimal and periodic and u is periodic (of the same period as x) and minimal subject to the conditions $u_0 = \xi$ and periodicity. In particular, we have

$$x_0 < u_a + D < U^{3B+1}x_0 < x_0 + \varepsilon , \quad (7.16a)$$

$$x_q < u_{a+q} + D < U^{3B+1}x_0 < x_0 + \varepsilon . \quad (7.16b)$$

Having established (7.15) and (7.16) we may argue just as before, and we obtain

$$|h(u_0,\ldots,u_q) + Y' - h(v_a,\ldots,v_{a+q})| < 23\theta\varepsilon , \quad (7.17)$$

where $Y' = \partial_1 h(x_0,x_1)(v_a - v_{a+q} + p)$.

Note that $Y - Y' = \partial_1 h(x_0,x_1)(v''_a - v''_{a+q} - v_a + v_{a+q})$, since $\xi_{i-} = v''_i$, for $i = a$ or $a + q$. We have $|h(v''_a,\ldots,v''_{a+q}) - h(v_a,\ldots,v_{a+q}) - Y + Y'| < |h(v''_a,v_{a+1}) - h(v_a,v_{a+1}) + h(v_{a+q-1},v''_{a+q}) - h(v_{a+q-1},v_{a+q}) - Y + Y'| < |\partial_1 h(v_a,v_{a+1})(v''_a - v_a) + \partial_2 h(v_{a+q-1},v_{a+q})(v''_{a+q} - v_{a+q}) - Y + Y'| + \theta([\varepsilon] + 1)\varepsilon < 3([\varepsilon] + 1)\varepsilon$. In the last inequality, we have used $|\partial_1 h(v_a,v_{a+1}) - \partial_1 h(x_0,x_1)| < [\varepsilon] + 1$, $|\partial_2 h(v_{a+q-1},v_{a+q}) - \partial_2 h(x_{q-1},x_q)| < [\varepsilon] + 1$, and $\partial_1 h(x_0,x_1) + \partial_2 h(x_{q-1},x_q) = 0$. The inequalities may be obtained in the same way as the corresponding inequalities in the proof of (7.14). Using $\varepsilon < 7$, we obtain

$$|h(v''_a,\ldots,v''_{a+q}) - h(v_a,\ldots,v_{a+q}) - Y + Y'| < 37\theta\varepsilon . \quad (7.18)$$

From (7.2), (7.3), (7.11), (7.12), (7.14), (7.17), and (7.18), we obtain

$$\begin{aligned}|P_\omega(\xi) - P_{p/q}(\xi)| &< |h(v''_a,\ldots,v''_{a+q}) - h(\xi_{a-},\ldots,\xi_{a+q-}) - \\ &\quad - h(u_0,\ldots,u_q) + h(x_0,\ldots,x_q)| + 14\theta\varepsilon \\ &< |h(x_0,\ldots,x_q) + Y - h(\xi_{a-},\ldots,\xi_{a+q-})| + \\ &\quad + |h(u_0,\ldots,u_q) + Y - h(v''_a,\ldots,v''_{a+q})| + \\ &\quad + 14\theta\varepsilon \\ &< 23\theta\varepsilon + 23\theta\varepsilon + 37\theta\varepsilon + 14\theta\varepsilon \\ &= (97)\theta\varepsilon < (100)\theta\varepsilon .\end{aligned}$$

We have proved this estimate for $p/q < \omega \le (p+1)/q$. As we remarked above, it follows that $|P_\omega(\xi) - P_{p/q}(\xi)| \le (200)\theta\varepsilon$ for $p/q < \omega$, and we may prove this estimate for $\omega < p/q$ in the same way. Since $\varepsilon \le 6(q^{-1} + |q\omega^* - p|)$, we have proved Theorem 7.1. □

§8. A MODULUS OF CONTINUITY

In [5], we proved that ΔW_ω is continuous at irrational ω. The quantity ΔW_ω is closely related to $P_\omega(\xi)$, and the methods of [5] also show that $\omega \to P_\omega(\xi)$ is continuous at irrational ω. Likewise, it should be possible to prove the estimate of Theorem 7.1 for ΔW_ω in place of $\omega \to P_\omega(\xi)$, by a slight modification of the proof given in the previous section.

In this section, we obtain a modulus of continuity for $\omega \to P_\omega(\xi)$ at any irrational number ω, as a consequence of Theorem 7.1. It should be possible to also obtain such a modulus of continuity by the methods of [5], but we were unable to obtain as sharp an estimate by the method of [5] as we obtain here. In particular, we show here that $\omega \to P_\omega(\xi)$ satisfies a Hölder condition at Diophantine ω. We have been unable to obtain such a result by the method of [5]. In proving the continuity of ΔW_ω in [5], we used the strict convexity [9] of Percival's Lagrangian. The estimates showing strict convexity which we have been able to obtain are not good enough to show that ΔW_ω satisfies a Hölder condition at Diophantine ω. In this sense, the methods of this paper are an improvement over the methods of [5].

The modulus of continuity for $\omega \to P_\omega(\xi)$ which we obtain depends on how well ω may be approximated by rational numbers. We review a few salient facts concerning such approximations. A rational number p/q is said to be a <u>best rational approximation</u> of an irrational number ω if $|\omega q - p|$ is smaller than $|\omega q' - p'|$ for any $p', q' \in \mathbf{Z}$ with $0 < q' \le q$. It is obvious that any irrational number admits an infinite number of best rational approximations. It is a well known and easy consequence of the pigeon hole principal that if p/q is a best rational approximation to an irrational number ω, then $|\omega q - p| < q^{-1}$. One says that ω satisfies a <u>Diophantine condition of order</u> α if there exists $\varepsilon > 0$ such that

$|q\omega - p| \ge \varepsilon q^{-\alpha}$ for all $p,q \in \mathbf{Z}$.

<u>Corollary 8.1. Let h be a continuous real valued function on $\mathbf{R}^2$ satisfying (H_1)-(H_5) and $(H_{6\theta})$. Let $P_\omega(\xi) = P_{\omega,h}(\xi)$ denote Peierls's barrier with respect to h. Let ω be an irrational number, p/q a best rational approximation of ω, and ρ a rotation symbol such that $|\omega - \rho^*| < q^{-2}$. Then</u>

$$|P_\omega(\xi) - P_\rho(\xi)| \le (6000)\theta/q .$$

<u>Proof</u>. Immediate from Theorem 7.1. □

This estimate gives the desired modulus of continuity. In particular, we have the following two results (as a consequence of Corollary 8.1):

Corollary 8.2. $\omega \to P_\omega(\xi)$ is continuous at any irrational number ω, uniformly in ξ. □

Corollary 8.3. If ω satisfies a Diophantine condition of order α, then $\omega \to P_\omega(\xi)$ satisfies a Hölder condition of order $1/2\alpha$ at ω, i.e. $|P_\omega(\xi) - P_\rho(\xi)| \leq \text{const.}|\omega - \rho^*|^{1/2\alpha}$, for $|\omega - \rho^*| \leq 1$.

Proof. Suppose $|q\omega - p| \geq \epsilon q^{-\alpha}$, for all $p,q \in \mathbf{Z}$. Let ρ be a rotation symbol with $|\omega - \rho^*| \leq 1$. Let p/q be the best rational approximation to ω with $q < |\omega - \rho^*|^{-1/2}$ and q as large as possible subject to this condition. Let P/Q be the best rational approximation to ω with $Q > q$ and Q as small as possible subject to this condition. Since $|q\omega - p| \geq \epsilon q^{-\alpha}$, we have $Q \leq q^\alpha/\epsilon$, by the pigeon hole principal. Since $Q \geq |\omega - \rho^*|^{-1/2}$, we have $q \geq (\epsilon|\omega - \rho^*|)^{-1/2\alpha}$. Then $|P_\omega(\xi) - P_\rho(\xi)| \leq (6000)\theta(\epsilon|\omega - \rho^*|)^{1/2\alpha}$, by Corollary 8.1. □

REFERENCES

[1] S. Aubry and P. Y. Le Daeron, 'The discrete Frenkel-Kontorova model and its extensions', Physica **8D** (1983), 381-422.

[2] S. Aubry, P. Y. Le Dearon, and G. André, 'Classical ground-states of a one-dimensional model for incommensurate structures'. Preprint (1982).

[3] V. Bangert, 'Mather sets for twist maps and geodesics on tori'. Preprint (1986), to appear in Dynamics Reported.

[4] A. Katok, 'More about Birkhoff periodic orbits and Mather sets for twist maps'. Preprint (1982).

[5] J. Mather, 'A criterion for the non-existence of invariant circles'. Publ. IHES (1986), 153-204.

[6] J. Mather, 'More Denjoy minimal sets for area preserving diffeomorphisms', Comment. Math. Helvetici **60** (1985), 508-557.

[7] J. Mather, 'Dynamics of area preserving mappings', to appear in Proceedings of ICM, 1986.

[8] J. Mather, 'Existence of quasi-periodic orbits for twist homeomorphisms of the annulus', Topology **21** (1982), 457-467.

[9] J. Mather, 'Concavity of the Lagrangian for quasi-periodic orbits', Comment. Math. Helvetici **57** (1982), 356-376.

CHAOTIC ORBITS IN THE THREE BODY PROBLEM

Richard Moeckel
School of Mathematics
University of Minnesota
Minneapolis, MN 55455

ABSTRACT. This paper describes the construction of a compact invariant set for the planar three body problem consisting of orbits which pass near the triple collision singularity. As they do so they exhibit chaotic changes of configuration. The invariant set is described via symbolic dynamics.

INTRODUCTION

The compact invariant set we will construct contains infinitely many new periodic orbits embedded in a rich network of homoclinic and heteroclinic connections. As these solutions pass near the singularity they can exhibit dramatic changes of configuration; for example, an orbit may approach collision with the three masses forming a nearly perfect equilateral triangle and emerge from a neighborhood of the singularity nearly collinear. The whole invariant set is described via symbolic dynamics, so by choosing "random" sequences of symbols we can also produce orbits with wildly varying configurations.

The research described here is presented with complete proofs in two papers [M1,M2]. It was partially supported by NSF grant, and by the Mathematics Research Institute at Berkeley. The author also wishes to thank the organizers of the Il Ciocco conference for inviting him to give the talk on which this paper is based.

1. PHASE SPACE

The planar three body problem concerns the motion of three point masses m_1, m_2, m_3 under the influence of their mutual gravitational attraction. It is a Hamiltonian dynamical system with Hamiltonian function:

$$H(p,q) = 1/2\ p^T A p - U(q)\ .$$

P. H. Rabinowitz et al. (eds.), Periodic Solutions of Hamiltonian Systems and Related Topics, 203–219.

Here $q = (q_1, q_2, q_3)$ and $p = (p_1, p_2, p_3)$ where q_j is the position of the jth mass and p_j is its momentum. Thus p and q are elements of R^6. A is the 6×6 diagonal matrix $\text{diag}(m_1, m_1, m_2, m_2, m_3, m_3)$ and $U(q)$ is minus the potential energy:

$$U(q) = \sum_{i<j} \frac{m_i m_j}{\|q_i - q_j\|} .$$

Note the singularities on the set $\Delta = \{q_i = q_j \text{ for some } i \neq j\}$.

This twelve-dimensional phase space can be reduced to only five dimensions by making use of the well known integrals of motion. We may assume without loss of generality that the total momentum $p_1 + p_2 + p_3 = 0$ and that the center of mass $m_1 q_1 + m_2 q_2 + m_3 q_3 = 0$. This reduces the number of dimensions by four. The angular momentum $p \times q = \Sigma\, p_j \times q_j$ is a constant, ω, and the Hamiltonian is invariant under a simultaneous rotation of all position and momentum variables in the plane. By fixing ω and passing to a quotient manifold under the symmetry we eliminate two more dimensions. Finally, the energy $H(p,q) = h$ can be fixed. We always choose $h < 0$. Let $M(h,\omega)$ denote the five-dimensional manifold obtained in this way. The topological structure of $M(h,\omega)$ depends in a complicated way on h, ω. However, we will always be concerned with the case of small angular momentum (so that close approaches to triple collision are possible). It turns out that for $h < 0$ and $\omega \neq 0$ sufficiently small $M(h,\omega)$ is homeomorphic to the Cartesian product of a two-sphere with three deleted points and a three-sphere. The first factor is the q space with the singular set Δ deleted and the circular symmetry quotiented out while the second represents the p space. It is interesting to note that the zero angular momentum manifold $M(h,0)$ has a different structure; it is homeomorphic to the product of the thrice punctured two-sphere and an open disk. This anomoly will disappear when appropriate normalized variables are introduced later.

The five famous periodic orbits of Euler and Lagrange provide a framework for the construction of our invariant set. These arise from the so-called relative equilibria of the three body problem. There are five special configurations of the three masses with the property that the gravitational force on each mass is directed toward the center of mass with magnitude proportional to the distance from it. It follows that this force could be balanced by a centrifugal force if the configuration were uniformly rotated at an appropriate rate. Thus each such configuration gives rise to a simple periodic orbit. For the three body problem the relative equilibrium configurations are the two rotationally distinct equilateral triangular ones and three collinear ones for which the exact spacing is determined by the masses and by which mass is in the middle. Figure 1 depicts the five relative equilibria for some choice of masses. It turns out that the circular orbits are part of a family of elliptical periodic orbits parametrized by angular momentum, ω (Figure 2). As the angular momentum tends to 0 the ellipses become more and more eccentric and in the limit we get a solution which begins and ends in triple collisions. Between the collisions the configuration is a homothetic expansion and contraction.

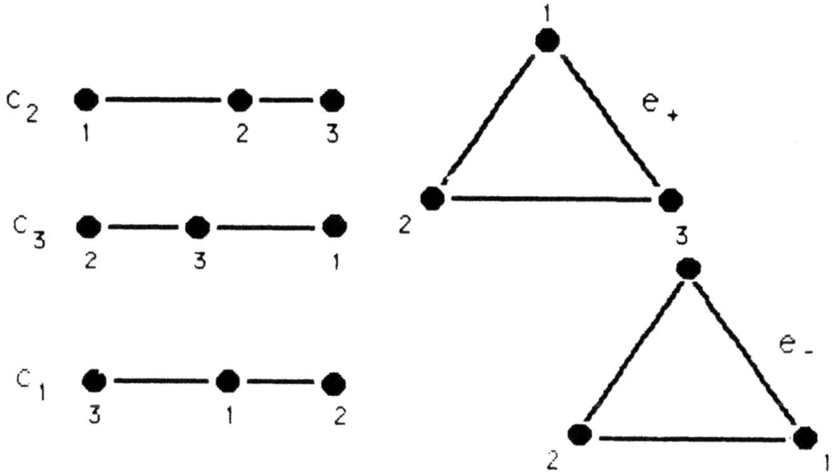

FIGURE 1

As noted above, we are primarily interested in the case of small angular momentum. It turns out that for ω sufficiently small the two equilateral periodic orbits are hyperbolic. Viewed in the five-dimensional manifold $M(h,\omega)$ they have three-dimensional stable and unstable manifolds. Thus the possibility of transverse homoclinic and/or heteroclinic orbits arises.

Theorem. For a nonempty open set of mass triples and for ω sufficiently small, there are orbits homoclinic to the equilateral solutions of Lagrange and orbits heteroclinic between the two of them.

The nonempty set of mass triples contains all triples m_1, m_2, m_3 such that two masses are sufficiently close to equal. However, only technicalities prevent the proof from being valid for **all** masses. The homoclinic and heteroclinic orbits are just a small part of the compact invariant set mentioned in the introduction; it is to this construction that the rest of the paper is devoted.

We study the case of small angular momentum by viewing it as a perturbation of the limiting case of zero angular momentum. We have already noted the anomolous behavior of the limiting process. The introduction of new coordinates eliminates this difficulty. The problem originates in the triple collision singularity. Triple collision can occur only when $\omega = 0$. If it does occur, it must be at the center of mass, i.e., at the origin. McGehee developed a system

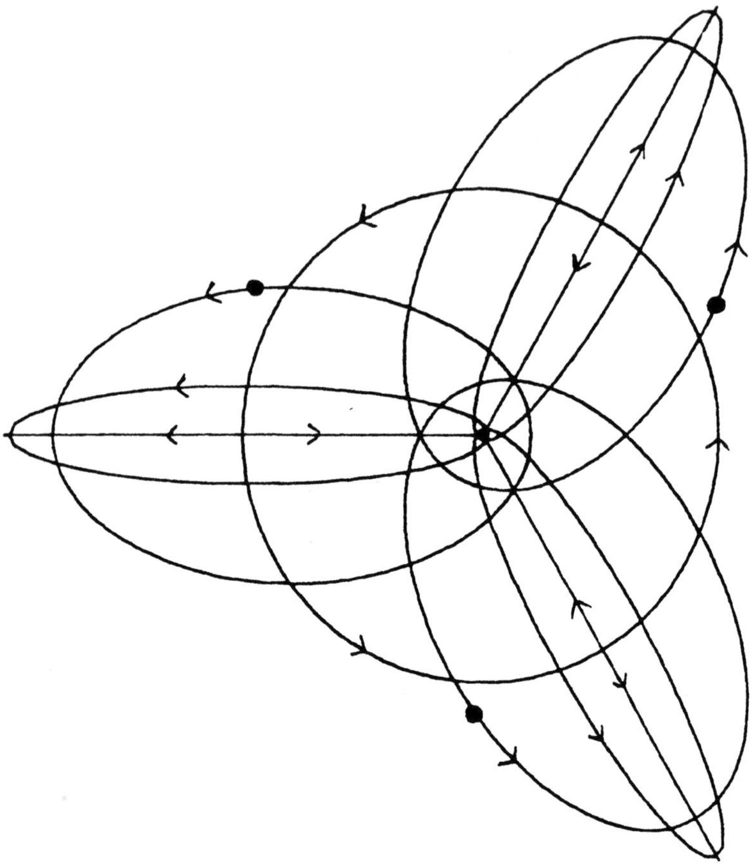

FIGURE 2

of "polar" coordinates in q space which "blow up" the origin. Together with scaling of the p variables and a change of timescale these coordinates replace the triple collision singularity by a smooth four-dimensional manifold which forms a boundary to $M(h,0)$. Moreover, the vectorfield extends smoothly to this collision manifold and the extension reflects the behavior of near collision orbits in $M(h,0)$ [Mc]. However, it does not reflect the limiting behavior of families of orbits of $M(h,\omega)$ as $\omega \to 0$. For this we consider the limit of the manifolds $M(h,\omega)$. In the space of the new variables, $M(h,\omega) \to M(h,0) \cup \partial M(h,0) \cup M_0$ where $\partial M(h,0)$ is the four-dimensional collision manifold mentioned above and M_0 is another five-dimensional manifold which fits together with $M(h,0)$ at a corner along $\partial M(h,0)$ as indicated schematically in Figure 3. This new component is fictitious in the sense that it lies entirely in the set blown up from the origin by the coordinate change. However, the convergence of $M(h,\omega)$ to its limit is continuous so the flow on M_0 is the limit of parts of the flows on $M(h,\omega)$. For example, the five families of periodic orbits have limits which are restpoint cycles consisting of two restpoints in $\partial M(h,0)$ connected in one direction by an orbit in $M(h,0)$ and in the other direction by an orbit in M_0. Figure 3 is also intended as a schematic depiction of this limiting process. The orbit segment in $M(h,0)$ is just the homothetic expansion and contraction described above as the limit of the elliptical orbits with eccentricity approaching 1. This orbit does not accurately reflect the behavior of the $\omega \neq 0$ orbits near collision; the homothetic orbit begins and ends in collision whereas the elliptical orbits spin around very near collision without colliding. It turns out that the segment of the limiting cycle in M_0 is the limit of this spinning behavior as $\omega \to 0$; the blown up configuration variables spin rigidly around by 360°. The limiting orbit fails to be periodic only because of the restpoints.

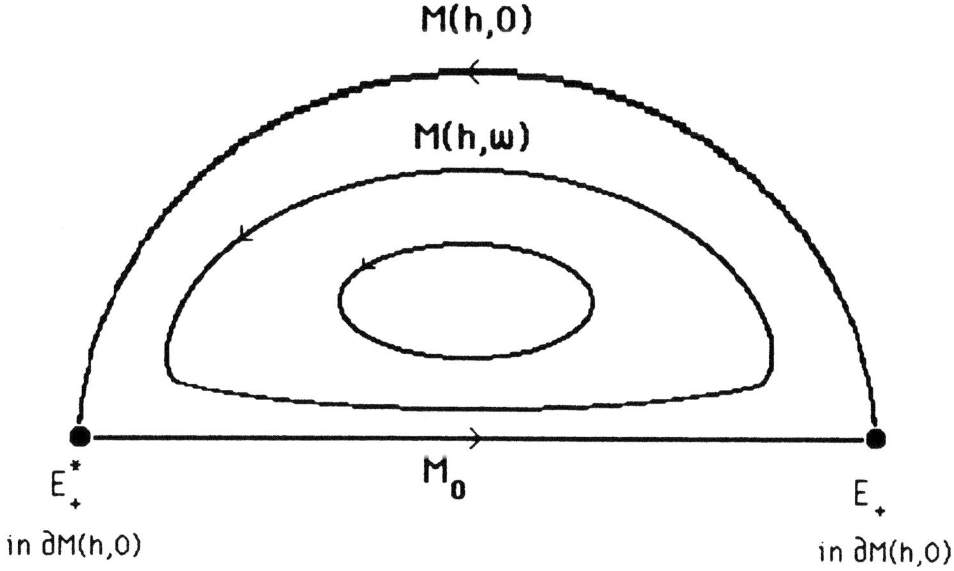

FIGURE 3

The restpoints and the connecting orbits play an important role in what follows. We will now describe some of their properties. There are exactly ten restpoints in $\partial M(h,0)$, two in each of the five restpoint cycles. All of the restpoints are hyperbolic in $\partial M(h,0)$ and the flow on this manifold is gradient-like, i.e., there is a Lyapunov function which decreases on all nonconstant solutions. The labelling of the restpoints is derived from that of the corresponding relative equilibria (Figure 1). For example, the two restpoints in the cycle corresponding to the equilateral configuration e_+ are called E_+ and E_+^* with the star on the one which has the higher value of the Lyapunov function. These equilateral restpoints and the corresponding ones with subscript $-$ have two-dimensional stable and unstable manifolds in $\partial M(h,0)$. Viewed in the two five-dimensional components $M(h,0)$ and M_0 they have three-dimensional stable and unstable manifolds. The unusual phenomenon of a hyperbolic restpoint in a five-dimensional space having both of its invariant manifolds three-dimensional is explained by the fact that $M(h,0)$ and M_0 meet at a corner near the restpoints. A similar thing happens to a saddle point in the plane if we restrict the flow to the positive x and y axes.

The equilateral restpoint cycle contains, besides the restpoints themselves, connecting orbits of the form $E_+ \to E_+^*$ in $M(h,0)$ and $E_+^* \to E_+$ in M_0. These orbits are intersections of the invariant manifolds $Un(E_+) \cap St(E_+^*)$ and $Un(E_+^*) \cap St(E_+)$ respectively. Both of these are transverse intersections of three-dimensional manifolds inside five-dimensional ones. Thus the restpoint cycle can be viewed as a kind of hyperbolic invariant set with three-dimensional stable and unstable manifolds just like the elliptical Lagrange orbits of which it is the limit. To produce the homoclinic orbits of the theorem we will look first for orbits homoclinic to the restpoint cycle and then perturb to nonzero angular momentum. The homoclinic orbits we will find will be further transverse intersections of $Un(E_{+,-})$ and $St(E_{+,-}^*)$ in $M(h,0)$.

The mechanism by which these further intersections are produced is best illustrated in the special case of two equal masses. In this case there is an invariant three-dimensional subset of $M(h,0) \cup \partial M(h,0) \cup M_0$ consisting of orbits whose configuration is always an isosceles triangle with the two equal masses symmetrically placed. This subsystem has been studied by several authors [D1,S-L,M3]. Figure 4 depicts the part of this three-dimensional subsystem near triple collision. The surface in the figure is the isosceles part of $\partial M(h,0)$. The exterior of the surface is in $M(h,0)$ while the interior is in M_0; the fact that they meet at a corner along the surface could unfortunately not be adequately sketched. We will only describe the part in $M(h,0)$ since this is where we are looking for the additional intersections of $Un(E_{+,-})$ and $St(E_{+,-}^*)$.

Three of the five relative equilibria are isosceles configurations: the two equilateral ones and the collinear one with the odd mass between the two equal ones. As a result the surface of Figure 4 contains six of the ten restpoints of the full system. Moreover, the homothetic orbits connecting the unstarred restpoints to the starred ones in $M(h,0)$ are isosceles and so are also shown in the figure. Now viewed from within the two-dimensional collision surface, the equilateral restpoints $E_{+,-}$ and $E_{+,-}^*$ are saddles while the collinear restpoints C and C^* are respectively a sink and a source. If the odd mass is not too large compared to the equal ones (mass ratio not more that 55/4) the eigenvalues of the linearized

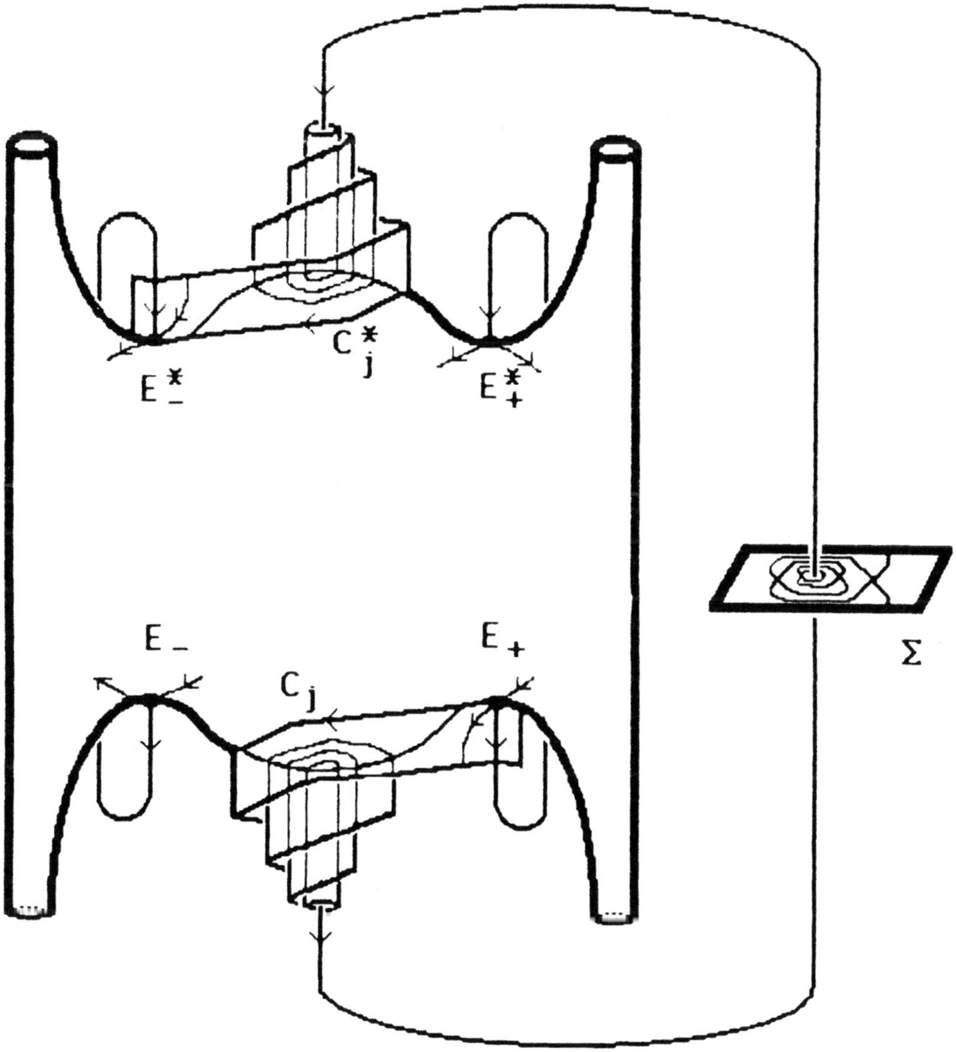

FIGURE 4

equations near the collinear restpoints are non-real complex conjugates and so nearby orbits spiral around them. If we now view the restpoints in $M(h,0)$ the starred ones get an extra stable eigenvalue while the unstarred ones get an extra unstable eigenvalue. Thus $Un(E_{+,-})$ and $St(E^*_{+,-})$ are both two-dimensional manifolds in the three-dimensional manifold $M(h,0)$. Note that within the collision surface certain branches of the invariant manifolds of the equilateral saddles fall into the spiraling sink and source. The effect of this on the larger invariant manifolds in $M(h,0)$ is amply illustrated in the figure. If $St(E^*_{+,-})$ is followed in backward time along the branch coming from C^* it coils around the collinear homothetic orbit like a scroll. Similarly if $Un(E_{+,-})$ is followed in forward time it coils around the same orbit. If we erect a cross section, Σ, to the flow along the collinear orbit the manifolds intersect it in curves which spiral in opposite senses converging to the point where the collinear orbit hits the section. As a result there are infinitely many crossings of the two manifolds in any neighborhood of this point. These crossings are at least topologically transverse since the manifolds involved are all real analytic and so have crossings of finite order.

This three-dimensional subsystem is present only for the case of two exactly equal masses. Furthermore, we have no information about the transversality of these intersections when they are viewed in the full five-dimensional manifold $M(h,0)$. However, the features of the flow which underlie the isosceles proof are present in the full problem as well even when no pair of masses are equal. We will briefly summarize what were the important features. First, there are complex eigenvalues at the collinear restpoints. Second, there are transverse connecting orbits of the form $C^* \dashrightarrow E^*_{+,-}$ and $E_{+,-} \dashrightarrow C$ within the collision manifold $\partial M(h,0)$. Third, curves which spiral in opposite directions around the same point tend to intersect. We will try to find analogous features in the five-dimensional setting.

In the full planar three body problem complex eigenvalues are present for every choice of the masses. Recall that there are three collinear restpoint pairs $\{C_j, C^*_j\}$ distinguished by which mass is between the other two. The nature of these restpoints and in fact of the whole flow depends on the mass ratios rather than on the mass values themselves. In the simplex of normalized mass triples, $m_1 + m_2 + m_3 = 1$, we have indicated in Figure 5 those triples for which each collinear restpoint has real eigenvalues. It can be seen that there is a large open set in which all three collinear restpoints exhibit spiraling and that always at least two of the three do so. Thus the first ingredient of the isosceles construction is present in abundance in the planar problem.

It is proved in [M_4] that the second ingredient, the connecting orbits $C^*_j \dashrightarrow E^*_{+,-}$ and $E_{+,-} \dashrightarrow C_j$, $j = 1,2,3$, are present in $M(h,0)$ for all mass triples. Furthermore they represent at least topologically transverse intersections of the corresponding invariant manifolds. For technical reasons we require C^1 transversality in order to study the way that the invariant manifolds coil up as they pass near the collinear restpoints. We have succeeded in proving C^1 transversality only for mass triples close to the isosceles case of two equal masses. It should be emphasized however, that the topologically transverse connections obtained in [M_4] are probably C^1 transverse for most if not all mass triples and that even if they are not the construction could probably be pushed through.

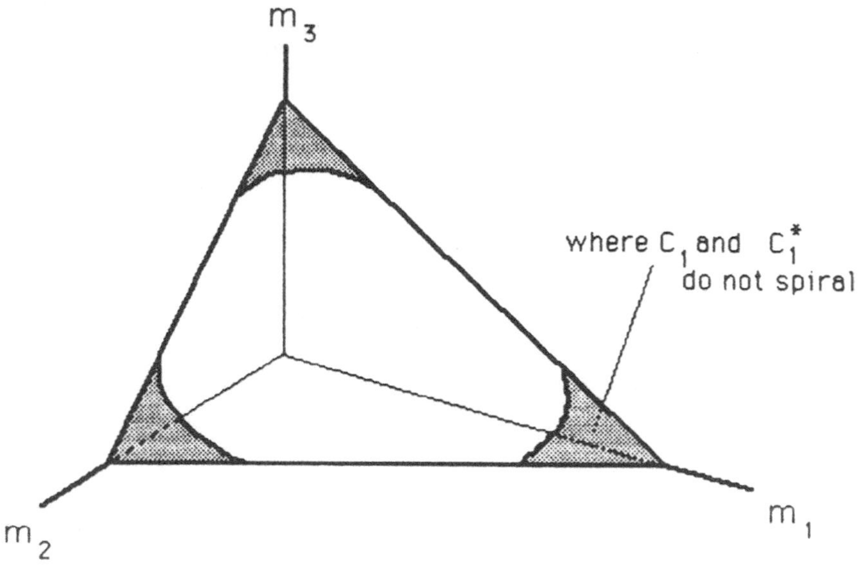

FIGURE 5

Now imagine following $Un(E_+)$, say, along a $E_+ \to C_j$ connecting orbit. Suppose that the masses have been chosen so that C_j is spiraling. What will happen to $Un(E_+)$ as it passes through a neighborhood of C_j? To properly answer this question requires some terminology to describe multidimensional spiraling manifolds. We will begin by stating what seem to be the key properties of "spirals" in dimensions two and three. The concept of a curve spiraling around a point in a plane is familiar enough; Figure 6 shows two possibilities for spiraling manifolds in $\mathbf{R}^3$. Both of them are spiraling around a circle, i.e., around a codimension-two submanifold of $\mathbf{R}^3$. This is also the case for a spiraling curve in $\mathbf{R}^2$. The presence of such a codimension-two manifold, C, to spiral around seems to be essential. Next, for convenience, we want spiraling manifolds to be parametrizable by angle at least near C. Here angle means the θ coordinate in the normal bundle to C; for a global θ to exist we assume that C is trivially embedded in the ambient space. Of course, if the spiraling manifold is of dimension greater than one other parameters will also be required in any parametrization. For example, in the figure, the spiraling surface can be viewed as a family of curves, one for each θ. The parameter describing the individual curves is essentially that of the central circle, C. Finally, again for convenience, we would like the family of manifolds of fixed θ to converge smoothly to a submanifold of C as $\theta \to \infty$ or as $\theta \to -\infty$. The submanifold of C to which a spiraling manifold converges is called its core. To summarize: a positive spiraling

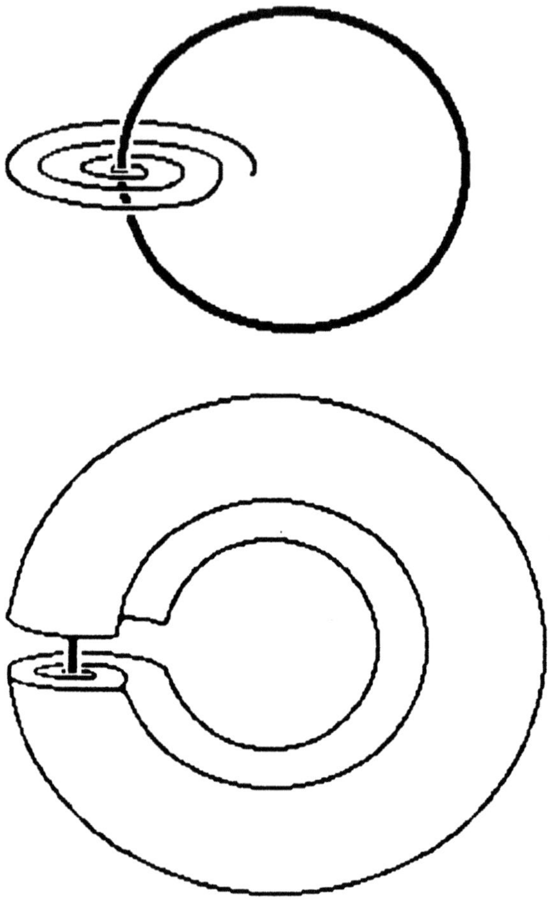

FIGURE 6

manifold, S, around a codimension-two manifold C consists of a core manifold S_∞ and a family of copies S_θ of S_∞ parametrized by angle around C and converging smoothly to S_∞ as $\theta \to \infty$. The definition of a negative spiraling manifold is obtained by changing ∞ to $-\infty$ throughout. It can be shown that a suitably technical version of this definition possesses coordinate invariance; in other words, applying a diffeomorphism to the ambient space takes a spiraling manifold around C to a spiraling manifold around the image of C. This property is important if we are to transport spirals from one part of phase space to another using the flow.

We now return to the three body problem where we were considering the fate of $Un(E_+)$ as it flows past C_j. $Un(E_+)$ and $St(C_j)$ are both three-dimensional so a transverse intersection between them consists of an isolated connecting orbit. Along this orbit we choose a two-dimensional disk in $Un(E_+)$ transverse to the flow (Figure 7). The central point of the disk lies on the connecting orbit but the other points on the disk will flow past C_j. The time required to flow past tends to ∞ as we approach the center of the disk. The claim is that the punctured disk emerges from a neighborhood of C_j as a spiraling surface.

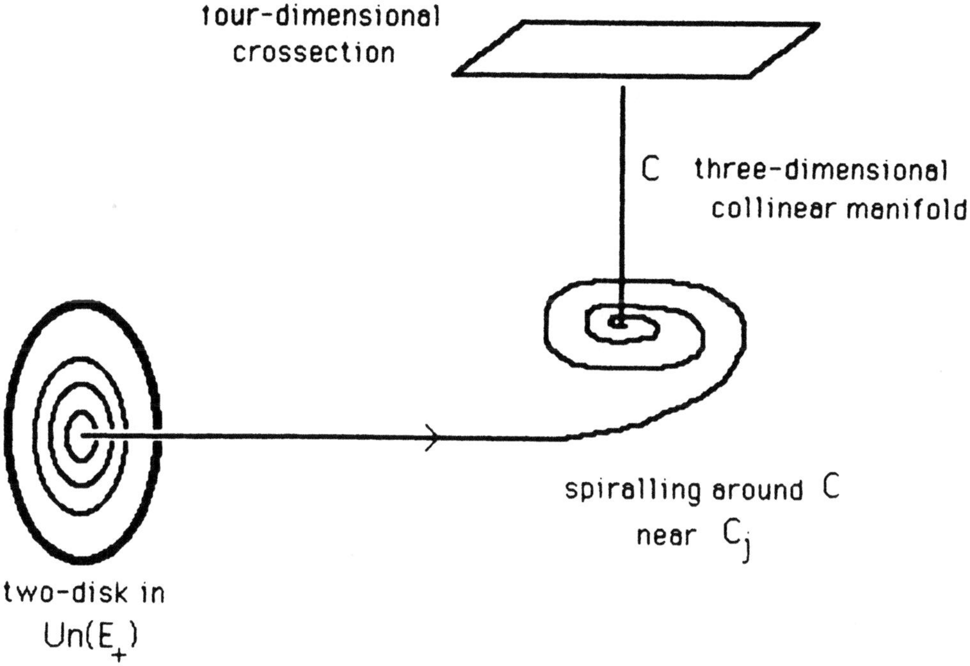

FIGURE 7

The first question to be addressed is: what does it spiral around? In the isosceles problem the collinear homothetic orbit was of codimension-two and the invariant manifolds spiraled around it. Now this orbit has codimension-four and cannot play the role of the central manifold. Fortunately, there is a natural codimension-two manifold to spiral around, the invariant collinear submanifold. If a collinear configuration is started with all three velocities along the line, then the configuration remains collinear for all time. The collection of all such collinear solutions forms a codimension-two subsystem, C, of M(h,0). Unlike the isosceles subsystem, the collinear subsystem in present for all choices of the mass triple. Of course, the collinear restpoints C_j, C_j^* and the collinear homothetic orbit lie inside C. Furthermore, the spiraling near C_j takes place in the two dimensions complementary to C. Now as our two-disk in $Un(E_+)$ flows past C_j the points closer to the center spend more time near the restpoint and so experience more spiraling and emerge with larger θ coordinates. More precisely, set up a four-dimensional cross section to the flow, Σ, along the collinear homothetic orbit near C_j. The center of this four-dimensional plane will be in the homothetic orbit. The manifold of collinear orbits, C, will intersect Σ in a two-dimensional surface. In the two complementary dimensions we set up a polar coordinate system. We follow points of the punctured disk in $Un(E_+)$ past C_j until they hit Σ. Then points starting closer to the center of the disk will emerge with larger θ coordinates; in fact it can be shown that the points of the punctured disk which will emerge with a given value of θ form a simple closed curve around the puncture as in Figure 7. Thus the image of this punctured disk in Σ can be viewed as a one parameter family of circles parametrized by θ and converging to some subset of $C \cap \Sigma$ as $\theta \to \infty$. In fact it is easy to see that this subset is precisely the unstable manifold of C_j intersected with Σ (which is indeed homeomorphic to a circle). In the terminology outlined above, the disk emerges as a surface in Σ around $C \cap \Sigma$ with core $Un(C_j)$. A similar thing happens to $St(E^*_{+,-})$ as it is followed in backward time past C^*_j. Because of the invariance of the concept of spiraling manifold under diffeomorphisms we may transport the spirals from $Un(E_{+,-})$ and those from $St(E^*_{+,-})$ to a common cross section Σ as in the isosceles problem. The question remains whether or not they must intersect.

To force intersections of multidimensional spirals, two things are necessary. First, as in the familiar cases, the manifolds must spiral in opposite senses. Second, we need a condition which guarantees that they will not pass through each other in the high dimensional ambient space. This will certainly be possible if the core manifolds do not intersect, since then the two spirals are converging to disjoint parts of C. For convenience we suppose that in fact the core manifolds intersect transversely in C. Then it can be shown that if the two spirals spiral in opposite senses, each point of intersection of the cores is a limit point of intersections of the spirals themselves [M2]. In our situation, the central manifold C is the manifold of collinear orbits and the core manifolds are $Un(C_j)$ and $St(C^*_j)$ which lie in C. These are each two-dimensional and so intersect Σ in curves; $C \cap \Sigma$ is a two-dimensional surface. It is a theorem of Devaney that these curves intersect transversely at the collinear homothetic orbit [D2]. It is not hard to show that the unstable spirals and stable spirals wind in opposite senses around C. In fact there is even a reflection symmetry carrying $St(E^*_{+,-})$ onto $Un(E_{+,-})$; such a symmetry reverses the direction of spiraling. It follows that each of $St(E^*_{+,-})$ intersects each of $Un(E_{+,-})$ infinitely often near the collinear homothetic orbit. Because of the real analyticity of the manifolds, there must in fact be infinitely many topologically transverse intersections.

In spite of the fact that Σ is four-dimensional it is possible and instructive to visualize the intersections of the spirals. Locally Σ will look like $\mathbf{R}^4$ and $C \cap \Sigma$ will look like $\mathbf{R}^2 \times \{(0,0)\} \subset \mathbf{R}^4$. Inside it, the core manifolds, $St(C_j^*)$ and $Un(C_j)$ will appear as transversely intersecting curves. We introduce polar coordinates in $\{(0,0)\} \times \mathbf{R}^2$. Fixing θ will produce a three-dimensional half-space as in Figure 8. Since the spiraling manifolds, $Un(E_+)$ and $St(E_-^*)$ say, are parametrizable by θ, they appear in each such half-space as curves. As we vary θ, these curves approach or recede from their respective core curves in $C \cap \Sigma$. For a given sense of changing θ, one will be approaching and one receding. From the figure, one can see how the combination of transversality of the core manifolds and opposite sense of spiraling produces transverse intersections of the spiraling manifolds near the point of intersection of the cores.

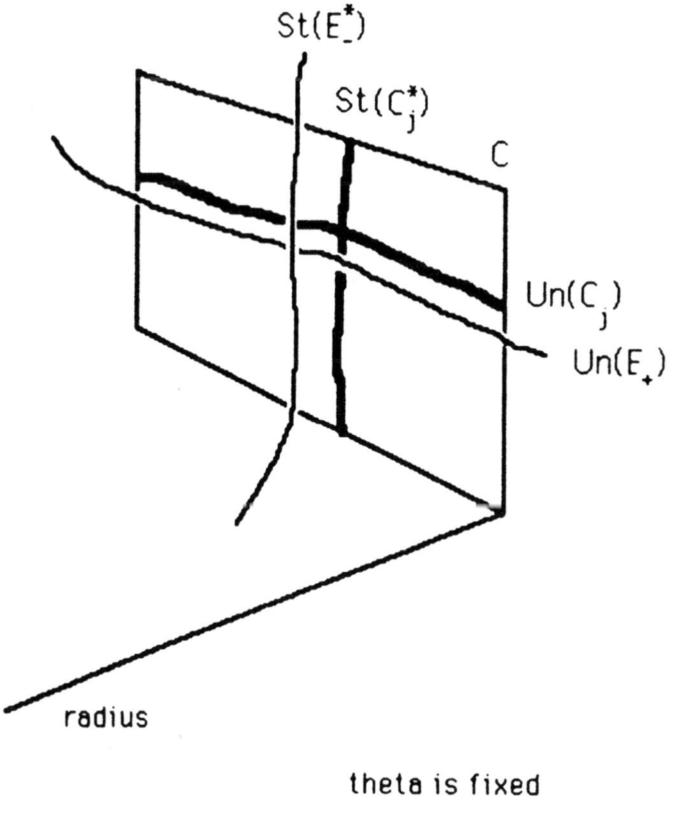

FIGURE 8

Having succeeded in finding many additional transverse connections $E_{+,-} \to E^*_{+,-}$ in $M(h,0)$ we now proceed to construct the promised invariant set for $\omega \neq 0$. We will use symbolic dynamics. Along each of the connecting orbits we construct a small four-dimensional box transverse to the flow; roughly speaking, two dimensions should be aligned with $Un(E_{+,-})$ and the other two with $St(E^*_{+,-})$. In addition, we construct a similar box along the $E^*_{+,-} \to E_{+,-}$ connecting orbit in M_0. We would like to say that there are Poincaré maps mapping these windows onto one another and stretching them in a favorable way. There is certainly abundant stretching taking place; the windows lie on orbits connecting hyperbolic restpoints so they flow forward up to these restpoints and are stretched out along the unstable manifolds. Recall however, that the invariant manifold $\partial M(h,0)$ prevents orbits in $M(h,0)$ from reaching M_0 and vice versa. This difficulty disappears as soon as we take $\omega \neq 0$. Then the restpoints vanish and the restpoint cycles become the equilateral periodic orbits of Lagrange. Now the stretched windows can flow on to cross the other windows. Specifically, the window set up along the $E^*_+ \to E_+$ connecting orbit in M_0 can be expected to flow to any window set up along $Un(E_+)$ in $M(h,0)$ as soon as we perturb to $\omega \neq 0$. In practice we can only get it to flow to finitely many of the infinitely many windows we set up on $Un(E_+)$.

A convenient device for organizing the information about which windows map to which under flow-defined Poincaré maps is a connection graph (Figure 9). The graph can be thought of in two ways. From one point of view, the vertices represent the four equilateral restpoints and the edges represent the connecting orbits we found. On the other hand we can think of the edges as representing the windows along these connecting orbits and a shared vertex as representing the existence of a Poincaré map between the corresponding windows for all sufficiently small nonzero ω.

We will formulate our main result in terms of this graph. One says that an orbit realizes a path in the graph if the orbit flows through the windows represented by the edges of the path in the order indicated by the path.

<u>Theorem</u>. Let any finite subgraph of the connection graph be given. Then for all sufficiently small nonzero ω, every path in the subgraph will be realized by at least one orbit of $M(h,\omega)$. Moreover, periodic paths are realized by at least one periodic orbit.

The proof uses the fact that the Poincaré maps stretch the boxes in a favorable way. It is similar to the more familiar symbolic dynamical arguments in dimension 2. Rather than dwell on it, we will describe some of the applications.

First we will locate the equilateral Lagrange orbits in the symbolic scheme. As $\omega \to 0$, we know that these orbits converge to restpoint cycles. Each of these cycles is represented in the graph by a periodic path consisting of two arrows: the arrow $E^*_{+,-} \to E_{+,-}$ and one of the infinitely many arrows in the other direction. It can

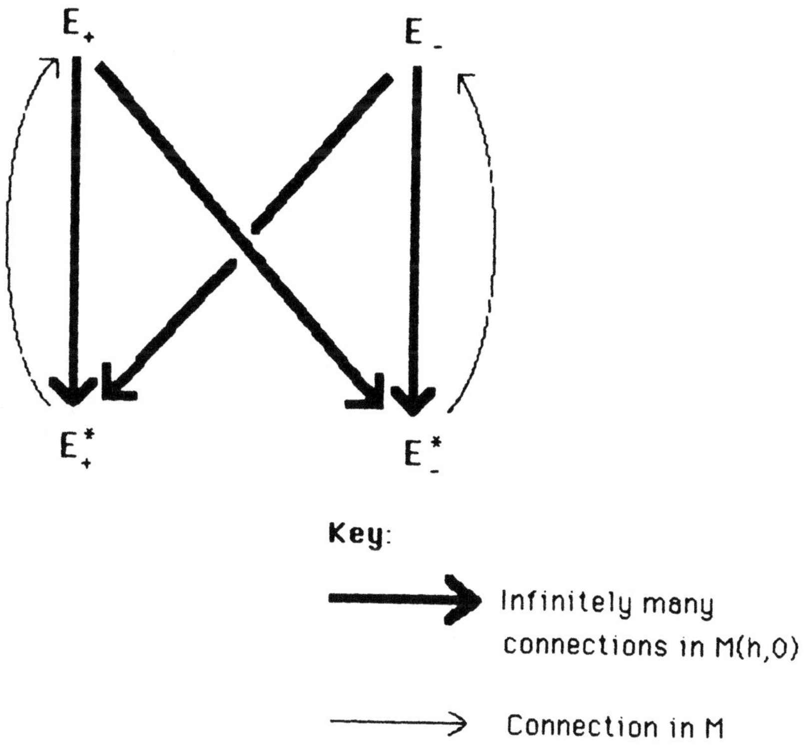

FIGURE 9

be shown that if the corresponding windows are sufficiently small, the equilateral Lagrange orbits are the only orbits which can realize this path. Now consider a path in the graph composed of an infinite repetition of the two Lagrange arrow followed by some other arrow, followed by another infinite repetition of the two Lagrange arrow. This orbit is not equal to the Lagrange orbit but tends to it in both forward and backward time, i.e., it is homoclinic to the Lagrange orbit. Similarly we can construct orbits heteroclinic between the two of them. This proves the previously mentioned theorem. Note that the behavior of these orbits while they are away from the Lagrange orbits is under our control to the extent that we can choose which window it will go through. If we choose a window very near to the collinear homothetic orbit then the qualitative behavior will be like that of the collinear homothetic orbit. So while the asymptotic behavior is equilateral, the intermediate behavior can be nearly collinear! By choosing periodic paths which incorporate both equilateral edges and nearly collinear ones we can construct periodic solutions of the three body problem which exhibit startling changes of configuration during their close approaches to triple collision; an example is shown in

Figure 10. Essentially anything is possible within the "selection rules" defined by the connection graph. Moreover by choosing aperiodic sequences we can produce seemingly random changes of configuration.

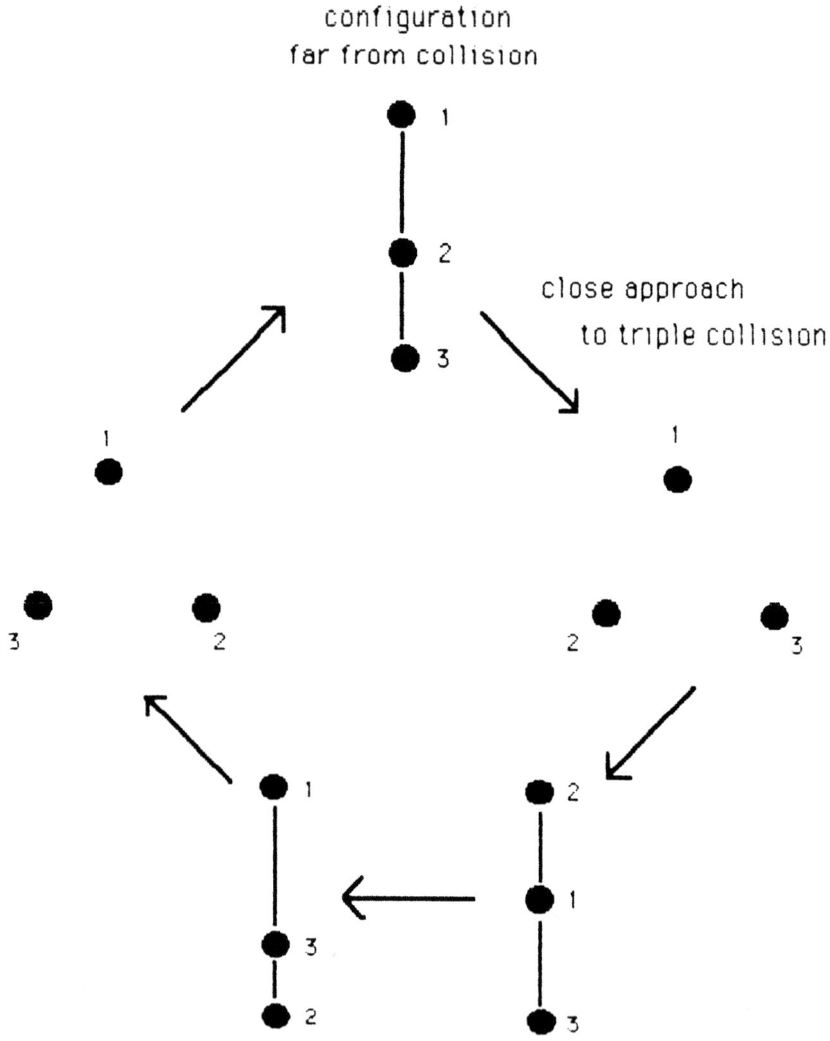

FIGURE 10

In summary, one can say that the stretching produced by the triple collision singularity combines with the recurrence of the classical orbits of Lagrange to produce chaotic behavior.

REFERENCES

D1 Robert L. Devaney, 'Triple collision in the isosceles three-body problem', Inv. Math. **60** (1980), 249-267.

D2 Robert L. Devaney, 'Structural stability of homothetic solutions of the collinear three-body problem', Cel. Mech. **19** (1979), 391-404.

M1 Richard B. Moeckel, 'Chaotic dynamics near triple collision', to appear in Arch. Rat. Mech.

M2 Richard B. Moeckel, 'Spiralling invariant manifolds', to appear in Jour. Diff. Eq.

M3 Richard B. Moeckel, 'Heteroclinic phenomena in the isosceles three-body problem', SIAM Jour. Math. Anal. **15** (1984), 857-876.

M4 Richard B. Moeckel, 'Orbits near triple collision in the three-body problem', Ind. Univ. Math. Jour. **32** (1983), 221-240.

Mc Richard McGehee, 'Singularities in classical celestial mechanics', Proc. Int. Cong. Math., Helsinki (1978), 827-834.

S-L C. Simo and E. Lacomba, 'Triple collision in the isosceles three-body problem', Bull. Amer. Math. Soc.

ON THE CONSTRUCTION OF INVARIANT CURVES AND MATHER SETS VIA A REGULARIZED VARIATIONAL PRINCIPLE

Jürgen Moser
Mathematik, ETH-Zentrum
8092 Zürich
Switzerland

ABSTRACT. In his work on monotone twist maps J. Mather constructed closed invariant sets with prescribed rotation number, using a degenerate variational principle. We propose a regularized variational principle with smooth solutions and show that they approximate the generally discontinuous solutions of Mather. This regularized variational principle involves the unknown function at a shifted argument. The Weierstrass excess function of calculus of variations is generalized to this situation.

1. INTRODUCTION

This note refers to monotone twist mappings of an annulus or a cylinder as they occur as section mappings in Hamiltonian systems of two degrees of freedom. For such mappings J. Mather developed an interesting theory establishing the existence of invariant sets for prescribed rotation numbers. These are, in general, Cantor subsets on closed Lipschitz curves, or, in favorable situations, invariant curves as they had been obtained previously by the so-called KAM theory under very restrictive assumptions. The orbits on these invariant sets are quasi-periodic, in a generalized sense, if the rotation number is irrational, and includes periodic orbits, if it is rational. For a recent survey article on Mather sets, see V. Bangert [1].

These Mather sets are rather complicated and they can be represented by discontinuous functions. It is the goal of this note to describe an alternate construction which yields approximations of these Cantor sets by smooth curves, which tend to the Cantor sets as the approximation parameter tends to zero. One may expect this approach to be useful for numerical computations since the discontinuities, annoying for calculations, are smoothed out by this procedure. The approach, to be described, is based on a regularized variational problem, which had been mentioned already in [3].

It is our aim to discuss more fully this variational problem which has the new feature of containing the unknown function at a shifted argument. In particular, we will extend the theory of

P. H. Rabinowitz et al. (eds.), Periodic Solutions of Hamiltonian Systems and Related Topics, 221–234.

extremal fields and the Weierstrass excess function to this situation. Moreover, we will see that the relevant functional has only a minimum as extremum and the solutions of the Euler equation with the corresponding boundary condition admits solutions which are unique up to a trivial phase shift.

In the last section we discuss a perturbation result in this context. It corresponds to the existence theorem of invariant curves as given by KAM theory, however, the proof has to be stripped of the use of repeated transformations usually employed. In fact, this led to a simpler proof*) of the existence proof for invariant tori avoiding the use of canonical transformation theory.

2. MATHER SETS

We restrict ourselves to the case of a cylinder which we represent by coordinates $(x,y) \in \mathbf{R}^2$ identifying points (x,y), (x',y) for which $x' - x$ are integers. The diffeomorphism $\Phi : \mathbf{R}^2 \to \mathbf{R}^2$ is given by

$$\Phi : (x_0,y_0) \to (x_1,y_1) = (f(x_0,y_0),g(x_0,y_0)) \tag{1}$$

where f,g are C^∞-functions satisfying

i) the periodicity condition:

$$f(x + 1,y) = f(x,y) + 1; \quad g(x + 1,y) = g(x,y) ,$$

ii) the condition to be exact symplectic:

$$\int y_0 dx_0 = \int y_1 dx_1$$

for any curve $y_0 = w(x_0) = w(x_0 + 1)$,

iii) and the condition of monotonicity

$$\frac{\partial f}{\partial y_0} > 0 .$$

We note that Mather's theory applies to homeomorphisms while we restrict ourselves to C^∞-diffeomorphisms.

We assume that, for each fixed $x_0, f(x_0,\cdot)$ maps $\mathbf{R}^1$ monotonically onto itself, so that we can represent y as a function of x and x_1. This leads to the representation of (1) by a generating function $h = h(x_0,x_1)$ in the form

$$y_0 = -h_1(x_0,x_1), \quad y_1 = h_2(x_0,x_1)$$

---

*)To be published in a paper by D. Salamon and E. Zehnder.

where h_1, h_2 denote the partial derivations of h with respect to the first or second argument. The conditions i)-iii) imply that

$$\left.\begin{aligned} &h(x_0 + 1, x_1 + 1) = h(x_0, x_1) \\ &-h_{12} > 0 \ . \end{aligned}\right\} \tag{2}$$

For symplicity we require that this condition holds uniformly, i.e. that

$$-h_{12} \geq \delta > 0 \quad \text{holds in} \quad \mathbb{R}^2 \ . \tag{3}$$

This condition implies that for fixed $y, x_0 \to -h_2(x_0, y)$ maps $\mathbb{R}^1$ monotonically onto $\mathbb{R}^1$ and that

$$h(x_0, x_1) \geq \frac{\delta}{2}(x_1 - x_0)^2 - c \tag{4}$$

with a constant c. This follows from (2), (3) and the identity

$$\begin{aligned} h(x_0, x_1) = h(x_0, x_0) &+ \int_{x_0}^{x_1} h_2(\eta, \eta) d\eta \\ &- \iint_{T(x_0, x_1)} h_{12}(\xi, \eta) d\xi dh \end{aligned} \tag{5}$$

where T is the right triangle two of whose sides lie on the vertical and the horizontal line through (x_0, x_1) and the third on the diagonal $(x_0 = x_1)$. In particular, h is bounded from below.

The problem is to find closed invariant curves $y = w(x)$ with a continuous w satisfying $w(x + 1) = w(x)$, or more generally invariant sets lying on such a curve, represented as a graph. In case of an invariant curve the mapping φ induces a circle mapping on $y = w(x)$ given by

$$x_0 \to f(x_0, w(x_0)) = x_1 \ ,$$

for which a rotation number α can be defined by

$$\alpha = \lim_{j \to \infty} \frac{x_j}{j} \ ,$$

x_j being the image of x_0 under the jth iterate of the above mapping.

In order to find an invariant set for a prescribed rotation number we represent this set (with Mather) parametrically as

$$x = u(\theta); \quad y = v(\theta) \tag{6}$$

where $u(\theta) - \theta$, $v(\theta)$ have period 1 and $u(\theta)$ is monotonically increasing, and such that the induced mapping is given by the rotation $\theta \to \theta + \alpha$. Analytically this requires the solution of the difference equation

$$\begin{aligned} u(\theta + \alpha) &= f(u(\theta), v(\theta)) \\ v(\theta + \alpha) &= g(u(\theta), v(\theta)) \end{aligned} \tag{7}$$

with the above specifications.

Under the above hypothesis, Mather's theorem [2] guarantees for any $\alpha \in \mathbf{R}$ the existence of a solution of the difference equation (7) for which

$$\begin{aligned} &\text{a)}\quad u(\theta) - \theta,\ v(\theta) \text{ have period } 1 \\ &\text{b)}\quad u(\theta) \text{ is strictly monotone increasing.} \end{aligned} \tag{8}$$

Such a monotone function u has at most countably many discontinuities. If it is continuous, then by (7) also v is continuous and represents an invariant curve. Its graph $y = w(x)$ is given by

$$w = v \circ u^{-1} . \tag{9}$$

If u has a jump at x_0^* then also at $x_0^* + j\alpha$, $j \in \mathbf{Z}$, and for irrational α the discontinuities are dense, $u(\theta)$ is the inverse of a Cantor function and the range of u is a one-dimensional Cantor set.

This paper is concerned with the solution of this difference equation (7) which with the generating function can be written in the equivalent form

$$v = -h_1(u, u^+), \quad v^+ = h_2(u, u^+)$$

where we define $u^{\pm}(\theta) = u(\theta \pm \alpha)$; $v^{\pm}(\theta) = v(\theta \pm \alpha)$. Eliminating v we obtain the difference equation

$$h_1(u, u^+) + h_2(u^-, u) = 0 , \tag{10}$$

which Mather solved with the help of the variational principle

$$\int_0^1 h(u, u^+)\, d\theta$$

devised by Percival. He obtained the solution of (10) by minimizing this functional over all functions satisfying (8). The existence proof of such a minimum is quite straightforward but the main difficulty is to verify that the "Euler equation" (10) holds for such a function class (8).

3. THE REGULARIZED VARIATIONAL PROBLEM

We propose to replace the above variational problem by

$$I_\nu^\alpha(u) = \int_0^1 \left(\frac{\nu}{2} u_\theta^2 + h(u,u^+)\right)d\theta; \quad u_\theta = \frac{\partial u}{\partial\theta}$$

by adding an artificial "viscosity term" with $\nu > 0$. We seek the minimum of this functional over the class X of u for which $\hat{u}(\theta) = u(\theta) - \theta$ belongs to the Sobolev space $H^1(S^1)$, i.e. $\hat{u}$ is required to have period 1 and its first derivative belongs to $L^2[0,1]$. No monotonicity is required. For h we have to require (2) and that it is bounded from below and smooth.

We summarize the main results in the following statements:

I) For $\nu > 0$ the above functional I_ν^α assumes its minimum in X and every minimal u of I_ν^α belongs to C^∞, i.e. $u - \theta \in C^\infty(S^1)$. Moreover, every minimal u satisfies the Euler equation

$$-\nu u_{\theta\theta} + h_1(u,u^+) + h_2(u^-,u) = 0 \tag{11}$$

where $u^\pm(\theta) = u(\theta \pm \alpha)$.

Every solution $u = u(\theta)$ of (11) with the boundary condition

$$u(\theta + 1) = u(\theta) + 1 \tag{12}$$

corresponds to an extremum of I_ν^α in X. The following argument shows that for fixed $\nu > 0$ this functional has in X a minimum as its only extremum, and the corresponding minimal $u = u(\theta)$ is unique up to the translation $\theta \to \theta + \text{const}$. We shall conclude this from the following statements, which are simple consequences of the maximum principle.

II) If $u_1, u_2 \in C^2(\mathbb{R})$ are any two solutions of (11) for $\nu > 0$ satisfying $u_1 \leq u_2$, then either $u_1 \equiv u_2$ or $u_1 < u_2$.

III) If $u_1, u_2 \in C^2(\mathbb{R})$ are any two solutions of (11) with $\nu > 0$ satisfying (12) then there exists a constant c such that

$$u_2(\theta) = u_1(\theta + c) ,$$

i.e. these solutions are unique up to the translation $\theta \to \theta + \text{const}$.

From this we conclude that for $\nu > 0$ the only extremum of I_ν^α is a minimum - whose existence was asserted in I).

IV) Any solution $u \in C^2(\mathbf{R})$ of (11) and (12) is strictly monotone and satisfies $u_\theta > 0$.

These results hold for $\nu > 0$. To study the limit $\nu \to 0$ we can normalize the minimal by the condition $u(0) = 0$ and denote this unique function by $u = u(\theta;\alpha,\nu)$.

V) There exists a sequence $\nu_k > 0$, $\nu_k \to 0$ such that $u_k(\theta) = u(\theta;\alpha,\nu_k)$ converges almost everywhere to a monotone function $u_*(\theta)$ satisfying at all points of continuity

$$h_1(u_*,u_*^+) + h_2(u_*^-,u_*) = 0, \quad u_*(\theta + 1) = u_*(\theta) + 1 .$$

This function u_* may have denumerably many discontinuities, giving rise to the gaps in the Mather set. This function is approximated by the smooth functions u_k.

One can use this smooth function $u(\theta;\alpha,\nu)$ to construct for each α a smooth curve

$$y = w(x;\alpha,\nu)$$

approximating the Mather set. For $\nu > 0$ these curves depend also smoothly on α and one could expect to get for fixed $\nu > 0$ a family curve covering the cylinder simply. This would require that $w = w(\theta;\alpha,\nu)$ depends monotonically on α, so that curves corresponding to different values of α do not intersect. We had hoped this to be the case but are not able to prove the monotonicity of w in α.

4. PROOFS

We give the proofs of the statements I)-V). To prove I) we note that h is bounded from below and therefore the functional I_ν^α is in X also bounded from below. Let $u_j = \theta + \hat{u}_j \in X$ be a minimizing sequence. From the form of I_ν^α it is clear that

$$\nu \int_0^1 \left(\frac{\partial u_j}{\partial\theta}\right)^2 d\theta$$

is bounded and since I_ν^α is invariant under the translation $u(\theta) \to u(\theta + \text{const})$ we can assume that

$$\int_0^1 \hat{u}_j d\theta = 0 .$$

Hence

$$\|\hat{u}_j\|_{H^1(S^1)}$$

is bounded and we can find a subsequence - called $\hat{u}_j$ again - which converges weakly in $H^1(S^1)$. Moreover, I_ν^α is lower semicontinuous with respect to weak convergence in $H^1(S^1)$, and thus the minimum of I_ν^α is attained by a function $u \in X$. Moreover, this minimal satisfies the integrated Euler equation

$$\nu u_\theta = \int_0^\theta (h_1(u,u^+) + h_2(u^-,u))d\theta + \text{const} .$$

From this one concludes that $u \in C^\infty$ and that u satisfies (11), (12). In other words, the usual direct methods of calculus of variations are applicable to this problem if $\nu > 0$.

To prove II) we set $z = u_2 - u_1 \geq 0$ and obtain from (11) a differential equation

$$-\nu z_{\theta\theta} + m(\theta)z + h_{12}(*)z^+ + h_{12}(**)z^- = 0$$

where the stars indicate some intermediate values and $m(\theta) = h_{11}(*) + h_{22}(**)$ a continuous function. From $h_{12} < 0$, $z^\pm \geq 0$ we obtain

$$-\nu z_{\theta\theta} + m(\theta)z \geq 0, \qquad z \geq 0$$

or

$$z_{\theta\theta} \leq \frac{m}{\nu} z, \qquad z \geq 0 .$$

From this inequality one concludes that either $z \equiv 0$ or $z > 0$, by the maximum principle. One can also proceed as follows. The set of zeroes of z is clearly a closed set on $\mathbf{R}$ and it suffices to show that it is open to see that it is empty or $\mathbf{R}$.

Now if this zero set is not empty and contains, say, $\theta = 0$ then we have $z(0) = 0$, $z'(0) = 0$ and therefore

$$0 \leq z(\theta) \leq \int_0^\theta (\theta - t) \frac{m(t)}{\nu} z(t)dt$$

hence, in some neighborhood $|\theta| < \delta$

$$z(\theta) \leq c\left|\int_0^\theta z(t)dt\right| .$$

By the Gronwall inequality this yields $z \equiv 0$ in $|\theta| < \delta$ proving that the zero set is open, hence $\mathbf{R}$.

The same argument gives III): We define for $c \in \mathbf{R}$

$$\mu(c) = \min_{\theta}(u_1(\theta + c) - u_2(\theta)) .$$

Since the function $u_1(\theta + c) - u_2(\theta)$ has the period 1, the minimum is taken on for some $\theta = \xi(c) \in [0,1]$. Also, from (12) we conclude that $\mu(c + 1) = \mu(c) + 1$ hence, if we establish the continuity of μ, we see that $\mu(c)$ takes on all values of $\mathbf{R}$, in particular, the value 0. If we pick c so that $\mu(c) = 0$ then $u_1(\theta + c) \geq u_2(\theta)$ with equality for $\theta = \xi(c)$ and, by II), we conclude $u_1(\theta + c) = u_2(\theta)$ for all θ.

We show the continuity of μ: Assume c_j is any sequence tending to c^* as $j \to \infty$, then

$$\mu(c_j) \leq u_1(\theta + c_j) - u_2(\theta)$$

implies for $j \to \infty$

$$\overline{\lim}\, \mu(c_j) \leq u_1(\theta + c^*) - u_2(\theta) \quad \text{for all} \quad \theta \in \mathbf{R} .$$

Thus $\overline{\lim}\, \mu(c_j) \leq \mu(c^*)$, i.e. μ is upper semicontinuous. To show $\underline{\lim}\, \mu(c_j) \geq \mu(c^*)$ we pick a subsequence c_j' so that $\underline{\lim}\, \mu(c_j) = \lim \mu(c_j')$ and a further subsequence c_j'' or c_j' such that $\xi(c_j'') \to \theta^*$. Then

$$\mu(c_j'') = u_1(\theta + c_j'') - u_2(\theta) \quad \text{for} \quad \theta = \xi(c_j'')$$

and $j \to \infty$ yields

$$\underline{\lim}\, \mu(c_j) = \lim \mu(c_j'') = u_1(\theta^* + c^*) - u_2(\theta^*) \geq \mu(c^*) .$$

Hence μ is continuous at c^*, hence in $\mathbf{R}$.

We turn to the proof of IV) and want to show that $u(\theta + c) - u(\theta) \geq 0$ for $c > 0$. Otherwise $\mu(c) = \min_{\theta}(u(\theta + c) - u(\theta))$ would be negative for some positive c, and since $\mu(c) \to +\infty$ for $c \to +\infty$ we can find a $c_0 > 0$ with $\mu(c_0) = 0$. Thus $u(\theta + c_0) \geq u(\theta)$ with equality for some θ. Thus by II) we have $u(\theta + c_0) = u(\theta)$ for all θ. This implies that $u(\theta)$ has the period c_0 hence is bounded contradicting the condition (12). Thus $u(\theta)$ is monotone and its derivative $u_\theta \geq 0$.

Differentiating (11) we obtain for $z = u_\theta$ the differential equation

$$Lz = -\nu z_{\theta\theta} + (h_{11}(u,u^+) + h_{22}(u^-,u))z + h_{12}(u,u^+)z^+ + \\ + h_{12}(u^-,u)z^- = 0 . \quad (13)$$

Since $z \geq 0$ we conclude from the maximum principle that $z \equiv 0$ or $z > 0$. The first case is excluded since it would give $u = \text{const}$; hence $u_\theta > 0$.

Finally, we establish V). We normalize $u = u(\theta;\alpha,\nu)$ by $u(0;\alpha,\nu) = 0$ and note that $u_\theta > 0$, $\int_0^1 u_\theta d\theta = 1$. By Helly's selection theorem there exists a sequence $\nu_k \to 0$ for which $u_k(\theta) = u(\theta;\alpha,\nu_k)$ converges almost everywhere to a monotone function $u_*(\theta)$. As a matter of fact $u_k(\theta)$ converges to $u_*(\theta)$ at every point of continuity of u_*. Moreover, $u_*(\theta + 1) = u_*(\theta) + 1$. The Euler equation (11) implies

$$\int_{\mathbf{R}} (-\nu\varphi_{\theta\theta}u + \varphi(h_1(u,u^+) + h_2(u^-,u)))d\theta = 0 \quad \text{for}$$

$$u = u_k, \ \nu = \nu_k$$

for all $\varphi \in C^2_{\text{comp}}(\mathbf{R})$. Hence, for $\nu = \nu_k \to 0$ we obtain via the theorem on bounded convergence

$$\int_{\mathbf{R}} \varphi(h_1(u_*,u_*^+) + h_2(u_*^-,u_*))d\theta = 0 \ .$$

From this relation follows the vanishing of the expression in parenthesis at all points of continuity of u_*.

5. THE WEIERSTRASS EXCESS FUNCTION

The arguments of the previous sections show that for the minimal u of I) constructed above one has

$$I^\alpha_\nu(v) \geq I^\alpha_\nu(u) \quad \text{for} \quad v \in X \ .$$

We want to establish this relation with the methods of calculus of variation using that the functions $u = u(x + c)$, $c \in \mathbf{R}$, form a "field of extremals".

We denote the integrand of I^α_ν by

$$F(x_0,x_1,p) = \frac{\nu}{2} p^2 + h(x_0,x_1) \tag{14}$$

where one has to replace x_0,x_1,p by $u(\theta)$, $u(\theta + \alpha)$ and $u_\theta(\theta)$, respectively. The unusual feature of this integrand is that F depends on the shifted argument $x_1 = u(\theta + \alpha)$. The Legendre condition corresponds to

$$F_{pp} = \nu > 0; \quad F_{x_0x_1} = h_{12} < 0 \tag{15}$$

in this case.

The fact that F does not depend on the independent variable θ implies that

$$I_\nu^\alpha(v) = \int_0^1 F(v,v^+,v_\theta)d\theta, \quad v \in X$$

is invariant under the translation $\theta \to \theta + \text{const}$.

The minimals $u(\theta + c)$, $c \in \mathbb{R}$, constructed in the previous section provide a field of extremals, i.e. the curves $x = u(\theta + c)$ cover the θ-x-plane in a simple fashion. Since $u_\theta > 0$ and u takes on all values of $\mathbb{R}$ we can define two functions ψ,χ by $\psi = u^+ \circ u^{-1}$, $\chi = u_\theta \circ u^{-1}$ so that

$$u^+ = \psi(u); \quad u_\theta = \chi(u)$$

where $\psi(x) - x$, $\chi(x)$ have period 1. Conversely, the family $u(\theta + c)$ is defined by the second relation. Note that $\psi'(x) > 0$.

We will define the function

$$E(x_0,x_1,p) = \frac{\nu}{2}(p - \chi(x_0))^2 - \iint_{D(x_0,x_1)} h_{12}(\xi,\eta)d\xi d\eta \geq 0 \quad (16)$$

where $D(x_0,x_1)$ is the domain bounded by horizontal and vertical segments through (x_0,x_1) and by part of the curve $x_1 = \psi(x_0)$. This function is the generalization of the Weierstrass excess function. For any $v \in X$ we have

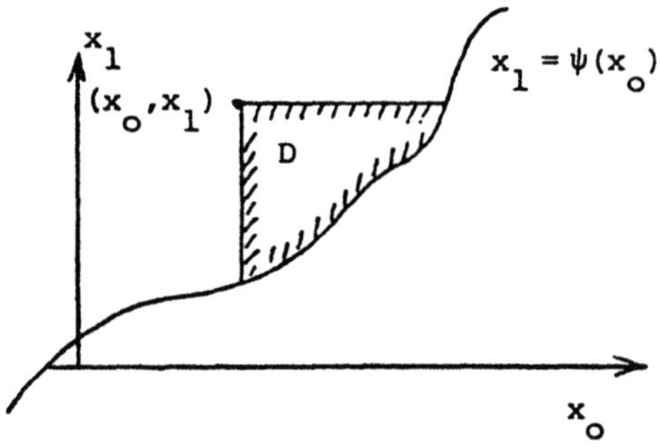

$$I_\nu^\alpha(v) = I_\nu^\alpha(u) + \int_0^1 E(v,v^+,v_\theta)d\theta$$

establishing the minimal character of u since $E \geq 0$.

The proof follows the standard procedure by subtracting from F a function

$$F^*(x_0,x_1,p) = f'(x_0)p + g(x_1) - g(x_0) + c_0 \quad (17)$$

where $f'(x),g'(x)$ have period 1 and c_0 is a constant. Then for any $v \in X$ the corresponding integral

$$\int_0^1 F^*(v,v^+,v_\theta)d\theta = \int_0^1 (f'(x) + \alpha g'(x))dx + c_0$$

is independent of v, and thus F and $F - F^*$ give rise to equivalent variational problems. We will choose f, g, c_0 so that the function

$$E(x_0,x_1,p) = F(x_0,x_1,p) - F^*(x_0,x_1,p) \geq 0$$

for all arguments, and it is equal to zero for $x_1 = \psi(x_0)$, $p = \chi(x_0)$. This requires

$$E_{x_0} = E_{x_1} = E_p = 0 \quad \text{for} \quad x_1 = \psi(x_0), \quad p = \chi(x_0) .$$

The second and third equation give

$$g'(x) = h_2(\psi^{-1}(x),x); \qquad f'(x) = \nu\chi(x) \tag{18}$$

while the first follows from these with the help of the Euler equation. Obviously, on the curve $x_1 = \psi(x_0)$, $p = \chi(x_0)$ the function E is a constant, which by appropriate choice of c_0 can be made equal to zero. This gives

$$c_0 = -\frac{\nu}{2}\chi^2 + h(x,\psi) - g(\psi) + g(x) . \tag{19}$$

Combining (17), (18) and (19) gives

$$F^* = \nu\chi p - \frac{\nu}{2}\chi^2 + h(x_0,\psi) + \int_{\psi(x_0)}^{x_1} h_2(\psi^{-1}(\eta),\eta)d\eta$$

and therefore

$$E = F - F^* = \nu\frac{p^2}{2} + h(x_0,x_1) - F^*$$

$$= \frac{\nu}{2}(p-\chi)^2 + h(x_0,x_1) - h(x_1,\psi) - \int_{\psi(x_0)}^{x_1} h_2(\psi^{-1}(\eta),\eta)d\eta .$$

Generalizing the formula (5) we can rewrite this as (16).

If we denote by $(I_\nu^\alpha)^*$ the integral corresponding to the integrand F^*, then this integral is independent of the choice of $v \in X$ by our previous consideration. Moreover, $E = 0$ for $x_0 = u$, $x_1 = u^+$, $p = u_\theta$. Therefore we have

$$I_\nu^\alpha(v) - (I_\nu^\alpha)^* = \int_0^1 E(v,v^+,v_\theta)d\theta$$

$$I_\nu^\alpha(u) - (I_\nu^\alpha)^* = 0$$

and eliminating $(I_\nu^\alpha)^*$

$$I_\nu^\alpha(v) - I_\nu^\alpha(u) = \int_0^1 E(v,v^+,v_\theta)d\theta > 0$$

as we claimed.

We remark that this argument can immediately be generalized to integrands $F(x_0,x_1,p)$ for which $F_{x_1p} \equiv 0$, $F_{pp} > 0$, $F_{x_0x_1} < 0$, i.e. for integrands of the form $F = a(x_0,p) + b(x_0,x_1)$.

6. PERTURBATION THEORY

In general the limit solution u_* of the minimal $u(\theta;\alpha,\nu)$ is discontinuous. However, if α is a badly approximable irrational number, i.e. satisfies with some constants c_0,τ the inequalities

$$|\alpha - \frac{p}{q}| > c_0q^{-\tau} \tag{20}$$

for all rational numbers, and if h is sufficiently close to $h^0(x_0,x_1)$ for which a smooth solution $u_0(\theta)$ of

$$h_1^0(u_0,u_0^+) + h_2^0(u_0^-,u_0) = 0; \quad u_0(\theta + 1) = u_0(\theta) + 1 \tag{21}$$

exists, then also (10) has a smooth solution. This corresponds to the invariant curve theorem, proven by an iteration process involving repeated transformations. It can be considered the simplest situation of KAM theory.

We want to point out that a similar result holds for the Euler equation (11): Its solution $u(\theta;\alpha,\nu)$ is a C^∞-function. However, under the above circumstances one can provide estimates $|u|_{C^{2,\beta}} < M$, $\beta \in (0,1)$, which are independent of ν and therefore give rise to a $C^{2,\beta}$-solution u_* for $\nu = 0$ corresponding to a twice differentiable invariant curve. Although the result is not surprising the proof of this result requires a different approach, independent of transformation theory, since the equation (11) expresses the invariance of the corresponding curves only for $\nu = 0$ but not for $\nu > 0$. One is led this way to a different, somewhat simpler, iteration procedure which actually is also applicable to higher dimensional problems.

We formulate the result. We assume that $u_0(\theta) = \theta$ is a solution of (21). This is not a restriction since we can achieve this by a transformation $\theta \to u_0(\theta)$. Thus we assume

$$h_1^0(\theta,\theta + \alpha) + h_2^0(\theta - \alpha,\theta) = 0 . \tag{22}$$

Moreover, we assume that $h^0(x_0,x_1)$ is C^∞ in a strip $|x_1 - x_0 - \alpha| < \rho$ and satisfies there the conditions (2). Next we consider a perturbed function

$$h(x_0,x_1,\lambda) = h^0(x_0,x_1) + \lambda h^1(x_0,x_1,\lambda)$$

with a C^∞-function h^1 satisfying the same periodicity condition as in (2). The problem is to estimate the solution $u = u(\theta;\alpha,\nu,\lambda)$ of the equation (11) for sufficiently small $|\lambda|$.

<u>Theorem</u>: Under the above assumptions for a given $\varepsilon > 0$ and $\beta \in (0,1)$ there exists a positive constant $\lambda^* = \lambda^*(\varepsilon)$ such that for $|\lambda| < \lambda^*$ the solution $u = u(\theta;\alpha,\nu,\lambda)$ normalized by $u = 0$ for $\theta = 0$ satisfies

$$|u - \theta|_{C^{2,\beta}} < \varepsilon$$

for all $\nu > 0$.

We will not give the proof here. It depends on a quadratically convergent iteration procedure but avoids transformation theory. We merely want to point out that the argument depends on the solution of the linear difference-differential equation

$$Lv = g \tag{23}$$

where L is the operator defined by (13). The right-hand side g is assumed to be smooth and of period 1, and also the solution v is required to have period 1. Note that $Lu_\theta = 0$, i.e. $u_\theta > 0$ is a solution of the homogeneous equation. From this one concludes that a necessary compatibility condition is that

$$\int_0^1 u_\theta g d\theta = 0 \ . \tag{24}$$

We want to indicate that the most general solution of $Lv = 0$ of period 1 is $v = \mu u_\theta$ with a constant μ, and that the equation (23) under the compatibility condition (24) possesses a smooth solution.

To prove the first statement we set $z = v/u_\theta$ and rewrite the difference-differential equation for z. The relevant formula is

$$u_\theta L(u_\theta z) = -\nu(u_\theta^2 z_\theta) + \nabla(h_{12}(u^-,u)u_\theta u_\theta^-(\nabla z^-)) \tag{25}$$

where the difference operator ∇ is defined by

$$\nabla\varphi = \varphi^+ - \varphi = \varphi(\theta + \alpha) - \varphi(\theta) \ .$$

Multiplying this expression by z and integrating we obtain

$$\int_0^1 vL(v)d\theta = \int_0^1 \{\nu u_\theta^2 z_\theta^2 - h_{12}(u^-,u)u_\theta u_\theta^-(\nabla z^-)^2\}d\theta \ . \qquad (26)$$

Since $h_{12} < 0$ we conclude from $Lv = 0$ that $z_\theta = 0$, i.e. $z = \text{const}$.

The same identity (25) can be used to obtain an a-priori estimate for the solution of (23), and thus can be used for an existence proof. We assume $c^{-1} < u_\theta < c$ and obtain from (26)

$$\int_0^1 vg d\theta = \int_0^1 vL(v)d\theta > \delta c^{-2} \int_0^1 |\nabla z^-|^2 d\theta$$

and because of $v = u_\theta z$

$$\int_0^1 |\nabla z^-|^2 d\theta < \delta^{-1}c^3 \int_0^1 zgd\theta = \delta^{-1}c^3 \left| \int_0^1 (\nabla z^-)(\nabla^{-1}g)d\theta \right|$$

or

$$\int_0^1 |\nabla z|^2 d\theta < \delta^{-2}c^6 \int_0^1 |\nabla^{-1}g|^2 d\theta \ .$$

With the help of such ν-independent estimates it is possible to establish the above theorem.

REFERENCES

Bangert, V., 'Mather sets for twist maps and geodesics on tori', preprint to appear in Dynamics Reported.

Mather, J., 'Existence of quasi-periodic orbits for twist homeomorphisms of the annulus', Topology **21** (1982), 457-467.

Moser, J., 'Recent developments in the theory of Hamiltonian systems', SIAM Reviews **28** (1986), 459-485.

THE OBSTRUCTION METHOD AND SOME NUMERICAL EXPERIMENTS RELATED TO THE STANDARD MAP

Arturo Olvera and Carles Simó
Departament d'Equacions Funcionals
Facultat de Matematiques
Universitat de Barcelona
Gran Via 585
Barcelona 08007, Spain

ABSTRACT. Invariant manifolds of hyperbolic periodic points can be used to show the non-existence of invariant rotational curves of one parameter families of area preserving twist maps on the cylinder. We studied the behavior of these periodic points concerning their eigenvalues, ordering and critical value of the parameter. The standard map was studied using numerical and perturbative methods.

1. INTRODUCTION

In this paper we present some results in a one parameter family of area preserving twist maps on the cylinder, the existence of invariant rotational curves (IRC) and self-similarity behavior.

In the first part of this paper we give a short description of the "obstruction method" for the destruction of invariant curves on the cylinder applied to a one parameter family of area preserving maps. Next, we study numerically some IRC of the standard map and their self-similarity behavior. The third part is devoted to compare the obstruction method and the existence of non-Birkhoff orbits on the symmetry lines. In the last part the behavior of the eigenvalues of the hyperbolic periodic points (HPP) on the symmetry line is studied when the value of the parameter is close to zero (using perturbation method) and when it is greater than the critical value.

2. OBSTRUCTION TO INVARIANT CURVES

In this section we give a short description of the method to find an upper bound of the critical value of the parameter such that a specific IRC disappears [1,2].

Let f_K be a one parameter family of area-preserving monotone diffeomorphisms on the cylinder $S^1 \times \mathbf{R}$, such that its lift on $\mathbf{R}^2$, F_K, has the form

P. H. Rabinowitz et al. (eds.), Periodic Solutions of Hamiltonian Systems and Related Topics, 235–244.

$$F_K(x,y) = (x + y + g_K(x),\ y + g_K(x))$$

where K is the parameter and $g_K(x)$ is a 1-periodic analytic function with zero average. F_K has the twist property.

Take two HPP $\mathbf{x}_0$ and $\mathbf{x}_1$ with rotation number ω_0 and ω_1 (resp.). Each HPP has homoclinic connection around the cylinder (the invariant manifolds of the images of $\mathbf{x}_0$ (resp. $\mathbf{x}_1$) have homoclinic intersection and it is possible to encircle the cylinder by segments of the stable and unstable manifolds of the images of $\mathbf{x}_0$ [1]). Then, if the unstable manifold of $\mathbf{x}_0$ ($\mathbf{W}^u\mathbf{x}_0$) and the stable manifold of $\mathbf{x}_1$ ($\mathbf{W}^s\mathbf{x}_1$) have heteroclinic points, there are not IRC with rotation number (RN) $\omega \in [\omega_0,\omega_1]$. Let K_0 the value of the parameter such that $\mathbf{x}_0$ and $\mathbf{x}_1$ have heteroclinic points for $K > K_0$ and they do not have them for $K < K_0$ (first heteroclinic tangency).

Let a sequence of HPP $\{\mathbf{x}_n\}$ defined like above such that the sequence of their RN $\{\omega_n\}$ has the following properties: i) $\omega_n = p_n/q_n$ with p_n, $q_n \in \mathbf{N}$ relative primes and q_n the period of $\mathbf{x}_n$; ii) $\lim \omega_n = \omega$; iii) $|\omega - \omega_{n+1}| < |\omega - \omega_n|$; iv) $(\omega - \omega_{n+1})(\omega - \omega_n) < 0$. Let K_n the values of K such that $\mathbf{x}_n$ and $\mathbf{x}_{n+1}$ have heteroclinic tangency. Then the sequence of $\{K_n\}$ is decreasing [1]. The limit value of the sequence, K_∞, is an upper bound of the critical value of the parameter (K_c) such that there is not IRC with RN ω for $K_0 > K > K_\infty$ and there is for $K_\infty > K > 0$. ω_n can be taken as the finite continued fraction $[a_1,a_2,\dots,a_n] = 1/(a_1 + 1/(a_2 + \dots + 1/a_n)\dots)$, if $\omega = [a_1,a_2,\dots]$ (see [3]).

The first heteroclinic tangency is impossible to detect because we should extend infinitely the invariant manifolds of the HPP. Hence we fix a particular tangency defined as follows:

Let $\mathbf{x}_0$ and $\mathbf{x}_1$ HPP, and $\mathbf{W}^u\mathbf{x}_0$ and $\mathbf{W}^s\mathbf{x}_1$ their invariant manifolds expanded in parametric form:

$$\mathbf{W}^u\mathbf{x}_0 = (\xi_0(s),\eta_0(s)) \qquad s,t \in \mathbf{R}^+ \cup \{0\}\ ,$$

$$\mathbf{W}^s\mathbf{x}_1 = (\xi_1(t),\eta_1(t))$$

such that $(\xi_0(0),\eta_0(0)) = \mathbf{x}_0$ and $(\xi_1(0),\eta_1(0)) = \mathbf{x}_1$.

We call heteroclinic tangency on the "first tongue" a tangency which happens for values of t_0 and s_0 such that:

i) $(\xi_0(s_0),\eta_0(s_0)) = (\xi_1(t_0),\eta_1(t_0))$

ii) $\left(\frac{d}{ds}\xi_0(s_0),\ \frac{d}{ds}\eta_0(s_0)\right) \cdot \left(\frac{d}{dt}\eta_1(t_0),\ -\frac{d}{dt}\xi_1(t_0)\right) = 0$

where $\cdot$ denotes the scalar product.

iii) The curvature of $\mathbf{W}^u\mathbf{x}_0$ (resp. $\mathbf{W}^s\mathbf{x}_1$) for $0 < t \leq t_0$ (resp. $0 < s \leq s_0$) has constant sign.

We take as k_n the minimal value of K for which i), ii) and iii) hold (see Figure 1). Then, for any n we have $K_n \leq k_n$ and hence the limit values satisfy $K_\infty \leq k_\infty$. We conjecture that $K_\infty = k_\infty$.

The numerical method for the computation of heteroclinic tangencies of the "first tongue" is described in [1,2].

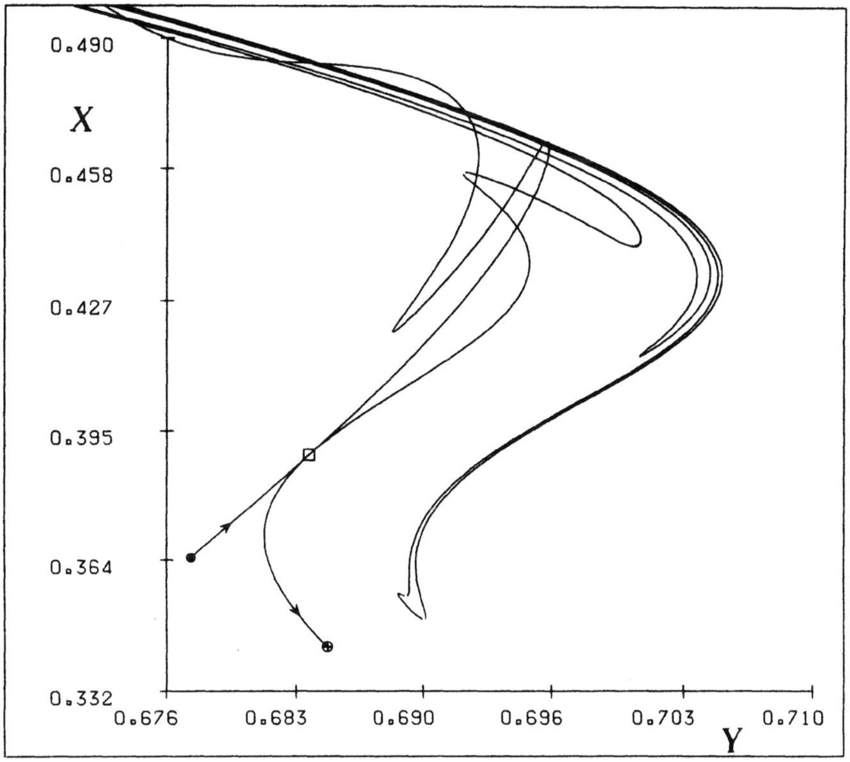

Figure 1. The stable invariant manifold of the HPP with RN $\omega = 5/8$ (⊕) has heteroclinic tangency (□) on the first tongue with the unstable invariant manifold of the HPP with RN $\omega = 8/13$ (•) at the parameter value $K = 1.085169$, for the standard map.

3. NUMERICAL RESULTS AND SELF-SIMILARITY

The obstruction method was used to study the behavior of the IRC on the standard map [4] (where $g_K(x) = (K/2\pi)\sin(2\pi x)$). This map has been studied for many years and it shows self-similarity behavior for the dynamics of a small neighborhood around noble-circles (i.e. IRC with RN $\omega = [a_1, a_2, a_3, \ldots, a_n, (1)^\infty]$). MacKay showed that the renormalization theory gives a nice way to understand the self-similarity behavior [5].

We studied different IRC in [1] and we obtained the following results:

a) When $\omega = (\sqrt{5} - 1)/2$, that is, when $\omega = [(1)^\infty]$ we computed the sequence of $\{k_n\}$ for $n = 2, \ldots, 15$. The values of k_n converge to $k_\infty \approx 0.971636$ in geometrical form with a ratio $\delta \approx 0.613818$, $\delta = \lim(k_{n+1} - k_n)/(k_n - k_{n-1})$. The value k_∞ is close to Greene's critical value for the IRC [6]. When the HPP $\mathbf{x}_n$ and $\mathbf{x}_{n+1}$ have heteroclinic tangency on the first tongue (for k_n) their residues ($R = (2 - \lambda - \lambda^{-1})/4$, λ being the eigenvalue) converge in geometrical form with a ratio δ to the value -0.648505 for $\mathbf{x}_n$ and -1.181067 for $\mathbf{x}_{n+1}$.

If we take other golden circles, that is those whose RN ω is of the form $\omega = [a_1,a_2,a_3,\ldots,a_n,(1)^\infty]$, the behavior is like above. Obviously, the limit value of $\mathbf{k}_\infty$ is different but the ratio δ and the limit value of the residues are the same.

b) Studying the IRC's with rotation numbers $\omega_a = [(a)^\infty]$ show phenomena similar to the golden case but the sequence of parameters converges to $\mathbf{k}_\infty$ with a ratio δ_a that is similar to the rotation number ($\delta_2 \approx 0.4087$ and $\delta_3 \approx 0.2951$). Also, the residues in each case converge (with a ratio δ_a) to a limit value.

c) Other kind of IRC, for which the RN has periodic continued fraction expansion $\omega = [(a_1,a_2,a_3,\ldots,a_i)^\infty]$ have been studied. In that case the self-similarity is observed every i steps of the process. This means that we should use $\delta_n^i = (k_{n+i} - k_n)/(k_n - k_{n-i})$ but it is still convergent when $n \to \infty$. δ_n^i is close to $\omega_{a_1}\omega_{a_2}\omega_{a_3}\cdots\omega_{a_i}$ (where $\omega_{a_j} = [(a_j)^\infty]$). The residues converge to i different limits.

d) Aperiodic continued fraction expansions of RN $\omega = [a_1,a_2,a_3,\ldots]$ do not show explicit self-similarity behavior but the ratio δ_n seems to depend on the last three elements of the expansion (a_{n+1},a_n,a_{n-1}) and the value of the residues on the last two terms (a_{n+1} and a_n).

4. NON-BIRKHOFF ORBITS

Let $\mathcal{O}(\mathbf{x})$ an extended orbit of $\mathbf{x}$ on $\mathbb{R}^2$ ($\mathcal{O}(\mathbf{x}) = \{F^i(\mathbf{x}) - (0,j) | i,j \in \mathbb{Z}\}$). Then $\mathcal{O}(\mathbf{x})$ is a Birkhoff orbit if $\mathbf{x}_0,\mathbf{x}_1 \in \mathcal{O}(\mathbf{x})$ and $\pi_1(\mathbf{x}_0) < \pi_1(\mathbf{x}_1)$ implies $\pi_1(F_K(\mathbf{x}_0)) < \pi_1(F_K(\mathbf{x}_1))$ (π_1 is the projection on the first variable).

Hall showed that in a twist map on the annulus there exist non-Birkhoff periodic orbits close to every IRC when it is destroyed [7]. These orbits belong to the instability Birkhoff zone created when the IRC disappears. Then, looking for non-Birkhoff orbits with RN close to some IRC is a method to detect the critical value of the parameter of this IRC.

We studied the behavior of the HPP's used in the obstruction method (along the symmetry lines on the standard map [6]) in regard of the loss of order in the orbit. We looked for a value of the parameter K such that a HPP with RN p/q (p,q relative primes and the orbit has a point in the symmetry lines) loses the order.

The HPP with RN $\omega_n = [a_1,a_2,\ldots,a_n]$ and $a_i = 1$ (the sequence of $\{\omega_n\}$ used in the golden mean circle) were studied until period 10946. For every $\omega_n = p_n/q_n$ we computed the scaled minimum distance

$$d(\omega_n,K) = q_n \min\{\pi_1(F_K(\mathbf{x}_i)) - \pi_1(F_K(\mathbf{x}_j)) \mid \mathbf{x}_i,\mathbf{x}_j \in O(\mathbf{x});$$
$$\mathbf{x}_j < \mathbf{x}_i; \quad i,j \in \mathbf{Z};\ i \neq j \text{ and } \mathbf{x} \text{ has RN } \omega_n\}$$

until values of K such that the magnitude of the residue of the HPP was greater than 3×10^5.

We did not find non-Birkhoff orbits, the value of $d(\omega_n,K)$ going to zero if $K > Kc$ (critical value) in exponential form. Figures 2 and 3 show the behavior of $\log(d(\omega_n,K))$ vs K when $\omega_n = 4181/6765$ and $\omega_n = 6765/10946$ (resp.).

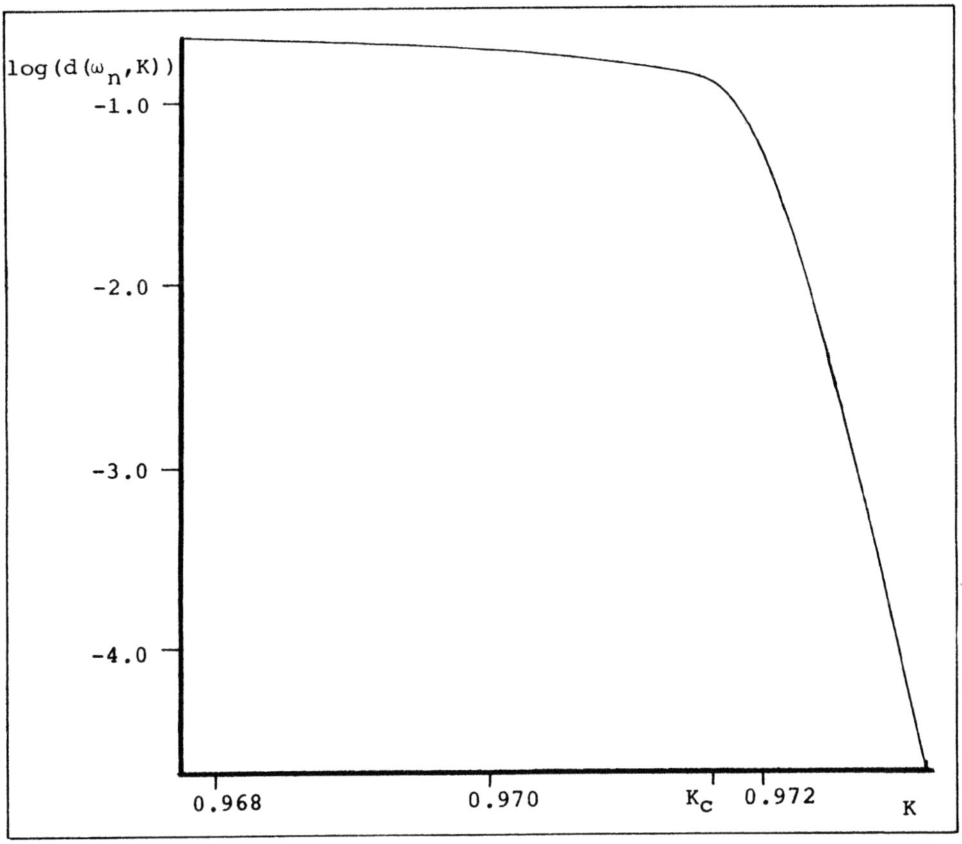

Figure 2. Graph of the function $\log(d(\omega_n,K))$ vs parameter K when the value of the RN is $\omega_n = 4181/6765$.

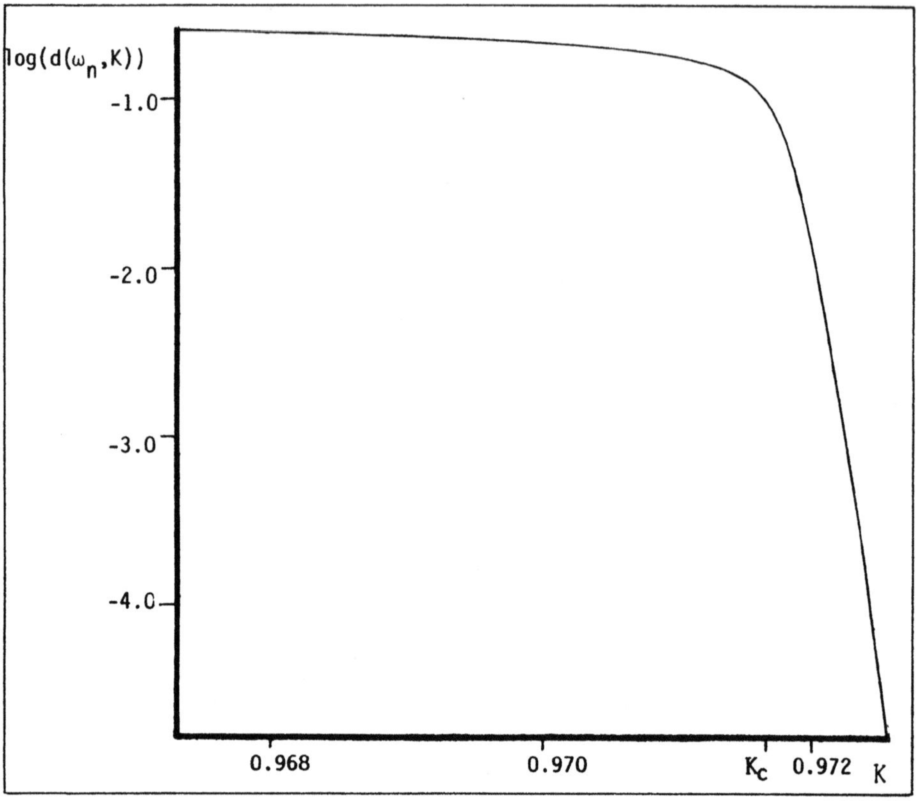

Figure 3. Graph of the function $\log(d(\omega_n,K))$ vs parameter K when the value of the RN is $\omega_n = 6765/10946$.

From this numerical experiment we conjecture that the orbits of HPP with RN p/q (where p and q are relative primes) having a point in any symmetry line are Birkhoff orbits for all values of K.

These experiments lead us to the following question: Do the non-Birkhoff orbits come from Birkhoff orbits? (i.e. is every non-Birkhoff orbit obtained by continuation of a Birkhoff orbit for some value of the parameter?).

It is interesting to see that the graphs of the function $\log(d(\omega_n,K))$ (Figures 2 and 3) change the slope around the critical value (of the golden mean circle). Perhaps this behavior is related to the creation of gaps between the destroyed IRC.

5. EIGENVALUES OF THE HPP

In section 2 we saw that the eigenvalues (or the residue) of the HPP have self-similarity behavior between the different RN $\{\omega_n\}$. We expect (following MacKay [5]) that the eigenvalues of two HPP with RN ω_n and ω_{n+1} allow for a renormalization relation. We investigate this possibility for the sequence of HPP with RN converging to the golden mean, when K is close to zero when $K > K_c$.

a) K close to zero (formal expansion)

When K is close to zero it is possible to obtain a formal expansion (via perturbations) of the eigenvalues of any HPP with RN p/q. But we must obtain first the position of the HPP as a function of its RN and K.

We looked for HPP with RN p/q on the symmetry line $y = 0$. Starting with $x_0 = p/q$ for $K = 0$ we obtained the first terms of the Taylor series of $\pi_1(F_K^q(x,0))$ following the method used in [2]. A fixed point of F_K^q is obtained as the zero of the function $\pi_1(F_K^q(x,0)) - x$. Using Newton formally we obtain an expression of $x_{p/q} = \Sigma a_i(p/q)K^i$ (the coefficients a_i only depend on p/q). We get this series for HPP with RN 1/2, 2/3, 3/5, 5/8, 8/13, 13/21, and 21/34 (some of them were expanded around the symmetry line $y = 1/2$). In all the cases, the expansion looks like an exponential series. Writing them in the form

$$a_0 K \exp\left(\sum_{i=1}^{2} a_i(p/q)K^i\right) + p/q \ ,$$

we obtained the values of a_i scaled by p/q, given in Table 1.

TABLE 1

ω_n	a_0/ω_n	δ_0	a_1/ω_n	δ_1	a_2/ω_n	δ_2
2/3	0.068911		0.111144		-0.017421	
3/5	0.043093		0.351251		0.061644	
5/8	0.052739	-2.6766	0.220975	-1.8430	0.136994	1.0493
8/13	0.049042	-2.6088	0.261887	-3.1842	0.108186	-2.6155
13/21	0.050451	-2.6232	0.245263	-2.4609	0.120538	-2.3321
21/34	0.049912	-2.6161	0.251428	-2.6966	0.114350	-1.9960

where $\delta_i = (a_i(\omega_n) - a_i(\omega_{n-1}))/(a_i(\omega_{n+1}) - a_i(\omega_n))$. From this result we can conclude that the HPP of RN p/q is located in $(x_f,0)$ where x_f is close to

$$(p/q)((K/20)\exp[(p/q)(K/4 + K^2/9)] + 1) .$$

The next step is to find an expression of the residue for this HPP. Using the Taylor expansion of F^q around $(x_f,0)$ we computed the Jacobian and its trace as functions of K. We obtained that the residue $(R = (2 - \text{trace})/4)$ has the form:

$$R(p/q) = 2 + b_q K^q + b_{q+1} K^{q+1} + b_{q+2} K^{q+2} + \ldots$$

at least for $p/q = 1/2$, $2/3$ and $3/5$. These results agree with Greene [6]. We have not been able to compute the residue of higher order HPP because of the complexity of the calculations. We obtain for these rotation numbers the results given in Table 2.

TABLE 2

p/q	b_q	b_{q+1}	b_{q+2}
1/2	1.0	0.0	0.010416
2/3	1.125	-0.00123	-0.06785
3/5	1.0822	-0.02045	

Checking the relation obtained for the position of HPP, it agrees with the numerical values of x_f for $K < 1/2$, with an error less than $q_n \times 10^{-5}$ when $K = 1/2$.

Finally, the expression of x_f agrees with the renormalization theory developed by MacKay [5] for K close to zero (when $|K| << 1$ taking $x_f(p_n/q_n)$, and applying the renormalization operator N_1 in a neighborhood of the simple fixed point we obtain $x_f(p_{n+1}/q_{n+1})$).

b) Eigenvalues of HPP for $K > K_c$.

We studied numerically the eigenvalues of HPP (on the symmetry line $y = 0$) when the parameter value is greater than the critical one. We took four rotation numbers, 3/5, 55/89, 4181/6765 and 6765/10946. Then we computed the λ and λ^{-1} of the related HPP for increasing values of K until $|\lambda + \lambda^{-1}| \approx 5 \times 10^5$.

We have the Greene's results [6] assuring that the residues of every HPP with RN ω_n and $\lim \omega_n \to (\sqrt{5} - 1)/2$ converge to 0.259.. around the critical value of the parameter. In order to reproduce the asymptotic behavior when K is close to zero and Greene's results we used the following form of the trace of the Jacobian:

$$\text{trace} = 2 + (K/K_f)^q \exp \sum_{i=1}^{3} a_i (K - K_f)^i)$$

where q is the period of the HPP and K_f is the parameter for which the residue is $-1/4$. The coefficients a_i and K_f are obtained using the least squares method. We show the value of these elements in Table 3.

There results do not show direct method to renormalize them. Perhaps the function of the trace must be more complicated.

TABLE 3

p/q	K_f	a_1	a_2	a_3
3/5	0.9739271	0.034921	0.007194	-0.000770
55/89	0.9705100	0.077493	0.003902	0.0000392
4181/6765	0.9716170	0.136974	0.004296	0.0000781
6765/10946	0.9716160	0.145385	0.003101	0.0000355

6. ACKNOWLEDGEMENTS

This work has been possible thanks to a Grant from CONACYT (México) to the first author. The work of the second author has been supported by CAICYT Grant 3534/83C3 (Spain). The computer facilities of the University of Barcelona were used. The authors are grateful to J. Mather for some helpful discussions at Il Ciocco.

7. REFERENCES

[1] A. Olvera and C. Simó, Physica **18**D (1987).

[2] A. Olvera and C. Simó, 'Heteroclinic Tangencies on the Standard Map', Communication to the VIII Congreso de Ecuaciones Diferenciales y Aplicaciones, Santander, Spain (1985).

[3] A. Ya. Khintchine, 'Continued Fractions', Noordhoff, Groningen (1963).

[4] B. V. Chirikov, Phys. Reports **52** (1979), 263.

[5] R. S. MacKay, 'Renormalisation in Area Preserving Maps', Thesis, Princeton, University Microfilms Int., Ann Arbor, Michigan (1982).

[6] J. M. Greene, J. Math. Phys. **20** (1979), 1183.

[7] P. L. Boyland and G. R Hall, 'Invariant Circles and the Order Structure of Periodic Orbits in Monotone Twist Maps', preprint, Boston University (1985).

ON A THEOREM OF HOFER AND ZEHNDER*

Paul H. Rabinowitz
Mathematics Department
University of Wisconsin-Madison
Madison, Wisconsin 53706

A major question of interest in the study of Hamiltonian systems is that of finding sufficient geometrical conditions for an energy surface of the system so that the system has a periodic orbit on the surface. The first result of a general nature in this direction is due to Seifert [1] and many subsequent contributions have been made. See e.g. [2-8]. At this conference H. Hofer and E. Zehnder [9] obtained a new kind of result in this spirit. It doesn't necessarily give periodic solutions on the prescribed energy surface, say $M \equiv H^{-1}(1)$ but gives such solutions on nearby surfaces. More precisely they proved:

Theorem 1: Suppose $H \in C^1(\mathbb{R}^{2n},\mathbb{R})$ and $M \equiv H^{-1}(1)$ is a compact hypersurface (and in particular $H_z \neq 0$ on M). Then there exists a sequence $\varepsilon_m \to 0$ such that $H^{-1}(1 + \varepsilon_m)$ contains a periodic solution z_m of the corresponding Hamiltonian system

$$\dot{z} = \mathcal{J}H_z(z) \ . \qquad \text{(HS)}$$

Actually this result also contains bounds on the action of the periodic solution. Theorem 1 implies that either (i) there is a periodic solution of (HS) on M or (ii) along any subsequence, the minimal period of z_m tends to infinity.

Sufficient conditions are known for (i) to occur which involve a priori bounds for the action in terms of the period. See e.g. [10] or [11]. It remains a major open question as to whether (ii) can occur.

*This research was sponsored in part by the National Science Foundation under Grant No. MCS-8110556 and by the United States Army under Contract No. DAAG29-80-C-0041. Reproduction in whole or in part is permitted for any purpose of the U. S. Government.

P. H. Rabinowitz et al. (eds.), Periodic Solutions of Hamiltonian Systems and Related Topics, 245–253.

The work of Hofer and Zehnder was inspired by a very recent result of Viterbo [12]. He settled a generalization of a conjecture of Weinstein [5] that was motivated by the results of [1-4].

The purpose of this note is twofold. First we will extend Theorem 1 slightly to show there is a somewhat richer structure of periodic solutions of (HS) near M.

Theorem 2: Under the hypotheses of Theorem 1, either (i) there is a sequence $\varepsilon_m \to 0$ such that for each m, $H^{-1}(1 + \varepsilon_m)$ contains uncountably many distinct periodic solutions of (HS) or (ii) there are uncountably many values of ε near 0 such that $H^{-1}(1 + \varepsilon)$ contains a periodic solution of (HS).

Our second goal concerns the existence of special kinds of periodic solutions of (HS) when $H(p,q)$ is also even in p. Then if there is a $T > 0$ and a solution $z = (p,q)$ of (HS) such that $p(0) = 0 = p(T)$, by extending p as an odd function of t about 0 and T and q as an even function of t about 0 on T, we get a $2T$ periodic solution of (HS). Such special solutions which bounce back and forth between the boundary and interior of a potential well associated with M have been obtained by several authors [1,2,6-8, 13-16]. In particular they have been called brake orbits [2,13]. We prefer to call them bouncing orbits. Our second result is

Theorem 3: If, under the hypotheses of Theorem 2, H is also even in p, the conclusions of Theorem 2 hold with respect to bouncing orbits.

The proofs of Theorems 2-3 mainly involve small modifications of the arguments of [9]. Therefore we will be sketchy in our presentation.

To prove Theorems 2-3, following [9], we begin by defining a new family of Hamiltonians. Since $H_z \neq 0$ on M, there is an $\bar{\varepsilon} > 0$ such that $H_z \neq 0$ on $H^{-1}([1 - \bar{\varepsilon}, 1 + \bar{\varepsilon}]) \equiv K$. Let B denote the bounded and U the unbounded component of $\mathbb{R}^{2n} \backslash K$. Then $0 \in B$ and we can assume $H^{-1}(1 + \bar{\varepsilon}) \subset \bar{U}$. Let γ denote the diameter of K and choose r and $b > 0$ such that

$$\text{(i)} \quad \gamma < r < 2\gamma \tag{4}$$

$$\text{(ii)} \quad \frac{3}{2}\pi r^2 < b < 2\pi r^2 .$$

For each $\delta \in (0,\bar{\varepsilon})$, let $f_\delta \in C^\infty(\mathbb{R},\mathbb{R})$ such that $f_\delta(s) = 0$ if $s \le 1 - \delta$, $f_\delta(s) = b$ if $s \ge 1 + \delta$, and $f_\delta'(s) > 0$ if $s \in (1 - \delta, 1 + \delta)$. Next define $g \in C^\infty(\mathbb{R},\mathbb{R})$ such that $g(s) = b$, $s \le r$; $0 < g'(s) \le 3\pi s$ if $s > r$, $g(s) \ge \frac{3}{2}\pi s^2$ if $s > r$, and $g(s) = \frac{3}{2}\pi s^2$ for large s.

With the aid of these new functions, we introduce a new family of Hamiltonians depending on δ. Set

$$\begin{aligned}\bar{H}(z) &= 0 \quad \text{if} \quad z \in B \cup H^{-1}([1-\varepsilon, 1-\delta]) \\ &= f_\delta(\varepsilon) \quad \text{if} \quad z \in H^{-1}(1+\varepsilon) \quad \text{for} \quad \varepsilon \in (-\delta,\delta) \\ &= b \quad \text{if} \quad z \in U \quad \text{and} \quad |z| < r \\ &= g(|z|) \quad \text{if} \quad |z| > r \,.\end{aligned}$$

It is easy to verify that $\bar{H} \in C^1(\mathbf{R}^{2n},\mathbf{R})$ and satisfies

$$-b + \frac{3}{2}\pi|z|^2 < \bar{H}(z) < \frac{3}{2}\pi|z|^2 + b \,. \tag{5}$$

Moreover if H is even in p, so is $\bar{H}$.

Our goal is to find 1-periodic solutions of

$$\dot{z} = \lambda J\bar{H}_z(z) \tag{6}$$

for λ near 1 and show how they lead to the desired family of solutions of (HS) near M. There are existence mechanisms from the calculus of variations that can be used to find 1-periodic solutions of (6). An appropriate functional framework to prove Theorem 2 will be introduced first. Later it will be indicated how a slight modification yields Theorem 3.

Let E denote the Hilbert space $W^{1/2,2}(S^1,\mathbf{R}^{2n})$ in the space of 2n-tuples of 1-periodic functions which possesses a "derivative of order $\frac{1}{2}$". In terms of Fourier series,

$$z = \sum_{k\in\mathbf{Z}} a_k e^{2\pi ikt} \quad \text{iff} \quad \sum_{k\in\mathbf{Z}} (1+|k|)|a_k|^2 < \infty \,.$$

See e.g. [9] or [17, Chapter 6] for a more detailed description of this space. E has a splitting into $E = E^+ \oplus E^0 \oplus E^-$ where E^+, E^0, E^- are the subspaces of E on which the action integral

$$A(z) = \int_0^1 p \cdot \dot{q}\,dt$$

is positive definite, null, or negative definite. Thus any $z \in E$ can be written as $z = z^+ + z^0 + z^- \in E^+ \oplus E^0 \oplus E^-$. Under an appropriate inner product on E, $A(z)$ takes the form

$$A(z) = \|z^+\|^2 - \|z^-\|^2 \,.$$

For $z \in E$, set

$$I_\lambda(z) \equiv A(z) - \lambda \int_0^1 \bar{H}(z)dt . \tag{7}$$

Aside from the additional parameter λ, this is the functional treated by Hofer and Zehnder in [9]. Since $\bar{H} \in C^1(\mathbb{R}^{2n},\mathbb{R})$ and is quadratic for large $|z|$, $I_\lambda \in C^1(E,\mathbb{R})$ [17, Prop. B39]. Moreover, any critical point of I_λ is a classical solution of (6) [17, Chapter 6].

A trivial modification of the proof of Lemma 1 of [9] (to reflect the dependence of I_λ on λ) gives

Lemma 8: If z is a solution of (6) and $I_\lambda(z) > 0$, then z lies in $H^{-1}(1 - \delta, 1 + \delta)$.

To find a positive critical value of I_λ, we will use Theorem 5.29 of [17], a version of an abstract critical point theorem in [18].

Proposition 9 [17]: Let E be a real Hilbert space with $E = E_1 \oplus E_2$, $E_2 = E_1^\perp$, and P_1, P_2 the corresponding orthogonal projectors. Suppose $I \in C^2(E,\mathbb{R})$, satisfies the Palais-Smale condition, and $I(u) = \frac{1}{2}(Lu,u) + b(u)$ with $Lu = L_1P_1u + L_2P_2u$ where $L_i : E_i \to E_i$ is bounded and self-adjoint. If further b' is compact and there is a subspace $\tilde{E} \subset E$ and sets $S \subset E$, $Q \subset \tilde{E}$ and constants $\alpha > w$ such that

$$\begin{array}{rl} \text{(i)} & S \subset E_1 \text{ and } I|_S \geq \alpha \\ \text{(ii)} & Q \text{ is bounded and } I|_{\partial Q} \leq w \\ \text{(iii)} & S \text{ and } \partial Q \text{ link ,} \end{array} \tag{10}$$

then I possesses a critical value $c \geq \alpha$.

Remark 11: The topological linking in (iii) is in the sense of [17-18]. Moreover, Proposition 9 gives a characterization of c as a minimax:

$$c = \inf_{h \in \Gamma} \sup_{u \in Q} I(h(u)) \tag{12}$$

where $\Gamma \subset C(Q,E)$ is a class of functions whose precise definition need not concern us here other than to note that $id \in \Gamma$.

To apply Proposition 9 to our setting, let $E_1 = E^+$ and $E_2 = E_0 \oplus E^-$. It is clear that for any $\lambda \in \mathbb{R}$, I_λ has the form required by the proposition with

$$b(z) = -\lambda \int_0^1 \bar{H}(z)dt .$$

The quadratic behavior of $\hat{H}$ for large $|z|$ and Proposition B.37 of [17] imply b' is compact. Since $\hat{H}(\rho) = o(|\rho|^2)$ as $\rho \to 0$ in $\mathbb{R}^{2n}$, $b(z) = o(\|z\|^2)$ as $z \to 0$ in E (see Lemma 6.16 of [17]). Let B_ρ denote a ball of radius ρ centered at 0. It then follows that if $S = \partial B_\rho \cap E_1$, for sufficiently small ρ, I_λ satisfies (10) (i) of Proposition 9 with e.g. $\alpha = \frac{1}{2}\rho^2$. (See also Lemma 2 of [9]). Indeed such a choice of ρ suffices for all λ in a compact neighborhood of 1. To define the set Q, let

$$\varphi(t) = \frac{1}{\sqrt{2\pi}}((\sin 2\pi t)e_1, -(\cos 2\pi t)e_1) \tag{13}$$

where e_1 is the unit n vector $(1,0,\ldots,0)$. Set $\tilde{E} = (\text{span } \varphi) \oplus E_2$. For $r_1 > \rho$ and $r_2 > 0$, set

$$Q = \{r\varphi \mid r \in [0,r_1)\} \oplus (B_{r_2} \cap E_2)$$

and let ∂Q denote its boundary in $\tilde{E}$. Then it follows from Example 5.26 of [17] that S and ∂Q link, i.e. (10) (iii) holds. Finally by Lemma 3 of [9], for $r_2 = r_1 = r$ sufficiently large and $\lambda > \frac{2}{3}$, (10) (ii) holds with $w = 0$.

All of the hypotheses of Proposition 9 have now been verified except for the Palais-Smale condition. As in Lemma 5 of [9], this is satisfied if

$$-J\dot{z} = 3\pi\lambda z$$

has no nontrivial periodic solutions. This will be the case if $3\pi\lambda \notin 2\pi\mathbf{Z}$. In particular if $\lambda \in (\frac{2}{3},\frac{4}{3})$, the Palais-Smale condition holds. It therefore follows that for each $\lambda \in [\frac{3}{4},\frac{5}{4}]$, I_λ has a critical value $c_\lambda > \alpha$ and a corresponding critical point z_λ which is a 1-periodic solution of (6).

Now we turn to the question of the relationship of z_λ to (HS). As a first step in this direction, let $j(\lambda)^{-1}$ be the minimal period of z_λ. Then $j(\lambda) \in \mathbf{N}$. Set

$$M = \sup\{j(\lambda) \mid \lambda \in [\tfrac{3}{4},\tfrac{5}{4}]\} .$$

Proposition 14: $M < \infty$.

Proof: If not, there is a sequence of $(\lambda_m) \subset [\frac{3}{4},\frac{5}{4}]$ such that $\lambda_m \to \lambda$ and $j(\lambda_m) \to \infty$. By Lemma 8, $z_{\lambda_m}(t) \subset H^{-1}(1 - \delta, 1 + \delta)$ which implies uniform L^∞ bounds for z_{λ_m}. Then (6) provides L^∞ bounds for $\dot{z}_{\lambda_m}$. Hence by the Arzela-Ascoli Theorem and (6), a subsequence of z_{λ_m} converges uniformly to a 1-periodic solution $w(t)$ of

$$\dot{z} = \lambda J\bar{H}_z(z) \ .$$

Since by above remarks, $I_{\lambda_m}(z_{\lambda_m}) \geq \alpha > 0$, $I_\lambda(w) \geq \alpha > 0$. On the other hand, $j(\lambda_m) \to \infty$ implies w has minimal period 0, i.e. $w \equiv$ constant. Consequently $I_\lambda(w) \leq 0$, a contradiction. Thus $M < \infty$.

Note that $M \geq 1$. Let $\rho = \min(\frac{1}{4}, \frac{1}{2M+1})$.

Proposition 15: $\{z_\lambda \mid |\lambda - 1| < \rho\}$ are geometrically distinct.

Proof: If $z_\lambda(t)$ and $z_\mu(t)$ represent the same trajectory, there is an $r(t) \in C^1$ such that $z_\lambda(t) = z_\mu(r(t))$. Therefore

$$\dot{z}_\lambda = \lambda J\bar{H}_z(z_\lambda) = \frac{d}{dt} z_\mu(r(t)) = \mu J\bar{H}_z(z_\mu(r(t)))\dot{r} \ .$$

Hence

$$\lambda = \mu\dot{r}$$

or

$$r(t) = \frac{\lambda}{\mu} t + \gamma \tag{16}$$

and

$$z_\lambda(t) = z_\mu(\frac{\lambda}{\mu} t + \gamma) \ . \tag{17}$$

By (17),

$$z_\lambda(t + j(\lambda)^{-1}) = z_\lambda(t) = z_\mu(\frac{\lambda}{\mu} t + \frac{\lambda}{\mu} j(\lambda)^{-1} + \gamma)$$
$$= z_\mu(\frac{\lambda}{\mu} t + \gamma) \ . \tag{18}$$

Therefore

$$\frac{\lambda}{\mu} j(\lambda)^{-1} \in j(\mu)^{-1}\mathbb{N} \ . \tag{19}$$

Similarly,

$$z_\mu(\frac{\lambda}{\mu} t + j(\mu)^{-1} + \gamma) = z_\mu(\frac{\lambda}{\mu} t + \gamma) = z_\lambda(t)$$
$$= z_\mu(\frac{\lambda}{\mu} (t + \frac{\mu}{\lambda} j(\mu)^{-1}) + \gamma) = z_\lambda(t + \frac{\mu}{\lambda} j(\mu)^{-1}) \tag{20}$$

so

$$\frac{\mu}{\lambda} j(\mu)^{-1} \in j(\lambda)^{-1}\mathbb{N} \ . \tag{21}$$

Combining (19) and (21) yields

$$\frac{\lambda}{\mu} = \frac{j(\lambda)}{j(\mu)} \; . \tag{22}$$

We can assume $\lambda > \mu$ and therefore $j(\lambda) > j(\mu)$. Thus

$$1 + \frac{2\rho}{1 - \rho} = \frac{1 + \rho}{1 - \rho} > \frac{\lambda}{\mu} = \frac{j(\lambda)}{j(\mu)} > 1 + \frac{1}{j(\mu)} > 1 + \frac{1}{M} \tag{23}$$

and

$$\rho > (2M + 1)^{-1} \; . \tag{24}$$

But (24) is contrary to our definition of ρ and the Proposition follows.

Proof of Theorem 2: The above results show for each $\lambda \in (1 - \rho, 1 + \rho)$, (6) possesses a 1-periodic solution z_λ and these functions are geometrically distinct. By Lemma 8, each z_λ lies on $H^{-1}(1 + \varepsilon)$ for some $|\varepsilon(\lambda)| < \delta$. By the definition of $\bar{H}$, the surfaces $H^{-1}(1 + \varepsilon)$ and $\bar{H}^{-1}(f_\delta(\varepsilon))$ are the same. A standard result - see e.g. Prop. 6.47 of [17] - then implies that a reparameterization ρ_λ of z_λ is a periodic solution of (HS). Observing that δ can be made as small as we please and that we have uncountably many distinct z 's and ρ_λ's, the conclusions of Theorem 2 follow.

Aside from working in a different function space, the proof of Theorem 3 is roughly the same as that of Theorem 2.

Proof of Theorem 3: Let

$$X = \{z = (p,q) \in E \,|\, p \text{ is odd about } 0 \text{ and } \tfrac{1}{2} \text{ and } q \text{ is even about } 0 \text{ and } \tfrac{1}{2}\} \; .$$

Let $X^0 = E^0 \cap X$ and $X^\pm = E^\pm \cap X$. As was noted earlier, if H is even in p, so is $\bar{H}$. Hence $\bar{H}_p, \bar{H}_q$ are then respectively odd and even in p. Consider I_λ as defined in (7) on X. Note that φ defined in (13) belongs to X^+. It then easily follows that the arguments of Theorem 2 carry over unchanged for Theorem 3. Note however, in Proposition 15, since $z(t) = (p(t),q(t))$ with $p(0) = 0$, $\lambda = 0$. But this does not otherwise effect the argument.

As a final remark, suppose for all $z \in M$, $p \cdot H_p(z) > 0$ if $p \neq 0$. Then it was shown in [10] that there are constants $\bar{a} > \underline{a} > 0$ such that

$$\underline{a} < T^{-1} \int_0^T p \cdot \dot{q} dt < \bar{a} \tag{25}$$

for any T periodic solution $z = (p,q)$ of (HS) on $H^{-1}(\ell)$ for any ℓ near 1 where $\underline{a}, \bar{a}$ are independent of ℓ. These bounds apply of course to bouncing orbits. Combining these bounds with Theorem 3, we get

Theorem 26: If $M = H^{-1}(1)$ is a compact hypersurface in $\mathbb{R}^{2n}$ and for all $z \in M$, $p \cdot H_p(z) > 0$ for $p \neq 0$, then M contains a bouncing orbit.

REFERENCES

[1] Seifert, H., 'Periodische Bewegungen mechanischen systeme', Math. Z. **51** (1948), 197-216.

[2] Weinstein, A., 'Periodic orbits for convex Hamiltonian systems', Ann. Math. **108** (1978), 507-518.

[3] Rabinowitz, P. H., 'Periodic solutions of Hamiltonian systems', Comm. Pure Appl. Math. **31** (1978), 157-184.

[4] Rabinowitz, P. H., 'Periodic solutions of a Hamiltonian system on a prescribed energy surface', J. Diff. Eq. **33** (1979), 336-352.

[5] Weinstein, A., 'On the hypotheses of Rabinowitz's periodic orbit theorems', J. Diff. Eq. **33** (1979), 353-358.

[6] Gluck, H. and W. Ziller, 'Existence of periodic solutions of conservative systems', Seminar on Minimal Submanifolds, Princeton University Press (1983), 65-98.

[7] Hayashi, K., 'Periodic solutions of classical Hamiltonian systems', Tokyo Univ. J. Math. **6** (1983), 473-486.

[8] Benci, V., 'Closed geodesics for the Jacobi metric and periodic solutions of prescribed energy of natural Hamiltonian systems', to appear Ann. Inst. H. Poincaré, Analyse Nonlineaire.

[9] Hofer, H. and E. Zehnder, 'Periodic solutions on hypersurfaces and a result by C. Viterbo', to appear in Inv. Math.

[10] Benci, V., H. Hofer, and P. H. Rabinowitz, 'A remark on a priori bounds and existence for periodic solutions of Hamiltonian systems', these proceedings.

[11] Benci, V. and P. H. Rabinowitz, 'A priori bounds for periodic solutions of Hamiltonian systems', to appear in Ergodic Theory and Dynamical Systems.

[12] Viterbo, C., 'A proof of the Weinstein conjecture in $\mathbb{R}^{2n}$', preprint, Sept. 1986.

[13] Ruiz, O. R., 'Existence of brake orbits in Finsler dynamical systems', Springer Lec. Notes in Math. **597** (1977), 542-567.

[14] van Groesen, E. W. C., 'Existence of multiple normal mode trajectories of even classical Hamiltonian systems', J. Diff. Eq. **57** (1985), 70-89.

[15] Rabinowitz, P. H., 'On the existence of periodic solutions of a class of symmetric Hamiltonian systems', to appear Nonlinear Analysis: T.M.A.

[16] Rabinowitz, P. H., 'On a theorem of Weinstein', to appear J. Diff. Eq.

[17] Rabinowitz, P. H., 'Minimax methods in critical point theory with applications to differential equations', C.B.M.S. Regional Conf. Ser. in Math. **65** (1986).

[18] Benci, V. and P. H. Rabinowitz, 'Critical point theorems for indefinite functionals', Inv. Math. **52** (1979), 241-273.

THE VALUE FUNCTION OF A MODIFIED JACOBI FUNCTIONAL

E. van Groesen
Department of Applied Mathematics
University of Twente
7500 AE Enschede, The Netherlands

ABSTRACT. Second order Hamiltonian systems with convex potentials are considered. To relate the value of the energy and the period for certain periodic motions, the value function of a modified Jacobi functional is investigated. A family of saddle points of this functional parameterized by the energy, provides periodic solutions for which the minimal period belongs to the generalized derivative of the value function.

1. INTRODUCTION

In this contribution we consider second order Hamiltonian systems

$$-\ddot{q} = V'(q), \quad q(t) \in \mathbb{R}^N , \tag{1.1}$$

with V the potential energy, a smooth function on $\mathbb{R}^N$ and V' its gradient. For periodic solutions of (1.1) we want to relate the period T to the value E of the total energy:

$$\frac{1}{2}\dot{q}^2 + V(q) = E . \tag{1.2}$$

In Gordon [6] and Lewis [11] there are some partial results in this direction, presupposing the existence of a smooth manifold of such periodic solutions; the requirement of smoothness seems to be difficult to verify in general. In this paper we will characterize for each value of E a periodic solution as a critical point of a functional J_E which is a modification of the usual Jacobi functional. This modified functional is easier to deal with from a functional analytic point of view; in particular, we obtain a one parameter family of periodic solutions as explicit saddle points of J_E for each E. This saddle point is in fact a minimizer of J_E on a naturally constrained subset N_E. The corresponding (mini-max) value of J_E defines a value function (see Clarke [5] and his contribution to these proceedings for other examples and applications of value functions). We will show that, under suitable conditions,

P. H. Rabinowitz et al. (eds.), Periodic Solutions of Hamiltonian Systems and Related Topics, 255–260.

the period of the solution belongs to the generalized derivative of this value function. This result can be used to further investigate the relation between E and T and to prove, for instance, a monotone dependence in specific cases. The relation between the period and the value function is not a consequence of standard results about value functions since, in our case, the critical value is a genuine saddle value or, expressed differently, both the functional J_E as well as the equality constraint in the definition of N_E depend on the parameter E.

For simplicity of exposition we will suppose from now on that V is an even function (see Remark 6 how the methods can be adapted to treat the general case). Then periodic solutions can be obtained by an appropriate continuation of the solution of a boundary value problem (cf. Rosenberg [14], Berger [3]). Specifically, on a normalized time interval, a solution x(t) of

$$\begin{aligned} -\ddot{x} &= T^2 V'(x) \\ x(0) &= \dot{x}(1) = 0 \end{aligned} \tag{1.3}$$

provides a so-called normal mode solution of period 4T by continuation of the function q(t) defined for a quarter period T by

$$q(t) = x(t/T) \ . \tag{1.4}$$

(So, normal modes are brake orbits that go through the origin of configuration space.) The transformation (1.4) relates the energy E and the 1/4-period T(> 0) like

$$\frac{1}{2}\frac{1}{T^2}\dot{x}^2 + V(x) = E \ . \tag{1.5}$$

2. THE MODIFIED JACOBI FUNCTIONAL J_E

On the space of functions $X := \{x \in H_1([0,1],\mathbb{R}^N) \mid x(0) = 0\}$ we consider the following functional $J_E : X \to \bar{\mathbb{R}} = \mathbb{R} \cup \{-\infty\}$:

$$J_E(x) = \begin{cases} 2[\int \frac{1}{2}\dot{x}^2 \cdot \{E - \int V(x)\}]^{1/2} & \text{if } \int V(x) < E \\ -\infty & \text{otherwise ;} \end{cases} \tag{2.1}$$

here $\int$ denotes integration over the interval (0,1). Note that J_E is a product of two functionals and that it differs in that respect essentially from the usual Jacobi functional

$$2 \int \sqrt{E - V(x)} \cdot \sqrt{\tfrac{1}{2}\dot{x}^2} \ .$$

We will call J_E a modified Jacobi functional since it results from the action functional on phase space in a way which is analogous to the way the usual Jacobi functional is obtained, the difference being that for the modification the energy constraint is prescribed in an integrated way instead of in a pointwise fashion. This integrated form of the energy constraint was introduced, and exploited in a different way, in the fundamental paper of Rabinowitz [13]. (See [7] for a historical account on the Euler-Maupertuis principle and details about the corresponding modification leading to (2.1).) The next result shows that J_E serves our purposes; the proof of it is straightforward.

Proposition 1. Let $x(\neq 0)$ be a critical point of J_E on X. Then x satisfies the boundary value problem (1.3) for a value of T given by

$$T = \frac{d}{dE} J_E(x), \quad \text{i.e.} \quad T^2 = \frac{\int \frac{1}{2} \dot{x}^2}{E - \int V(x)} . \tag{2.2}$$

The corresponding normal mode defined by (1.4) has T as its 1/4-period and has energy E.

3. MINIMIZING J_E ON NATURALLY CONSTRAINED SETS N_E

It is obvious that J_E is neither bounded from below nor from above on X. For rather general potentials V it is possible to prove the existence of at least one critical point using the mountainpass lemma of Ambrosetti and Rabinowitz [2]. Here we shall require V to be strictly convex which will enable us to provide a very explicit characterization of a saddle point of J_E. It is no restriction to assume that V is normalized such that $V(0) = 0$ and $V(x) > 0$ for all $x \in \mathbb{R}^N$.

Let us first introduce the natural constraints for the functional J_E (the idea goes back to Nehari [12]; Berger and Schechter [4] have considered such constraints in some generality; for an application of the idea in the context of Hamiltonian systems see also Berger [3] and Ambrosetti and Mancini [1]). For $E > 0$ let N_E be the subset

$$N_E := \{x \in X | E = \int V(x) + \frac{1}{2} V'(x) \cdot x\} . \tag{3.1}$$

Proposition 2. For each $E > 0$ the set N_E is a natural constraint for the functional J_E in the following sense:

(i) any critical point of J_E belongs to N_E, and

(ii) critical points of J_E on N_E are also critical points of J_E on all of X.

Proof. (i) Multiply the equation (1.3) by x and integrate by parts and use (2.2). (ii) By convexity of V, the set N_E is a regular manifold and Lagrange's multiplier rule is applicable. A critical point x of J_E on N_E thus satisfies

$$-\ddot{x} = T^2 V'(x) + \lambda\left(\frac{3}{2} V'(x) + \frac{1}{2} V''(x) \cdot x\right)$$

for T^2 given by (2.2) and some $\lambda \in \mathbf{R}$. To get the result, show that $\lambda = 0$ by multiplying this equation by x and integrating by parts, and using $x \in N_E$ and the convexity of V.

(The vanishing of the multiplier λ that results from the restriction to N_E explains the name "natural constraint".) □

As a consequence of this proposition, we may just as well look for critical points of J_E on N_E instead of on all of X. But on N_E, the functional J_E is strictly positive and we may look for a minimizer. The next result can be established (see [9]).

Proposition 3. For any $E > 0$ there exists at least one solution of the constrained minimization problem

$$\inf\{J_E(x) \mid x \in N_E\} \; . \tag{3.2}$$

Moreover, for such a solution, the corresponding normal mode has T given by (2.2) as its minimal 1/4-period.

Remark 4. By convexity of V, on each ray in X through the origin the functional J_E takes its maximum value at a unique point. The set N_E is precisely the collection of all these points and is strictly starshaped with respect to the origin. Consequently, introducing polar coordinates in X, with S the unit sphere $\{y \in X \mid \int \dot{y}^2 = 1\}$, the constrained minimization problem (3.2) is an explicit characterization of the following saddle point formulation:

$$\inf\{J_E(x) \mid x \in N_E\} = \inf_{y \in S} \max_{\rho > 0} J_E(\rho y) \; . \tag{3.3}$$

4. THE VALUE FUNCTION AND ITS RELATION TO THE PERIOD

We introduce the value function as the value of J_E at the critical point characterized by (3.2):

$$j(E) := \inf\{J_E(x) \mid x \in N_E\} \; . \tag{4.1}$$

Proposition 5. For $E > 0$, the value function $j(E)$ is continuous and monotonically increasing. If the right and left-hand side derivatives are denoted by $j'_+(E)$ and $j'_-(E)$ respectively, then if $\hat{x}$ is any solution of the constrained minimization problem (3.2), the minimal 1/4-period $\hat{T}$ satisfies

$$j'_+(E) \le \hat{T} \le j'_-(E) \; . \tag{4.2}$$

Proof. The monotonicity is an immediate consequence of the monotonicity of the function $E \to J_E(x)$ for each fixed x, and of the explicit mini-max characterization (3.3). To prove the first inequality of (4.2), the second one is analogous, let for $e > 0$ a number $\rho = \rho(e) \in \mathbb{R}$ be defined such that $\rho(e) \cdot \hat{x} \in N_{E+e}$. Then ρ is uniquely determined and $\rho(0) = 1$ and $\rho'(0) = [\int V'(\hat{x}) \cdot \hat{x} + \frac{1}{2} V''(\hat{x}) \cdot \hat{x} \cdot \hat{x}]^{-1}$ is finite. Since $j(E + e) \leq J_{E+e}(\rho\hat{x})$ by definition of j, the result follows from the fact that the first variation $\delta J_E(\hat{x};\cdot)$ vanishes and from (2.2):

$$\dot{j}'_+(E) \leq \lim_{e \to 0} \frac{1}{e} [J_{E+e}(\rho\hat{x}) - J_E(\hat{x})]$$

$$= \frac{d}{dE} J_E(\hat{x}) + \delta J_E(\hat{x};\hat{x}) \cdot \rho'(0) = \hat{T} . \qquad \square$$

In [10] this value function is exploited to find relations between T and E. For instance, for a specific class of subquadratic potentials it is shown that $\dot{j}$ is a convex function and hence, because of (4.2), is differentiable. Moreover, all solutions of (3.2) then have the same minimal 1/4-period $T = j'(E)$, and T runs from $j'_T(0)$ to ∞ if E runs from 0 to ∞; at a given E, all solutions of (3.2) are also minimizers on X of the usual Lagrange functional for solutions of prescribed period $T = j'(E)$:

$$\frac{1}{T} \int \frac{1}{2} \dot{x}^2 - T \int V(x) .$$

See [10] for further results and for superquadratic potentials.

Remark 6. In case V is convex but not necessarily even, one can look for brake-orbits by considering J_E on all of $H_1([0,1],\mathbb{R}^N)$. Then $\dot{x}(0) = \dot{x}(1) = 0$ result as natural boundary conditions and T is half of the period. Then analogous results can be obtained if one uses as natural constraints the sets

$$M_E := \{x | \int V'(x) = 0, \quad E = \int V(x) + \frac{1}{2} V'(x) \cdot x\}$$

(of codimension $N + 1$, in this case). See also Berger [3] for the additional constraint, and [9] for the existence result in this specific case.

REFERENCES

1. Ambrosetti, A. and G. Mancini, 'Solutions of minimal period for a class of convex Hamiltonian systems', Math. Analen **255** (1981), 405-421.

2. Ambrosetti, A. and P. H. Rabinowitz, 'Dual variational methods in critical point theory and applications', J. Funct. Anal. **14** (1973), 349-381.
3. Berger, M. S., 'Periodic solutions of second order dynamical systems and isoperimetric variational problems', Amer. J. Math. **93** (1971), 1-10.
4. Berger, M. S. and M. Schechter, 'On the solvability of semilinear gradient operator equations', Adv. in Math. **25** (1977), 97-132.
5. Clarke, F. H., Optimization and nonsmooth analysis, Wiley Interscience, New York, 1983.
6. Gordon, W. B., 'On the relation between period and energy in periodic dynamical systems', J. Math. Mech. **19** (1969), 111-114.
7. van Groesen, E. W. C., Hamiltonian flow on an energy surface: 240 years after the Euler-Manpertuis principle, in R. Martini (ed.), L. N. Physics, vol. 239, Springer, 1985.
8. van Groesen, E. W. C., 'On small period, large amplitude normal modes of natural Hamiltonian systems', Nonlin. Anal. T.M.A. **10** (1986), 41-53.
9. van Groesen, E. W. C., 'Analytical mini-max methods of Hamiltonian brake orbits of prescribed energy', J. Math. Anal. Appl., in press.
10. van Groesen, E., 'Duality between period and energy of certain periodic Hamiltonian motions', J. London Math. Soc., in press.
11. Lewis, D. C., 'Families of periodic solutions of systems having relatively invariant line integrals', Proc. Amer. Math. Soc. **6** (1955), 181-185.
12. Nehari, Z., 'On a class of nonlinear second order differential equations', Trans. Amer. Math. Soc. **95** (1960), 101-123.
13. Rabinowitz, P. H., 'Periodic solutions of Hamiltonian systems', Comm. Pure Appl. Math. **31** (1978), 157-184.
14. Rosenberg, R. M., 'Normal modes of non-linear dual-mode systems', J. Appl. Math. **27** (1960), 263-268.

PERTURBATIONS OF NONDEGENERATE PERIODIC ORBITS OF HAMILTONIAN SYSTEMS

Michel Willem
Institut Mathematique
2 Ch. du Cyclotron
B-1348 Louvain-la-Neuve
Belgium

INTRODUCTION

This work consists of three parts. In the first part we compute the Morse index and the nullity of the periodic solutions of equations of the form $\ddot{u} + g(u) = 0$.

The second part is devoted to a local perturbation theorem for nondegenerate periodic orbits of Hamiltonian systems. Using only the implicit function theorem, we generalize a recent theorem of Ambrosetti, Coti-Zelati and Ekeland [2]. An application is given to the forced pendulum equation.

The third part contains a global perturbation theorem for superlinear differential equations. The basic tool is Morse theory.

We shall use the following notations:

$$H_T^1 = \{u \in H^1(0,T;\mathbb{R}) : u(0) = u(T)\} ,$$

$$C_T^1 = \{u \in C^1(0,T;\mathbb{R}^{2N}) : u(0) = u(T)\} .$$

1. SECOND ORDER AUTONOMOUS DIFFERENTIAL EQUATIONS

This section is devoted to the study of the autonomous problem

$$\ddot{u}(t) + g(u(t)) = 0 , \tag{1}$$

$$u(0) - u(T) = \dot{u}(0) - \dot{u}(T) = 0 ,$$

where $g \in C^1(]-\ell,\ell[,\mathbb{R})$ for some $\ell \in]0,+\infty]$. We assume that the following condition holds:

$$g(-u) = -g(u), \quad 0 < g(u) \cdot u \quad \text{for} \quad 0 < |u| < \ell , \tag{A_0}$$

and that g satisfies one of the following growth assumptions:

P. H. Rabinowitz et al. (eds.), Periodic Solutions of Hamiltonian Systems and Related Topics, 261–265.

$$u^{-1}g(u) < g'(u) \quad \text{for} \quad 0 < |u| < \ell , \tag{A_1}$$

$$g'(u) < u^{-1}g(u) \quad \text{for} \quad 0 < |u| < \ell . \tag{A_1'}$$

Since the energy $(1/2)\dot{u}^2 + G(u)$, with $G(u) = \int_0^u g(s)ds$, is a first integral of (1), the initial conditions $u(0) = a \in]0,\ell[$, $\dot{u}(0) = 0$ provide a periodic solution of (1) with minimal period

$$P(a) = 2\sqrt{2} \int_0^a \frac{dx}{\sqrt{G(a) - G(2)}} .$$

The periodic solutions of (1), with period T, are the critical points of the functional φ defined on H_T^1 by

$$\varphi(u) = \int_0^T [\frac{1}{2} \dot{u}^2(t) - G(u(t))]dt .$$

The following theorem gives the Morse index $j(u,T)$ and the nullity $n(u,T)$ of a critical point u of φ with minimal period T/k.

Theorem 1. [6]. Under assumptions (A_0) and (A_1) (resp. (A_1')), if u is a critical point of φ with minimal period T/k and such that $u(0) \in]0,\ell[$, $\dot{u}(0) = 0$, then

$$j(u,T) = 2k, \qquad n(u,T) = 1 ,$$

$$\text{(resp.} \quad j(u,T) = 2k - 1, \qquad n(u,T) = 1) .$$

2. LOCAL PERTURBATIONS OF NONDEGENERATE PERIODIC ORBITS

In this section, we consider the existence of T-periodic solutions of the forced Hamiltonian system

$$J\dot{u} + \nabla H(u) = \varepsilon f(t,u,\varepsilon) \tag{2}$$

near a T-periodic solution v_0 of the autonomous system

$$J\dot{u} + \nabla H(u) = 0 .$$

We assume that

$$H \in C^2(\mathbb{R}^{2N},\mathbb{R}) . \tag{B_1}$$

If h is a T-periodic solution of the linearized system

$$J\dot{h} + H''(v_0)h = 0$$

then h is proportional to $\dot{v}_0$. (B_2)

$f \in C^2(\mathbf{R} \times \mathbf{R}^{2N} \times \mathbf{R}, \mathbf{R}^{2N})$, f is T-periodic in t. (B_3)

Let us define $M : C_T^1(0,T;\mathbf{R}^{2N}) \times \mathbf{R} \to C(0,T;\mathbf{R}^{2N})$ by

$$M(u,\varepsilon) = J\dot{u} + \nabla H(u) - \varepsilon f(\cdot,u,\varepsilon) .$$

Let us also define the projector P_λ on $C(0,T;\mathbf{R}^{2N})$ by

$$P_\lambda u = \int_0^T (u(t),\dot{v}_\lambda(t))dt\ \dot{v}_\lambda$$

where $v_\lambda(t) = v_0(t + \lambda)$. By the implicit function theorem (see [4]) the system

$$(I - P_\lambda)M(v_\lambda + w,\varepsilon) = 0, \quad P_\lambda w = 0$$

defines near $\varepsilon = 0$ and $Z = \{v_\lambda : \lambda \in \mathbf{R}\}$ a function $w = w^*(\lambda,\varepsilon)$. Clearly $w^*(\lambda,0) = 0$ and w^* is T-periodic in λ. Equation $M(u,\varepsilon) = 0$ is then equivalent to $N(\lambda,\varepsilon) = 0$ where

$$N(\lambda,\varepsilon) = \int_0^T (M(v_\lambda + w^*(\lambda,\varepsilon),\varepsilon),\dot{v}_\lambda)dt .$$

Since $N(\lambda,0) = 0$, this is a bifurcation problem. Moreover N is T-periodic in λ.

<u>Theorem 2</u>. Under assumptions (B_1) to (B_3), if $f(t,u,\varepsilon) = \nabla_u F(t,u,\varepsilon)$ then, for $|\varepsilon|$ small enough, system (2) has at least 2 T-periodic solutions near Z.

<u>Proof</u>. Let us define φ_ε on $C_T^1(0,T;\mathbf{R}^{2N})$ by

$$\varphi_\varepsilon(u) = \int_0^T \left[\frac{1}{2} (J\dot{u},u) + H(u) - \varepsilon F(t,u,\varepsilon)\right]dt$$

and ψ_ε on $\mathbf{R}$ by

$$\psi_\varepsilon(\lambda) = \varphi_\varepsilon(v_\lambda + w^*(\lambda,\varepsilon)) .$$

Since ψ_ε is T-periodic, this function has at least a maximum and a minimum on $[0,T[$. It suffices then to verify that, near $\varepsilon = 0$, if $\psi_\varepsilon'(\lambda) = 0$ then $N(\lambda,\varepsilon) = 0$. □

We define now

$$h(\lambda) = \int_0^T (f(t,v_\lambda(t),0),\dot{v}_\lambda(t))dt .$$

<u>Theorem 3</u>. Under assumptions (B_1) to (B_3), if $h(\lambda_0) \neq 0$, there is a neighborhood V of $(0, v_{\lambda_0})$ in $\mathbf{R} \times C_T^1$ such that, if

$(\varepsilon, u) \in V$ is a solution of (2), then $\varepsilon = 0$ and $u \in Z$.

<u>Proof</u>. Since $D_\varepsilon N(\lambda_0, 0) = h(\lambda_0) \neq 0$, it suffices to apply the implicit function theorem. □

<u>Theorem 4</u>. Under assumptions (B_1) to (B_3), if λ_0 is a simple zero of h, there exists a continuous function $u^* : [-\bar{\varepsilon}, \bar{\varepsilon}] \to C_T^1$ such that $u^*(0) = v_{\lambda_0}$ and $u^*(\varepsilon)$ is a solution of (2). Moreover there is a neighborhood V of $(0, v_{\lambda_0})$ in $\mathbf{R} \times C_T^1$ such that if $(\varepsilon, u) \in V$ is a solution of (2) then either $u = u^*(\varepsilon)$ or $\varepsilon = 0$ and $u \in Z$.

<u>Proof</u>. Since $D_\varepsilon N(\lambda_0, 0) = h(\lambda_0) = 0$ and $D_{\lambda\varepsilon} N(\lambda_0, 0) = h(\lambda_0) \neq 0$, it suffices to apply the Crandall-Rabinowitz bifurcation theorem in the simple case of a real function of two real variables. □

<u>Remark</u>. Theorem 2 generalizes, by a direct approach, theorem 3.6 of [2]. Theorem 3 generalizes theorem 1 of [1] and theorem 4.2 of [2]. Theorem 4 generalizes, with a simpler proof, theorem 1 of [1].

<u>Example</u>. Let $f \in C(\mathbf{R}, \mathbf{R})$ be a τ-periodic function and consider the problem

$$\ddot{u}(t) + \sin u(t) = \varepsilon f(t) \tag{3}$$

$$u(j\tau) - u(0) = \dot{u}(j\tau) - \dot{u}(0) = 0 ,$$

where j is a positive integer. Assume that $k^{-1} j\tau > 2\pi$ and let v_0 be the solution of

$$\ddot{v} + \sin v = 0$$

$$v(0) = a,\ P(a) = k^{-1} j\tau,\ \dot{v}(0) = 0 .$$

By Theorems 1 and 2, for $|\varepsilon|$ small enough, problem (3) has at least two solutions near $Z = \{v_0(t + \lambda) : \lambda \in \mathbf{R}\}$. (See [5] for the case when $f(t) = \cos t$ or $f(t) = \sin t$.)

3. GLOBAL PERTURBATIONS OF NONDEGENERATE PERIODIC ORBITS

Let us consider the problem

$$\ddot{u}(t) + |u(t)|^{P-2} u(t) = f(t) \tag{4}$$

$$u(0) - u(T) = \dot{u}(0) - \dot{u}(T) = 0$$

where $P > 2$, $f \in L^q(0, T; \mathbf{R})$ and $\frac{1}{P} + \frac{1}{q} = 1$. Let u_k be the solution of

$$\ddot{u}(t) + |u(t)|^{P-2} u(t) = 0$$

$$u(0) = a, \ P(a) = a^{1-P/2}P(1) = T/k, \ \dot{u}(0) = 0$$

and let $Z_k = \{u_k(t + \lambda); \lambda \in \mathbf{R}\}$.

Theorem 5. There exists $k_0 \in \mathbf{N}^*$ such that, if $k > k_0$, each Z_k generates two solutions of (4). In particular, problem (4) has infinitely many solutions. If the solutions of (4) are nondegenerate, then, for $k > k_0$, (4) has at least one solution with Morse index k.

Sketch of proof. Let us define φ on H_T^1 by

$\varphi(u) = \int_0^T [1/2\dot{u}(t)^2 - |u(t)|^P]dt$. According to Theorem 1, Z_k is a nondegenerate critical manifold of φ and the corresponding critical groups over $\mathbf{Z}_2$ are given by

$$C_n(\varphi,Z_k) \simeq \mathbf{Z}_2 \qquad n = 2k \text{ or } n = 2k + 1 ,$$

$$\simeq \{0\} \qquad n \leq 2k - 1, \ n \geq 2k + 2 .$$

Thus, for $\varepsilon = \varepsilon(k) > 0$, the Betti numbers of φ satisfies

$$B_n(\varphi^{c+\varepsilon},\varphi^{c-\varepsilon}) = 1, \quad n = 2k, \quad n = 2k + 1 ,$$

where $c = c(k) = \varphi(uk)$. Since the action corresponding to (4) is, in some sense, a perturbation of φ, it suffices then to use a truncation procedure and classical stability properties of the Betti numbers in MORSE theory (see [6]). □

Remarks. 1. The existence of infinitely many solutions of (4) is obtained by contradiction in [3].

2. The same method is applicable to more general nonlinearities [6].

REFERENCES

[1] A. Albizzati, 'Sélection de phase par un terme d'excitation pour les solutions periodiques de certaines équations différentielles', C. R. Acad. Sci. Paris +.996 (1983), 259-262.

[2] A. Ambrosetti, V. Coti-Zelati and I. Ekeland, 'Symmetry breaking in critical point theory and applications', preprint, 1985.

[3] A. Bahri and H. Berestycki, 'Existence of forced oscillations for some nonlinear differential equations', Comm. Pure Appl. Math. 37 (1984), 403-442.

[4] J. Hale, 'Bifurcation near families of solutions', in Differential Equations, Acta Universitatis Upsaliensis, Upsala, 1977.

[5] B. V. Schmitt and N. Sari, 'Sur la structure de l'équation du pendule forcé', preprint.

[6] M. Willem, 'Perturbation des variétés critiques nondégénerées et oscillations nonlineaires forces', preprint.

REMARKS ON PERIODIC SOLUTIONS ON HYPERSURFACES

E. Zehnder*)
Mathematical Institute
Ruhr University
Bochum, Federal Republic of Germany

In this note we shall briefly describe some well-known and some more recent results and open questions related to the existence problem of periodic orbits on energy surfaces. We start with the Hamiltonian equation

$$\dot{x} = J\nabla H(x) \equiv X_H(x) \quad \text{on} \quad x \in \mathbb{R}^{2n} , \tag{1}$$

$H \in C^\infty(\mathbb{R}^{2n})$ being a smooth function, and J being the nondegenerate skew symmetric matrix

$$J = \begin{pmatrix} 0 & -I \\ I & 0 \end{pmatrix} \in L(\mathbb{R}^{2n}) . \tag{2}$$

With ω we shall denote in the following the associated symplectic form

$$\omega(X,Y) = \langle JX,Y\rangle, \qquad X,Y \in \mathbb{R}^{2n} , \tag{3}$$

it is an exact and nondegenerate 2-form on $\mathbb{R}^{2n}$. Assume now that for some constant $E \in \mathbb{R}$ the set

$$S := \{x \in \mathbb{R}^{2n} | H(x) = E\} \tag{4}$$

is compact and a regular energy surface, i.e. $\nabla H(x) \neq 0$ for $x \in S$. Then S is a smooth compact hypersurface and the Hamiltonian vector field $X = X_H$ is tangential to S since $\langle \nabla H(x), X\rangle = 0$.

The flow of the vector field X_H on S is, in general, very intricate. In the exceptional case of an integrable Hamiltonian system near S, one knows, of course, that most solutions on S are

*)Research supported by the Stiftung Volkswagenwerk.

P. H. Rabinowitz et al. (eds.), Periodic Solutions of Hamiltonian Systems and Related Topics, 267–279.

quasiperiodic having n rationally independent frequencies and lying on invariant n-dimensional tori, moreover, the periodic solutions are dense on S. Under a small perturbation however this orbit structure changes dramatically, still many of the invariant tori have a continuation by the K.A.M. theory, and these tori lie in the closure of the set of periodic orbits. With increasing perturbation the invariance properties breakdown and very little is known about the orbit structure. Instead of studying the hopeless initial value problem for the flow one can try to find invariant subsystems, which, in general, leads to boundary value problems, and then try to study the flow nearby by perturbation methods. The simplest subsystems are the periodic solutions which will be considered in the following. There is a well-known variational principle on the loop space of S for which the periodic solutions are the critical points. It is the advantage of the variational principle that it singles out precisely the periodic solutions neglecting the complexity of the orbit structure of the flow.

In order to first formulate the existence problem of periodic solutions of X_H on S more geometrically we recall that the hypersurface S together with the symplectic 2-form ω determine the flow on S. Indeed, if S is a compact and smooth hypersurface in $\mathbb{R}^{2n}$ there is a smooth function $H \in C^\infty(\mathbb{R}^{2n})$ satisfying (4) and moreover $\nabla H(x) \neq 0$ for $x \in S$, see [16]. Therefore S is a regular energy surface for X_H. If S is also a regular energy surface for a second Hamiltonian vector field X_G, i.e. $S = \{x \in \mathbb{R}^{2n} | G(x) = \text{const}\}$ then there exists a smooth and nonvanishing function ρ on S such that

$$X_H = \rho X_G \quad \text{on} \quad S . \tag{5}$$

The two Hamiltonian systems have therefore, on S, the same orbits up to reparameterization governed by ρ. In particular they have the same periodic orbits and we can ask for periodic orbits on S independent of the Hamiltonian function chosen. In more abstract terms, the kernel of $\omega|S$ is 1-dimensional, indeed it is spanned by $JN(x) \in T_xS$, where $N(x)$ is the outer normal at S in X. We therefore have a line bundle $L_S \subset TS$ which has the direction of every Hamiltonian vector field X_H having S as regular energy surface. A periodic orbit on S is then simply a one-dimensional submanifold $P \subset S$ diffeomorphic to the circle S^1 and satisfying $TP = L_S|P$. We can therefore ask the question:

<u>Does a smooth compact hypersurface</u> $S \subset \mathbb{R}^{2n}$ <u>carry a periodic orbit?</u>

Although partial results mentioned below do exist, this question with respect to the distinguished symplectic form ω (3) is still open. In contrast, it is easy to construct hypersurfaces and symplectic structures different from ω for which the answer is no, as we shall see below. It should be emphasized that a very restricted class of vector fields on S is considered. An arbitrary vector field need not possess a periodic orbit. For example, using a result by Denjoy, P. A. Schweitzer [15] constructed on $S = S^3$, the 3-dimensional sphere, a C^1-vector field which has no periodic orbit. It is not known whether such a vector field exists in the class of analytic vector fields, for example. On higher dimensional spheres S^{2n-1}, $n > 3$, T. W. Wilson [14] found C^∞-vector fields without periodic orbits. It would be interesting to know, whether such examples do exist in the more restricted class of vector fields which preserve a volume element. For the torus instead of the sphere irrational translations have clearly no periodic orbits.

It is, of course, well-known that the flow of a Hamiltonian vector field X_H on S preserves a volume element. Consequently every point $x \in S$ is a nonwandering point, if S is compact. This fact was used by C. Pugh and C. Robinson [19] in order to conclude from the so-called C^1-Closing-Lemma the following statement which is of generic nature.

Theorem 1 [19]. C^1-generically the periodic orbits are dense on a compact energy surface.

This statement claims in particular that given a Hamiltonian system X_{H_1} near a regular compact energy surface, then there exists in every C^1-neighborhood of X_H a (in general different) Hamiltonian vector field X_G which has an energy surface on which the periodic orbits are dense. We point out that it is an important open question whether the result holds true for C^k with $k > 1$. Theorem 1 is related to a question asked by H. Poincaré. In his book "Les méthodes nouvelles de la mécanique céleste" (Tome 1, 1892, Chap. III, §36, p. 82) H. Poincaré expresses his strong belief that in the restricted class of Hamiltonian vector fields which are very close to integrable systems, the periodic orbits are dense. Although such systems have been studied extensively this question remained unanswered up to now. It can be proved, however, that, in general, the closure of the set of periodic orbits of nearly integrable systems is of positive measure on S, see for example [20].

In sharp contrast to the generic statement above we shall be concerned with existence results for a given Hamiltonian system. The break through in the global problem of existence of a periodic orbit on a hypersurface is due to P. Rabinowitz [5] and to A. Weinstein [6]. In 1978 they proved that a star like hypersurface, resp. a convex hypersurface always carries at least one periodic orbit. For the special situation of a convex hypersurface F. Clark and I. Ekeland [9] introduced the so-called dual action variational principle which not only allowed an easy existence proof but lead to multiplicity results, to a Morse theory for periodic solutions on a convex hypersurface and, most recently, to a symplectic invariant of a convex hypersurface, [10-13,29]. By this invariant one concludes in particular, that a convex hypersurface carries always at least 2 periodic orbits, a result which is optimal for $n = 2$. The dual action principle is not suitable in the search for periodic solutions on general hypersurfaces. Changing the Hamiltonian function in an appropriate way, P. Rabinowitz [5] reduced the problem of finding periodic solutions on a star like surface to the problem of finding appropriate periodic solutions of (1) in $\mathbb{R}^{2n}$ having a prescribed period, say $T = 1$ instead. These solutions are the critical points of the functional

$$\Phi(x) := \int_0^1 \left\{\frac{1}{2} \langle -J\dot{x},x\rangle - H(x)\right\}dt \tag{6}$$

defined on the loop space of periodic loops $x(0) = x(1)$ in $\mathbb{R}^{2n}$. In contrast to the dual action principle this functional is degenerate, it is bounded neither from above nor from below, so that standard variational techniques do not apply directly. But Rabinowitz demonstrated using minimax arguments that this degenerate variational principle can be used effectively for existence proofs, which lead to a deeper understanding of critical point theory for strongly indefinite variational functionals, [4,7,18]. The variational principle (6) in connection with C. Conley's homotopy index theory (see Benci [30]) turned out to be useful also in finding fixed points of Hamiltonian maps on the torus T^{2n} [21]; the Morse-theory for forced oscillations on T^{2n} [22] relates the critical points of (6) in the time periodic case to the intrinsic winding numbers of the corresponding forced oscillations. For related problems in symplectic geometry, M. Chaperon [23] and J. C. Sikorav [24] introduced different and more geometric variational approaches, which remind of the generating function techniques, see also [25-27]. For still another variational principle designed in order to find minimal periodic orbits and also Mather sets for monotone twist-mappings in the plane we refer to J. Moser's note in these proceedings.

In 1979 A. Weinstein [2] conjectured that hypersurfaces which are of contact type as explained further on carry at least one periodic orbit. Indeed, most recently C. Viterbo [1] succeeded in proving this conjecture. Subsequently some of his ideas were used by H. Hofer and E. Zehnder [28] to prove that every slightly thickened compact smooth hypersurface admits a periodic orbit. To make this statement precise we first give a

<u>Definition</u>. A parameterized family of compact hypersurfaces in $\mathbb{R}^{2n}$ modelled on a compact hypersurface S is a diffeomorphism $\psi : (-1,1) \times S \to \mathbb{R}^{2n}$ onto an open and bounded neighborhood of S such that $\psi(0,x) = x$ for $x \in S$. In the following we shall abbreviate $S_\varepsilon = \psi(\{\varepsilon\} \times S)$.

Observe that every compact smooth hypersurface $S \subset \mathbb{R}^{2n}$ belongs to such a family. For example, take a smooth function $H \in C^\infty(\mathbb{R}^n)$ such that $S = H^{-1}(1)$ and $\nabla H(x) \neq 0$ for $x \in S$. Then the vector field $X = \lambda \cdot \nabla H(x)$ with $\lambda(x)^{-1} = |\nabla H(x)|^2$ defined near S is transverse to S. Its flow φ^t with $\varphi^0 = \mathrm{id}$ satisfies $H(\varphi^t(x)) = 1 + t$ for $x \in S$ and $|t|$ small and can be used to define a parameterized family $\psi : (-1,1) \times S \to \mathbb{R}^{2n}$.

We shall denote the action of a loop x in $\mathbb{R}^{2n}$ having period T by

$$A(x) := \frac{1}{2}\int_0^T \langle -J\dot{x},x\rangle dt \ , \tag{7}$$

where $X : [0,T] \to \mathbb{R}^{2n}$ satisfies $x(0) = x(T)$.

Theorem 2 [28]. Let $S \subset \mathbb{R}^{2n}$ be an arbitrary compact smooth hypersurface and let ψ be a parameterized family of hypersurfaces modelled on S. Then there exists a constant $d = d(\psi) > 0$ such that for every $0 < \delta < 1$ there is an ε in $|\varepsilon| < \delta$ for which the hypersurface S_ε carries a periodic orbit x satisfying the estimate:

$$0 < A(x) < d \ .$$

Observe that there are no assumptions on the compact hypersurface S. The theorem does not claim a periodic orbit on S itself. But it shows for example for a family $S_E = \{x \in \mathbb{R}^{2n} | H(x) = E\}$, $E \in I$, of compact regular energy surfaces that for a dense set in the interval I the corresponding surfaces S_E have a periodic orbit. The following simple example shows that the statement does not hold true for every hypersurface $S \subset \mathbb{R}^n$ if we take on $\psi\{(-1,1) \times S\}$ a symplectic structure different from ω.

Consider the 4-dimensional manifold $T^3 \times I$, with I being an open interval, and with coordinates $\xi_1,\xi_2,\xi_3(\text{mod } 1)$ and $\varepsilon \in I$. Define the Hamiltonian function $H \in C^\infty(T^3 \times I)$ by $H(\xi,\varepsilon) = \varepsilon$. We look for a symplectic structure defined by a skew symmetric nondegenerate matrix J such that the Hamiltonian vector field is given by

$$\frac{d}{dt}\binom{\xi}{\varepsilon} = J\nabla H = \binom{\alpha}{0} \tag{8}$$

for a constant and rationally independent vector $\alpha = (\alpha_1,\alpha_2,\alpha_3) \in \mathbb{R}^3$. For example, we can choose

$$J = \begin{pmatrix} 0 & 1 & 0 & \alpha_1 \\ -1 & 0 & 0 & \alpha_2 \\ 0 & 0 & 0 & \alpha_3 \\ -\alpha_1 & -\alpha_2 & -\alpha_3 & 0 \end{pmatrix} . \tag{9}$$

Then define for $r^2,s^2 > 0$ and $0 < \varepsilon < \min\{r^2,s^2\}$ the diffeomorphism $\psi : T^3 \times I \to \mathbb{R}^4$ by

$$\psi(\xi,\varepsilon) = \begin{pmatrix} (r^2 + \varepsilon \cos^2 \xi_3)^{\frac{1}{2}} \cos \xi_1 \\ (r^2 + \varepsilon \cos^2 \xi_3)^{\frac{1}{2}} \sin \xi_1 \\ (s^2 + \varepsilon \sin^2 \xi_3)^{\frac{1}{2}} \cos \xi_2 \\ (s^2 + \varepsilon \sin^2 \xi_3)^{\frac{1}{2}} \sin \xi_2 \end{pmatrix} . \tag{10}$$

The mapping ψ defines a parameterized family over T^3, the hypersurfaces being given by $H(x) = \varepsilon$, with $H = F^2 + G^2$, where $F = x_1^2 + x_2^2 - r^2$ and $G = x_3^2 + x_4^2 - s^2$. Clearly, none of the hypersurfaces in this family carries a periodic orbit since α is an irrational vector. The induced symplectic form on $\psi(T^3 \times I) \subset \mathbb{R}^4$, is, however, not equivalent to ω, indeed it is not an exact 2-form.

In view of this example one could ask for conditions on S and the symplectic structure on $\psi\{(-1,1) \times S\}$ which enable to conclude the statement of Theorem 2. Is it, for example, necessary that the symplectic structure can be extended to a symplectic structure on $\mathbb{R}^{2n}$ as indicated by the proof of Theorem 2 which we sketch at the end of this note?

We first illustrate Theorem 2 by some applications. In order to find periodic solutions not only near S but on the given hypersurface we shall require additional properties for S. We first assume that there exists a vector field X in a neighborhood of S which is transverse to S and which satisfies, in addition

$$L_X \omega = \omega . \tag{11}$$

Such distinguished hypersurfaces are called of contact type, see A. Weinstein [2]. Examples are the star like hypersurfaces, indeed $X(x) = \frac{1}{2} x$ satisfies (11), and a hypersurface $S \subset \mathbb{R}^{2n}$ is transverse to X precisely if it is star like with respect to the origin. Now, the flow of a vector field X with (11) satisfies

$$(\varphi^t)^* \omega = e^t \omega , \tag{12}$$

and allows to define a very special parameterized family ψ of hypersurfaces modelled on S. It has the property that if $y(t)$ is a periodic orbit on S_ε for some ε then $x(t) := \varphi^{-\varepsilon}(y(t))$ is a periodic orbit on the given hypersurface $S = S_0$ as one verifies readily. We therefore conclude from Theorem 2 the following existence result due to C. Viterbo [1]. It was conjectured by A. Weinstein [2] under the additional assumption that $H^1(S) = 0$.

Theorem 3 (C. Viterbo [1]). *Every compact smooth hypersurface of contact type carries at least one periodic orbit.*

For the next application of Theorem 2 we need an additional metric structure; if x is a periodic orbit of period T we denote its length by

$$\ell(x) := \int_0^T |\dot{x}(t)|dt \ , \tag{13}$$

where we have used the Euclidean metric. We begin with a simple remark.

Let ψ be a parameterized family modelled on S and set $U = \psi((-1,1) \times S)$. Define the Hamiltonian function $H : U \to \mathbb{R}$ by setting $H(\psi(\varepsilon,x)) = \varepsilon$. By Theorem 2 there is a sequence $\varepsilon_j \to 0$ such that S_{ε_j} carries the periodic solution x_j. With T_j we denote its period. On $\psi([-\delta,\delta] \times S)$, $0 < \delta < 1$ we have the estimate $M^{-1} \leq |\nabla H(x)| \leq M$ for some constant $M > 0$. From the Hamiltonian equation $-J\dot{x}_j = \nabla H(x_j)$ we therefore conclude that for all $j \geq 1$

$$M^{-1}T_j \leq \ell(x_j) \leq M \cdot T_j \ . \tag{14}$$

Assuming that $\ell(x_j)$ or equivalently that the periods T_j are bounded for $j \geq 1$ one concludes readily by the Ascoli-Arzela theorem that a subsequence of x_j converges to a periodic solution x^* on $S = S_0$ as $j \to \infty$.

In view of this observation the following a priori statement is an immediate consequence of Theorem 2.

Theorem 4 [28]. *Let* $S \subset \mathbb{R}^{2n}$ *be a smooth compact hypersurface. Assume there is a constant* $K > 0$ *such that*

$$\frac{1}{K}\,\ell(x) \leq |A(x)| \leq K\ell(x) \tag{15}$$

for all periodic solutions) in a parameterized family* ψ *modelled on* S. *Then* S *carries at least one periodic orbit.*

Note that the second estimate always holds true. For applications of this a priori result we refer to V. Benci, H. Hofer and P. Rabinowitz in these proceedings, where the a priori bounds (15) are verified for a large class of Hamiltonian functions containing the classical Hamiltonian systems of the form

$$H(x) = \frac{1}{2}\,|p|^2 + V(q) \ , \tag{16}$$

---

*) The existence is, of course, not assumed.

with $x = (p,q) \in \mathbf{R}^n \times \mathbf{R}^n$. The question arises here, whether the periodic orbits guaranteed by Theorem 4 are so-called brake orbits (see these proceedings).

In view of Theorem 2 and the remark prior to Theorem 4 a hypersurface which does not admit a periodic orbit gives rise to an abundance of periodic solutions nearby having large periods; more precisely

Proposition 1. If $S \subset \mathbf{R}^{2n}$ is a compact smooth hypersurface without any periodic solution, then for every Hamiltonian vector field X_H admitting S as regular energy surface the following holds true: Given an open neighborhood U of S and a constant $K > 0$, then X_H possesses a periodic orbit x which is contained in U and which has length $\ell(x) > K$.

Unfortunately we do not know of any hypersurface which meets the assumption of the proposition (the symplectic form is ω in (3)). In view of the statement it might not be so easy to construct such an example. The tempting idea to use the translation on a torus fails according to the following observation due to J. Moser.

Proposition 2. There is no hypersurface $S \subset \mathbf{R}^4$ diffeomorphic to T^3 which is a regular energy surface for a Hamiltonian vector field whose flow on S is equivalent to the translation on T^3 by an irrational vector.

The statement holds true for every T^{2n-1} and, moreover, in every 2n-dimensional symplectic manifold provided the symplectic 2-form is an exact form. This follows from the proof.

Proof. We shall prove the statement with respect to any exact symplectic structure, and consider the manifold $M = T^3 \times I$ with the exact symplectic form $d\lambda$, with the 1-form

$$\lambda = \sum_{j=1}^{4} a_j(x)dx_j \quad \text{and} \quad a_j \in C^\infty(M) \ , \tag{17}$$

where $x_1, x_2, x_3 (\text{mod } 1)$ and $x_4 \in I$. Assume there is a Hamiltonian vector field $X = X_H$,

$$d\lambda \lrcorner X = -dH \ , \tag{18}$$

for which $T^3 \times \{0\} = S$ is an energy surface, such that, moreover,

$$X = \sum_{j=1}^{3} \alpha_j \frac{\partial}{\partial x_j} \quad \text{on} \quad S \tag{19}$$

for a constant and rationally independent vector $\alpha = (\alpha_1, \alpha_2, \alpha_3)$. The flow φ^t of X leaves $d\lambda$ invariant, i.e. $(\varphi^t)^* d\lambda = d\lambda$; moreover $\varphi^t(S) = S$, so that

$$(\varphi^t)^* j^* d\lambda = j^* d\lambda \ , \tag{20}$$

the map $j : S \to M$ denoting the injection map. On S the flow is given by

$$\varphi^t(x,0) = (x + \alpha t, 0), \qquad x \in T^3 \ . \tag{21}$$

The coefficients of $d\lambda$ are

$$c_{ij} := \frac{\partial}{\partial x_j} a_i - \frac{\partial}{\partial x_i} a_j, \qquad 1 \leq i,j \leq 4 \ . \tag{22}$$

By (20) and (21) the c_{ij} are, on S, independent of $x \in T^3$, if $i,j \leq 3$. Integration over the torus then shows that $c_{ij} = 0$, if $i,j \leq 3$. Therefore the matrix $C = (c_{ij})$, $1 \leq i,j \leq 4$ is not regular on S, so that $d\lambda$ cannot be a symplectic 2-form, as we wanted to prove.

One could ask whether the statement of the proposition holds true under the weaker assumption that the flow on T^3 is merely ergodic. The next proposition excludes also the torus T^3 which is foliated into invariant 2-dimensional tori T^2 on which the flow is a translation by an irrational vector in $\mathbb{R}^2$.

<u>Proposition 3. There is no hypersurface $S \subset \mathbb{R}^4$ diffeomorphic to T^3 which is a regular energy surface for a Hamiltonian vector field whose flow on S is equivalent to the translation on T^3 given by $(\alpha_1, \alpha_2, 0)$ with α_1, α_2 rationally independent.</u>

<u>Proof</u>. We use the same set up and notation as in the previous proof replacing, however, (19) by

$$X = \sum_{r=1}^{2} \alpha_j \frac{\partial}{\partial x_j} \quad \text{on} \quad S \ , \tag{23}$$

where $S = T^3 \times \{0\} \subset M = T^3 \times I$. Consider a torus T^2 in S defined by $x_3 = \text{const}$, and let $i : T^2 \to M$ be the injection mapping. We claim that T^2 is a Lagrangian submanifold, i.e. $i^* d\lambda = 0$. Indeed, since $\varphi^t(T^2) = T^2$ we conclude that $(\varphi^t)^* i^* d\lambda = i^* d\lambda$. Moreover, since $\alpha = (\alpha_1, \alpha_2)$ is an irrational vector, the 2-form $i^* d\lambda$ on T^2 is constant. From $i^* d\lambda = d(i^* \lambda)$ we therefore conclude by integration over T^2 that $i^* d\lambda = 0$ as claimed. Consequently

$$c_{ij} = 0 \quad \text{on} \quad S, \qquad 1 \leq i,j \leq 2 \ . \tag{24}$$

As in the previous proof one shows that, moreover,

$$c_{ij} = c_{ij}(x_3) \quad \text{on} \quad S, \qquad 1 \leq i,j \leq 3 \ . \tag{25}$$

The aim is to find a point x_3^* such that $c_{ij}(x_3^*) = 0$ for $1 \leq i,j \leq 3$. It then follows that the matrix C is not a regular matrix so that $d\lambda$ is not a symplectic 2-form. For this purpose we first pick a simpler 1-form $\bar{\lambda}$ on M for which

$$j^*d\lambda = j^*d\bar{\lambda} , \tag{26}$$

where $j : S \to M$ denotes the injection map. Define

$$\bar{\lambda} = \sum_{r=1}^{4} \bar{a}_r dx_r , \tag{27}$$

with

$$\bar{a}_r = \int_{T^2} a_r dx_1 dx_2 .$$

Then, by (24) and (25)

$$d(j^*\bar{\lambda} - j^*\lambda) = 0 , \tag{28}$$

so that

$$j^*\bar{\lambda} = j^*\lambda + df , \tag{29}$$

for a function $f(x)$ defined on $x \in \mathbf{R}^3$. In order to prove that f is a periodic function we observe that $f(x) = \langle\gamma,x\rangle + \hat{f}(x)$, $\hat{f}$ periodic. Since $\bar{a}_r = a_r + f_{x_r}$ on S, for $1 \leq r \leq 3$, integration over T^3 shows that indeed $\gamma = 0$, so that $\bar{\lambda}$ satisfies (26). Replacing λ by $\bar{\lambda}$ we can therefore assume that

$$\frac{\partial}{\partial x_1} a_j = \frac{\partial}{\partial x_2} a_j = 0 \quad \text{on} \quad S, \qquad 1 \leq j \leq 3 . \tag{30}$$

From (18) we conclude

$$j^*(d\lambda \lrcorner X) = 0 , \tag{31}$$

which gives, in view of (22), (23) and (30) the relation:

$$\frac{\partial}{\partial x_3} a_1 \cdot \alpha_1 + \frac{\partial}{\partial x_3} a_2 \cdot \alpha_2 = 0 , \tag{32}$$

on S. On the other hand, the periodic function $a_1(x_3)\alpha_2 - a_2(x_3)\alpha_1$ on S, has a critical point, x_3^*, at which

$$\frac{\partial}{\partial x_3} a_1 \cdot \alpha_2 - \frac{\partial}{\partial x_3} a_2 \cdot \alpha_1 = 0 . \tag{33}$$

Since $\alpha_1^2 + \alpha_2^2 \neq 0$, we get by (32) and (33)

$$\frac{\partial}{\partial x_3} a_1 = \frac{\partial}{\partial x_3} a_2 = 0 . \tag{34}$$

Summarizing there is a point on S where $c_{ij} = 0$ for $1 \leq i,j \leq 3$, as we wanted to prove.

We finally sketch the ideas of the proof of Theorem 2. Let $\psi : (-1,1) \times S \to \mathbb{R}^{2n}$ be the parameterized family of compact hypersurfaces modelled on S, take $0 < \delta < 1$, and define a very special Hamiltonian H setting

$$H(x) = f(\varepsilon) \quad \text{if} \quad x = \psi(\varepsilon,y) ,$$

$y \in S$ and $|\varepsilon| \leq \delta$ for an appropriate function f. Extending H to a convenient function on $\mathbb{R}^{2n}$ we look for 1-periodic solutions of $\dot{x} = J\nabla H(x)$ on $\mathbb{R}^{2n}$ which are the critical points of the functional Φ defined in (6). It turns out that there are plenty of periodic solutions which unfortunately do not lie on the parameterized family and whose Conley-Morse-index is not easy to compute. The function H however is constructed in such a way that, a priori those critical points x of Φ which represent periodic solutions on S_ε for some $|\varepsilon| < \delta$ satisfy $\Phi(x) > 0$. Such a critical point can be found by a mini-max argument for Φ defined on the space $E = H^{1/2}(S^1,\mathbb{R}^{2n})$. The underlying idea is to construct two bounded sets Σ and Γ with the properties that

$$\Phi|\Gamma \geq \beta > 0$$

and

$$\Sigma \cdot s \cap \Gamma \neq \emptyset, \quad s \geq 0 .$$

The dot denotes the gradient flow of $\frac{d}{ds} x = -\nabla\Phi(x)$ on E. Using the P.S.-condition for Φ one concludes that

$$c := \inf_{s \geq 0} \sup_{x \in \Sigma} \Phi(x \cdot s) \geq \beta \tag{35}$$

is indeed a critical value for Φ. The critical points correspond to the required periodic orbits on S_ε for $|\varepsilon| < \delta$. For the details we refer to [28].

It has already been pointed out that the importance of periodic solutions lies in the fact that they provide a handle to study the very intricate flow nearby using perturbation arguments. For example, if the Floquet-multiplier of a periodic solution are on $|\lambda| = 1$ and meet finitely many nonresonance conditions and if, moreover, the nonlinear Birkhoff invariants of lower order meet a nondegeneracy condition, then one knows that in a small neighborhood of the periodic solution the system is close to an integrable system. Using K.A.M.

theory one therefore finds an abundance of quasiperiodic solutions and also an abundance of other periodic solutions of large period surrounding the given periodic solution, [20]. The global variational principles mentioned above are a powerful tool to establish the existence of global periodic solutions but are not suited to give such information about the nature of these solutions.

We would like to thank J. Moser for valuable discussions.

REFERENCES

[1] C. Viterbo, 'A proof of the Weinstein conjecture in $\mathbb{R}^{2n}$', preprint, September 1986.

[2] A. Weinstein, 'On the hypotheses of Rabinowitz's periodic orbit theorems', J. Diff. Eq. **33** (1979), 353-358.

[3] V. Benci and P. Rabinowitz, 'A priori bounds for periodic solutions of a class of Hamiltonian systems', preprint (1986).

[4] V. Benci and P. Rabinowitz, 'Critical point theorems for indefinite functionals', Inventiones Math. **52** (1979), 241-273.

[5] P. Rabinowitz, 'Periodic solutions of Hamiltonian systems', Comm. Pure Appl. Math. **31** (1978), 157-184.

[6] A. Weinstein, 'Periodic orbits for convex Hamiltonian systems', Ann. Math. **108** (1978), 507-518.

[7] H. Hofer, 'On strongly indefinite functionals with applications', Trans. Amer. Math. Soc. **275** (1983), 185-214.

[8] E. H. Spanier, Algebraic Topology, Mc Graw Hill: New York (1966).

[9] F. Clarke and I. Ekeland, 'Hamiltonian trajectories having prescribed minimal period', Comm. Pure Appl. Math. **33** (1980), 103-116.

[10] I. Ekeland and J. M. Lasry, 'On the number of periodic trajectories for a Hamiltonian flow on a convex energy surface', Ann. Math. **112** (1980), 283-319.

[11] I. Ekeland and H. Hofer, 'Periodic solutions with prescribed minimal period for convex autonomous Hamiltonian systems', Invent. Math. **81** (1985), 155-188.

[12] I. Ekeland, 'Une théorie de Morse pour les systèmes hamiltoniens convexes', Ann. Inst. Henri Poincaré, Annalyse non linéaires, **1** (1984), 19-78.

[13] I. Ekeland and H. Hofer, 'Convex Hamiltonian energy surfaces and their periodic trajectories', preprint, September 1986.

[14] T. W. Wilson, 'On the minimal sets of nonsingular vectorfields', Ann. Math. **84** (1966), 529-536.

[15] P. A. Schweitzer, 'Counterexamples to the Seifert conjecture and opening closed leaves of foliations', Ann. Math. **100** (1974), 386-400.

[16] E. Lima, 'Orientability of smooth hypersurfaces and the Jordan-Brouwer separation theorem', Expositiones mathematicae, to appear.

[17] V. Benci, H. Hofer and P. Rabinowitz, 'A priori bounds for periodic solutions on hypersurfaces', these proceedings.

[18] P. Rabinowitz, 'Minimal methods in critical point theory with applications to differential equations', CBMS Lectures, Regional Conference Series in Math. Nr. **65** (1986).

[19] C. C. Pugh and R. C. Robinson, 'The C^1 closing lemma, including Hamiltonians', Ergodic Theory and Dynamical Systems **3** (1983), S. 261-313.

[20] C. C. Conley and E. Zehnder, 'An index theory for periodic solutions of a Hamiltonian system, geometric dynamics', Springer Lecture Notes in Math. 1007, 122-145.

[21] C. C. Conley and E. Zehnder, 'The Birkhoff-Lewis fixed point theorem and a conjecture of V. I. Arnold', Invent. Math. **73** (1983), 33-49.

[22] C. C. Conley and E. Zehnder, 'A global fixed point theorem for symplectic maps and subharmonic solutions of Hamiltonian equations on tori', AMS Proceedings of Symposia in Pure Mathematics, **45**, Part 1 (1986), 283-299.

[23] M. Chaperon, 'Quelques questions de géométrie symplectique', Séminaire Bourbaki 1982-83, No. 610, Astérisque **105-106** (1983), 231-249.

[24] J. C. Sikorav, 'Problémes d'intersections et de points fixes en géométrie Hamiltonienne', preprint Univ. Paris Sud (1986).

[25] A. Floer, 'Proof of the Arnold conjecture for surfaces and generalizations for certain Kähler-manifolds', Duke Math. J. **3** (1986), 1-32.

[26] F. Laudenbach and J. C. Sikorav, 'Persistence d'intersections avec la section nulle un cours d'une isotopie Hamiltonienne dans un fibre cotangent', Invent. Math. **82** (1985), 349-357.

[27] H. Hofer, 'Lagrangian embeddings and critical point theory', Ann. Inst. Henri Poincaré, Analyse non linéaire **2** (1985), 407-462.

[28] H. Hofer and E. Zehnder, 'Periodic solutions on hypersurfaces and a result by C. Viterbo', preprint RUB, November 86, to appear in Inventiones Math.

[29] I. Ekeland and H. Hofer, 'Global and local invariants for convex energy surfaces and their periodic trajectories: a survey', these proceedings.

[30] V. Benci, 'Some application of the Morse-Conley theory to the study of periodic solutions of second order conservative systems', these proceedings.

INDEX

GPSR Compliance

The European Union's (EU) General Product Safety Regulation (GPSR) is a set of rules that requires consumer products to be safe and our obligations to ensure this.

If you have any concerns about our products, you can contact us on ProductSafety@springernature.com

In case Publisher is established outside the EU, the EU authorized representative is:

Springer Nature Customer Service Center GmbH
Europaplatz 3
69115 Heidelberg, Germany

Batch number: 10607214

Printed by Printforce, the Netherlands